解读传统建筑

Reading the Traditional Architecture

周晶 李天 编著

图书在版编目(CIP)数据

解读传统建筑:Reading the Traditional Architecture:汉英对照/周晶,李天编著.—西安:西安交通大学出版社,2018.2
ISBN 978-7-5605-7500-1

Ⅰ.①解… Ⅱ.①周… ②李… Ⅲ.①古建筑-建筑艺术-中国-汉、英
Ⅳ.①TU-092.2

中国版本图书馆 CIP 数据核字(2018)第 010164 号

书　　名 解读传统建筑
编　　著 周　晶　李　天
策划编辑 屈晓燕
文字编辑 季苏平

出版发行 西安交通大学出版社
(西安市兴庆南路 10 号　邮政编码 710049)
网　　址 http://www.xjtupress.com
电　　话 (029)82668357　82667874(发行部)
(029)82668315(总编办)
印　　刷 西安日报社印务中心

开　　本 720 mm×1000 mm　1/16　**印张** 17　**字数** 412 千字
版次印次 2018 年 8 月第 1 版　2018 年 8 月第 1 次印刷
书　　号 ISBN 978-7-5605-7500-1
定　　价 40.00 元

读者购书、书店添货或发现印装质量问题,请与本社营销中心联系、调换。
订购热线:(029)82665248　(029)82665249
投稿热线:(029)82665355
读者信箱:lg_book@163.com

前言

双语教材《解读传统建筑》(*Reading the Traditional Architecture*)较为广泛地选取古今中外文明进程中产生的典型传统建筑类型,使读者通过阅读各种建筑类型的代表性建筑,了解中外传统建筑的发展历程、基本构成要素与地域特征和时代特点,以及民族文化与宗教信仰在建筑艺术上的表现,对建筑艺术在人类文明发展史上的重要意义有进一步认识。

教材由十章组成,每章介绍一种传统建筑类型。每章由引言、四篇阅读文章以及知识点三个环节构成。引言为典型建筑类型的简短介绍;阅读文章选取世界范围内代表性的传统建筑案例,为章节核心内容;知识点则对各民族文化中独特的建筑文化现象进行简短介绍,目的是为有深度阅读意向的读者提供兴趣点。

针对建筑艺术的特点,教材配有大量手绘图,多数为作者自绘,部分采自往昔绘画作品,以增加一般读者阅读兴趣和对建筑实例的感性认识;平面图与剖视图则为专业阅读者提供深入了解建筑布局与空间特征的选择。每篇阅读材料还提供了一定数量的注释,辅助阅读者了解建筑的历史背景与文化气氛。

教材采用正文以英文为主,辅以汉语注释的形式。鉴于传统建筑涵盖多种文化背景,涉及专有词汇较多,有些词汇用法较为特殊,教材较为详细地编列了词汇表,偏重建筑艺术表达法,以方便读者查阅。

编著者

2018 年 2 月

Contents

目录

Chapter One

Vernacular House

Vernacular house is a broad, grassroots concept which encompasses fields of architectural study including aboriginal, indigenous, ancestral, rural, and ethnic architecture and is contrasted with the more intellectual architecture called polite, formal, or academic architecture just as folk art is contrasted with fine art.

Vernacular house is designed based on local needs, availability of construction materials and reflecting local traditions. At least originally, vernacular house did not use formally-schooled architects, but relied on the design skills and tradition of local builders. It tends to evolve over time to reflect the environmental, cultural, technological, economic, and historical context in which it exists.

Vernacular house is influenced by a great range of different aspects of human behavior and environment, leading to different building forms for almost every different context; even neighboring villages may have subtly different approaches to the construction and use of their dwellings, even if they at first appear the same. Despite these variations, every building is subject to the same laws of physics, and hence will demonstrate significant similarities in structural forms. The Encyclopedia defines vernacular architecture as... comprising the dwellings and all other buildings of the people. Related to their environmental contexts and available resources they are customarily owner or community-built, utilizing traditional technologies. All forms of vernacular architecture are built to meet specific needs, accommodating the values, economies and ways of life of the cultures that produce them.

Frank Lloyd Wright described vernacular architecture as "Folk building growing in response to actual needs, fitted into environment by people who knew no better than to fit them with native feeling".

Stilt Houses for Chinese Southern Ethnics[1]

A stilt house is a raised structure that is most commonly built above water, although it also may be built over dirt or sand. It is sometimes called a pile dwelling because it is supported by large stakes, known as piles, that are driven directly into the water or into the shoreline. These structures typically rest 3.5-4 meters off the ground to allow for high tide, and are designed to avoid flooding and water damage. Created from bamboo or other water-resistant timber and reinforced with deck boards and sometimes concrete, the stilt house dates back to prehistoric times, but it is still commonly found around the world, especially in places prone to flooding.

1. Bamboo House(竹楼)[2]

The traditional Thai houses usually feature a bamboo or wooden structure, raised on stilts and topped with a steep gabled roof. The area beneath the house is used for storage, crafts, lounging in the daytime, and sometimes for livestock. The houses are raised due to heavy flooding during monsoon season each year, and in more ancient times for predators.

Thai houses are made from a variety of wood and are often built in just a day as prefabricated wood panels are built ahead of time and put together on site by a master builder. More houses are built with bamboo, a material that is easily constructed and does not require professional builders. Most homes start out as a single family home and when a daughter gets married, an additional house is built on site to accommodate her new family.

A traditional Thai house is usually built as a cluster of physically separate rooms arranged around a large central terrace. The terrace is the largest singular part of the home as it makes up to 40% of the square footage, and up to 60% if the veranda is included. An area in the middle of the terrace is often left open to allow the growth of a tree through the structure, providing welcome shade. The tree chosen is often flowering or scented. (See Fig. 1. 1)

Fig. 1. 1 Thai Bamboo house

Thai building and living habits are often based on superstitious and religious beliefs. Traditional Thai houses are built in accordance with three ancient principles: "material preparation, construction, and dwelling". Materials, including site and orientation, the taste and smell of soil, and the names of trees that will be used to build houses and so on, will be carefully chosen. Second, the construction must be done mindfully. For instance, only a person of acknowledged spiritual power is allowed to perform a ritual when the first column is put into the ground. The time allotted for that ritual needs to be precisely calculated and fixed. Similarly, construction of the guardian spirit house and proper conduct of the housewarming ceremony are also essential. The third principle is proper behavior in the completed house. For example, the threshold is believed to be inhabited by "a household guardian spirit", therefore, stepping on it is prohibited. If residents of the house do not follow this precept, spiritual protection will disappear. Another example is that if someone sleeps under the girders, it is believed that ghosts will cause him difficulty in breathing. All of these observances serve the purpose of making houses sacred places and pleasing "good" spirits in order to receive their protection against "bad" spirits.

It is important for the Thai people to draw in their natural surroundings by placing potted plants around the terrace. In the past there were strict taboos regarding which plants could be placed directly around the house. The level of the floor changes as one moves from room to terrace, providing a wide variety of positions for sitting or lounging around the living areas.

Furniture in Thai house is sparse and only includes a bed platform, dining table, and loose cushions for sitting. Sleeping areas are set up so that the beds are aligned with the shorter end of the room as sleeping parallel with the length is similar to lying in a coffin . The direction that the head points towards can never be the west as that is the position bodies are laid in before cremation.

2. Diaojiao Stilt House(吊脚楼)[3]

The west area of Hu'nan(湖南) and east of Guizhou(贵州) Province of China is a mountainous area where many ethnic groups live, including Tujia(土家), Miao(苗), Yao(瑶), Dong(侗) and Zhuang(壮), etc. In this area, the chain of mountains runs up and down, rivers densely distributed, and trees are abundant. The unique geographic environment has given birth to the stilt house which is unique, elegant, close to rivers and mountains. The integration of cultures of various ethnic groups has endowed the stilt house with rich and diversified cultural connotations: the stilt house of Tujia people embodies the universal view of the coexistence of human and God, while the Dong people like to beautify the house by carving the capitals into bamboo shape, and carving flower patterns on the banisters of the corridor.

The Diaojiao stilt house is usually built beside the river or in steep mountain areas in different heights, it is usually divided into the front part and the rear part. The front part is a storied building built on the ground and similar to the fence-style building, while the rear part is a bungalow. In front of the house is a corridor formed by wood balusters with a height of about 1 meter. Suspended windows are equipped in some houses for rest and enjoying cool. Sundries can be stored under the floor, and by this, the space is utilized smartly. The space of the stilt house is arranged facilely in various shapes. The shortage of construction land is relatively relieved, for the stilt house makes use of the hillside or river, which is originally not suitable for building houses. Furthermore, it has the advantages of aeration, moisture-proof, viper and wild animal defense. (See Fig. 1. 2)

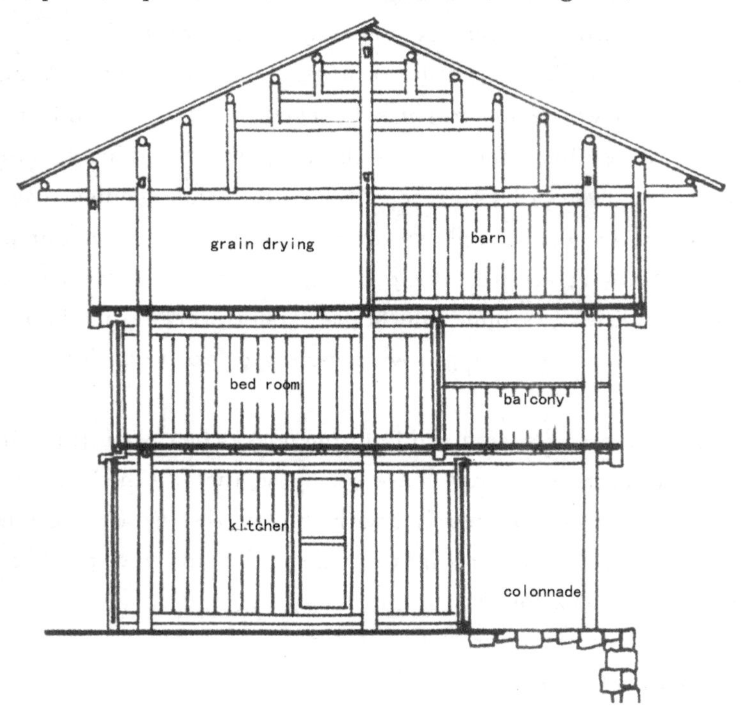

Fig. 1. 2 Section of Diaojiao Stilt House

Residing in the high and cold mountainous area with quite wet and foggy climate which is not suitable for accommodation, Miao people constructed a kind of wooden-house above the ground along the mountain ridge with good ventilation so as to avoid the constructing restrictions, such as steep slopes and wetness.

Situated at the foot of the Leigong Mountain along Bai River in Leishan County of southwestern Guizhou Province, Xijiang Miao Village with a population of 1,000 people is definitely a wonderful destination to appreciate the unique Miao-style wooden stilt house. The stilt houses in this village are normally constructed on a 30-70 degree mountain slope. The construction techniques adopted by Xijiang Miao people have a long history over 7,000 years. Raised above the soil with a length of 7 to 14 meters, the wooden stilt house is usually built with three or four storeys. Each storey has three to five rooms. The upper floor is used to store the provisions; the second floor for people living and the first floor for sundries or feeding livestock. These Miao-style houses are scattered along the mountain ridges which is recognized as the most representative look of Miao village.

Along with changes and innovations, modern wooden stilt house is now commonly built with solid and moisture-proof material which helps local residents to save their farming land and forest.

Unfortunately, as more and more young people left village for big city, there are fewer people living in the Miao-style house and learning how to build it.

3. Kam Village[4] (侗寨)

The Kam(Dong 侗族)live mostly in eastern Guizhou, western Hu'nan, and northern Guangxi, China and small numbers of Kam speakers are found in Vietnam. Dong villages are traditionally built near rivers, or at the foot of mountains. Dong villages have a uniqueness that distinguishes them from those of neighboring ethic groups such as the Zhuang, Miao and Yao people—the drum tower which dominates the landscape of a typical Dong village, with its formation of tiered, multistory roofs. (See Fig. 1. 3)

Fig. 1. 3 Dong people village

Usually, an average-size Kam village has 200-300 homes, although the smallest ones have only 10-20 and the largest ones have more than 1, 000 homes. Besides the drum tower, typical Kam villages also have stilt wooden houses , ancient and sacred trees, covered bridges ("wind-and-rain bridge"), wayside pavilions with wooden or stone benches, bullfighting arenas, wells surrounded by stone rims near sacred trees, traditionally communally owned fish-ponds, racks for drying grain and granaries, village entrances—to protect against intruders, where "blocking the way" ceremonies are held, and altars to Sa Sui which the main deity Kam believe in.

3.1 Drum Tower

Dong legend has it that the drum tower was built in accordance with the sample of the "king of cedar", and the overall outline of the Dong drum tower really looks like a cedar, embodying the concept of worshipping big trees. The drum tower, which is the tallest and most revered structure in the village and constructed without nails, is a testimony to the architectural skills of the Dong people.

There are two types of drum towers—pagoda and hall type, while the pagoda types is more popular. Inside the tower, four big pillars stand erect, and benches between the pillars encircle the central fire place. A long drum made of a hollowed tree trunk, hanging within the tower formerly served as a warning device against invaders. On the top rises a multi-eaved, octagonal finial pavilion. The eave angle is upturned, and the over all outline is changeable.

The drum towers provide venues for the villagers to discuss and settle important matters, hold important festivals or entertainments such as singing and the playing wind instruments. When there are important things to talk about, the drum is beaten by a respected villager to summon the villagers. The fire on the fire place burns almost all year round.

3.2 Roofed Bridge

As a Dong saying goes, where there is a river, there must be a bridge. In fact, whenever the Dongs live in compact communities, there will be various bridges. The one that best reflects the Dongs' superb bridge-building techniques is the stone-based woodroofed bridge named Fengyu (wind-and-rain bridge). It has several pavilions built onto it. The bridge is tile-roofed and flanked by wooden railings and stools, so it looks like a long corridor. Not a single nail or screw is used in the construction of the bridge; instead, the whole wooden structure is held together by mortise and tenon joints.

Similar to the drum tower, the roofed bridge is also the essence of Dong village. When a family member goes to travel or comes back, the roofed bridge will be the place for seeing off and welcoming. Most of the roofed bridges are set in the front or at the back of the village with benches on both sides, in summer, people will enjoy the cool here.

Notes:

1. Stilt Houses：干栏式民居是中国西南少数民族的居住建筑样式，以竹木为主要建筑材料，通常为两层，下层豢养动物和堆放杂物，上层住人。干栏式建筑适合那些雨水多、比较潮湿的地方，如我国广西中西部、云南东南部、贵州西南部和越南北部等。东南亚一带盛行的栅居，也属于干栏式建筑，以适应潮湿多雨的需要。干栏式房屋构筑办法是用竖立的木桩为基础，其上架设竹、木质小龙骨作为承托地板悬空的基座，基座上再立木柱和架横梁，构筑成框架状的墙围和屋盖，柱、梁之间用树皮茅草或竹条板块或用草泥填实。
2. Bamboo House：傣家竹楼属干栏式建筑，为上下两层的高脚楼房，整个竹楼的梁、柱、墙及附件都用竹子制成。竹楼底层一般不住人，是饲养家禽的地方。上层为人们的居住空间。竹楼室内的布局很简单，一般分为堂屋和卧室两部分。堂屋设在木梯进门的地方，堂屋内设有火塘，是烧饭做菜的地方。堂屋的外部设有阳台和走廊，是傣家妇女做针线活的地方。竹楼内空间较大，少遮挡物，通风条件好，非常适宜于潮湿多雨的气候条件。竹楼房顶呈"人"字型，易于排水，不会造成积水。竹楼下层高约七八尺，四无遮栏，牛马拴束于柱上。
3. Diaojiao Stilt House：吊脚楼为半干栏式建筑，渝东南及桂北、湘西、鄂西、黔东南地区的吊脚楼特别多，是苗族、壮族、布依族、侗族、水族、土家族等民族的传统民居。吊脚楼多依山靠河而建，呈虎坐形，以"左青龙，右白虎，前朱雀，后玄武"为最佳屋场，讲究朝向，或坐西向东，或坐东向西。吊脚楼的正屋多建在实地上，厢房除一边靠在实地和正房相连，其余三边皆悬空，靠柱子支撑。吊脚楼高悬地面既通风干燥，又能防毒蛇、野兽，楼板下放置杂物。
4. Kam Village：侗寨大多就其地形，依山傍水，整体布局大致分为平坝型、山麓型、山脊型和山谷型。一般侗寨由风雨桥、鼓楼、凉亭、寨门、吊脚楼、井亭、晾天架、萨殿组成。侗寨中央或寨边巍峨的建筑是侗寨的独特标志——鼓楼，大多数侗寨的鼓楼与戏台、芦笙坪三者一体，是村民集会、娱乐、议事的重要场所。侗寨民居吊脚楼很有特色，它既保留古越人干栏式建筑风格，又吸收中原建筑的特点。

Earthen Dwellings Tulou[1] in Fujian

Tulou buildings (土楼) are rural dwellings unique to the Hakka[2] in the mountainous areas in southeastern Fujian, China. A Tulou is usually a large, enclosed and fortified earth building, most commonly rectangular or circular in configuration, with very thick load-bearing rammed earth walls about three to five stories high. Most of the Tulou are found in a relatively small geographical area, straddling the boundary between the Yongding and Nanjing counties, Fujian province. They were mostly built between the 12th and the first half of 20th centuries. A total of 46 Tulou sites were inscribed as World Heritage Site. (See Fig. 1. 4)

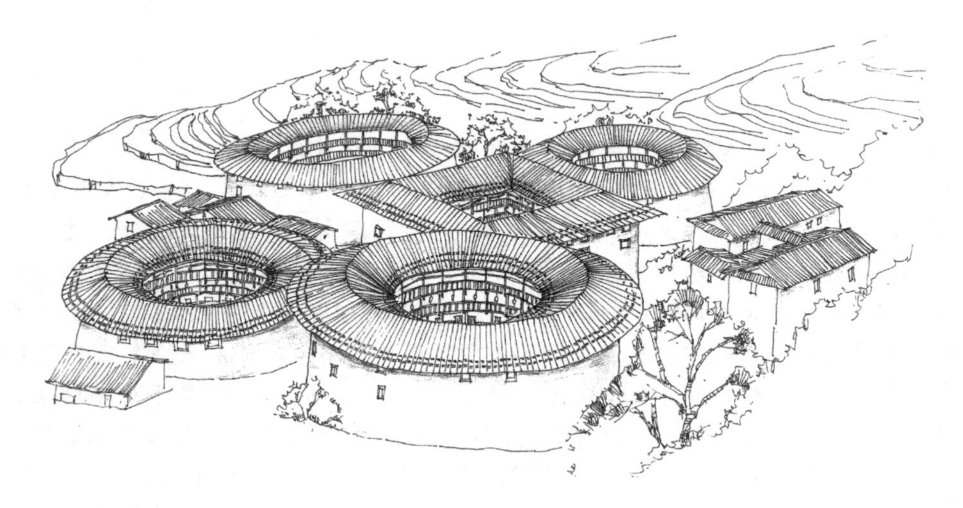

Fig. 1. 4 Bird view of a Hakka village

1. The Description of Tulou Buildings

The Tulou buildings have the appearance of forts, with large overhanging eaves, small windows set high into the massive outer walls on the third and fourth floors. The form and scale and the building material of a Tulou reinforce the sense of complete contextual synthesis in surrounding landscape. A Tulou is like a village since each Tulou contains many of the functions normally necessary for a village in many respects.

1.1 Layout of a Tulou

The layout of Tulou followed Chinese traditional dwelling principles: "closed outside, open inside" with an enclosure wall with living quarters around the peripheral and a common courtyard at the centre. The ground floor plan of a Tulou building can be in circle, semicircle, oval, square, rectangle, even irregular pentagon form. A small building at the center served as ancestral hall[3] for ancestry worshipping, festivals, meetings, weddings, funerals and other ceremonial functions.

The courtyard is used for drying clothes and rice, communal activities, and for children's play.

The Tulou is dense and compact, with up to 250 small uniform rooms, constructed in two-or three-story wooden structures. They are placed around the buildings periphery and arranged symmetrically around the Tulou's central axis. Through identical galleries all of the rooms look out onto the open courtyard, as is normally the case in most Chinese traditional houses. (See Fig. 1. 5)

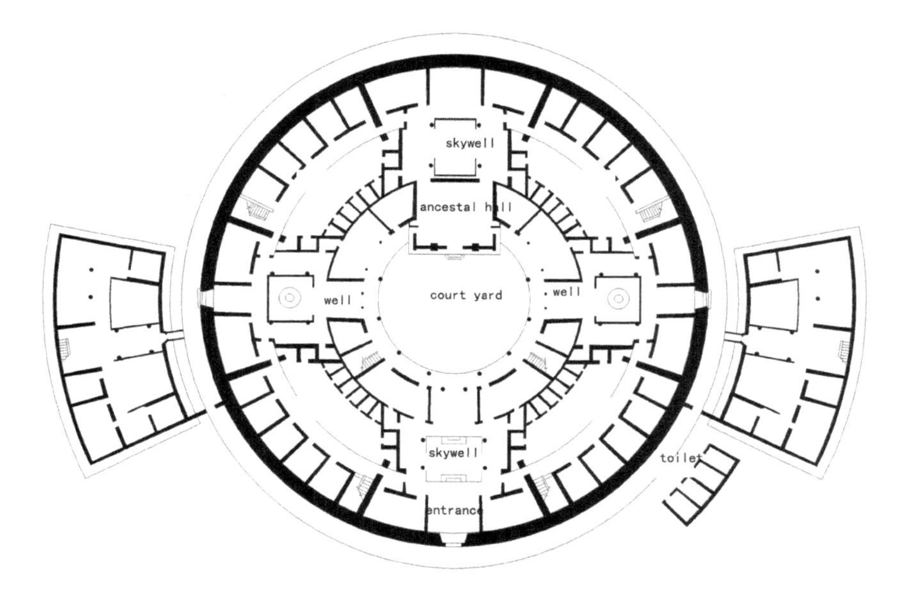

Fig. 1. 5 The layout of a round Tulou

1. 2 The Allotment of the Rooms of Tulou

The rooms of a Tulou are shared among its inhabitants in such a way that a single family unit uses two or three rooms on each floor, in a vertical segment of the building. One room on the ground floor is the kitchen and another is used as dining and living room. The preparation of meat and vegetables is done in the courtyard immediately in front of the kitchen where the oven and firewood is to be found. The stove in the kitchen is vented to the outside through small openings in the outer wall. Steep stairs lead to the verandas that ring the upper levels. The sleeping quarters are on the first and second floor, the food, clothes and valuables are stored on the top floor, but in some round Tulou such as Zhenchenglou(振成楼), the bedrooms are found on the upper floors.

2. The Construction Features of Tulou

Tulou is probably the largest, and most defensively advanced traditional village residences known in China. The Tulou building was fortification like, with outer walls of stamped clay up to 1. 5 meters thick and 18 meters high, an iron-clad portal, weapon slits under the eaves of the large overhanging roof, and a connecting gallery that enabled rapid movement of people and weaponry.

2. 1 The Features of Wall Construction

The foundation of Tulou building was built with paved stones on top of compacted earth ground, in two to three tiers. There is a circular drain around the top tier foundation to prevent rainwater from damaging the outer wall. In most cases, the weight bearing outer wall of Tulou consists of two sections, the lower section is built from cut stone blocks or river cobbles held together with a lime, sand and clay mixture to a height of about one or two meters, depending upon the regional flood water level. The compacted earth wall stacked on top of the stone section. The construction of earth wall from compacted earth mixed with sticky rice and reinforced with horizontal bamboo sticks was described first in Song dynasty building standard *The Yingzao Fashi*(营造法式)[4].

The walls were built inclined toward the centre, such that the natural force of gravity pushes the wall together. The thickness of the Tulou wall decreases with height as specified in *The Yingzao Fashi*. The bottom two storeys of Tulou are solid with no window nor gun hole, windows are open only from the third to fifth storeys, because rooms at the bottom storey served as family storage rooms and the upper storeys were living quarters.

The rooftops were covered with baked clay tiles, arranged in radial form; insertion technique was used at regular intervals to compensate for larger circumference at the outside. Majority of roof tiles were laid from top to bottom, the gap caused by radial layout was compensated by small sections of tiles laid in λ shape inserts. This technique allowed the tiles to be laid in radial form without visible gaps, and without the use of small tiles at top, larger tiles at bottom. The eaves usually extend about two meters, protecting the earth wall from damage by rainwater pouring from the eaves.

Circular corridors from 2nd to uppermost level were made of wood boards laid on horizontal wooden beams with one end inserted into the earth wall. The corridors are protected with a circle of wooden railings. Stairwells are distributed evenly around the corridors, four sets of stairwells being the usual number. Each stairwell leads from ground floor to the highest floor.

Public water wells in groups of two or three are usually located at the centre court; more luxurious Tulous have in-house water well for each household in the kitchen. Most Tulous have built water pipes to offer protection to the upper wooden floors against fire.

2.2 The Advantages of Round Tulou

Among the over 400 existing Tulou buildings, the round Tulou is exceeded by the square one in numbers. The round Tulou, range widely in size, some exceed 70 meters in diameter and there are several less than 20 meters in diameter, the largest is *Zaitianlou*(在田楼) in Zhaoan(兆安) with 2.4 m thick walls and a diameter of 91 meters, while the smallest round Tulou is with a diameter of 17 meters.

The circular form has several advantages: first, a circular form is easier to build because of the identical cross-section throughout and without the need for complex roof and wall corner construction; Second, the circular form allows more economic use of material. Wood is more expensive than clay to obtain, transport and work. For each bay[5] the outer rim of clay is longer than that of wood which faces the courtyard. Further, a given amount of material gives a 41% larger courtyard and approximately a 13% larger building area in the circular than in the rectangular Tulou.

Since the portal is the most vulnerable point of attack, it is therefore protected by an ingenious fire-dowsing system with an internal gutter above which is connected to a water tank situated on the second floor. The animal pens, a water well and food stockpiles in the courtyard provided for a lengthy conflict.

3. Social features of Tulou building

Unlike other housing structures around the world with architecture illustrating social hierarchy, Tulou exhibits its unique characteristic as a model of community housing for equals. All rooms were built the same size with the same grade of material, same exterior decoration, same style of windows and doors, and there was no "penthouse" for "higher echelons"; a small family owned a vertical set from ground floor to "penthouse" floor, while a larger family would own two or three vertical sets.

Tulous were usually occupied by one large family clan of several generations; some larger Tulou had more than one family clan. Besides the building itself, many facilities such as water wells, ceremonial hall, bathrooms, wash rooms, and weaponry were shared property. Even the surrounding land and farmland, fruit trees etc. were shared. The residents of Tulou farmed communally. Each small family has its own private property, and every family branch enjoys its privacy behind closed doors.

In the old days, the allotment of housing was based on family male branch; each son was counted as one branch. Public duties such as organization of festivals, cleaning of public areas, opening and closing of the main gate, etc., was also assigned to a family branch on a rotational basis.

All branches of a family clan shared a single roof, symbolizing unity and protection under a clan; all the family houses face the central ancestral hall, symbolizing worship of ancestry and solidarity of the clan. When a clan grew, the housing expanded in radial form by adding another outer concentric ring, or by building another Tulou close by, in a cluster. Thus, a clan stayed together. We take Chengqilou(承启楼)[6] as example.

Chengqilou, "called the king of Tulou", built in 1709, is a massive round Tulou with four concentric rings surrounding an ancestral hall at the center. The outer ring is 62.6 meters in diameter and four storeys tall, with 288 rooms, saying 72 rooms on each level, the circular corridor on 2nd to 4th floor, with four sets of staircases at cardinal points connecting ground to top floors. A large outward extending roof covers the main ring. The ground floor rooms are kitchens for family branches, the second level rooms are grain storage rooms, and the 3rd and 4th floor rooms are living quarters and bedrooms. The second ring is two storeys high with 80 rooms, saying 40 rooms on each level, the third ring is one storey with 32 rooms serving as community library; The 4th ring is in the very center of Tulou as the ancestral hall with circular covered corridor surrounding . There are 370 rooms in all, if a person stayed for one night in each room, it would take him more than a year to go through all the rooms.

Chengqilou has two main gates and two side gates. 15th generation Jiang clan with 57 families and 300 people live here. It is said that at its heyday, there were more than 80 family branches living in Chengqilou.

Notes:

1. Tulou：土楼，是分布在中国东南部的福建、江西、广东三省的客家地区，以生土为主要建筑材料、生土与木结构相结合，并不同程度地使用石材的大型居民建筑。其中，分布最广、数量最多、品类最丰富、保存最完好的是福建土楼，主要分布在福建省龙岩、永定、漳州南靖县和华安县，现存有3000余座。土楼的兴建高潮是在中国动乱与客家族群由中原向南方迁移之际，主要时期包含唐末黄巢之乱、南宋政权南移与明末清初。
2. Hakka：客家，是中国南方广东、江西、福建、广西、台湾等本地族群的主要组成部分，是世界上分布范围广阔、影响深远的汉族民系。客家先民始于秦朝征岭南融百越时期，历经西晋永嘉之乱、东晋五胡乱华，唐末黄巢之乱，宋室南渡。彼时，中原汉族大举南迁，陆续迁入南方各省，在与外界相对隔绝的状态下，经过千年演化，最迟在南宋已逐渐形成一支具有独特方言、风俗习惯及文化形态的汉族民系。客家以粤赣闽为基地，大量外迁到中国各地，含港澳台地区以及南洋乃至世界各地。客家文化既继承了古代正统汉族文化，又融合了南方土著文化，有古汉文化活化石之誉，耕读传家是客家文化的特点。
3. ancestral hall：宗祠，又称祠堂，主要用于祭祀祖先，也可以作为家族的社交场所，各房子孙办理婚、丧、寿、喜的场所。族亲们商议族内的重要事务也常在祠堂进行。有些祠堂附设学校，族人子弟就在这里上学。祠堂建筑一般都比民宅规模大、质量好，精致雕饰、用材上等。祠堂多数都有堂号，制成金字匾高挂于正厅。祠堂一姓一祠，在旧时代，外姓、族内妇女或未成年儿童不许擅自入内。
4. 《营造法式》：《营造法式》是北宋时期李诫在两浙工匠喻皓的《木经》基础上编成的，是北宋官方颁布的一部建筑设计、施工的规范书，是我国古代最完整的建筑技术书籍，标志着中国古代建筑已经发展到了较高阶段。全书34卷，357篇，3555条，宋崇宁二年(1103年)出版。《营造法式》是当时建筑设计与施工经验的集合与总结，对后世产生深远影响。
5. bay：间。在中国传统建筑中，四根木柱围成的空间称为“间”。建筑的迎面间数称为“开间”，或称“面阔”。建筑的纵深间数称“进深”。中国古代以奇数为吉祥数字，所以平面组合中绝大多数的开间为单数；而且开间越多，等级越高。北京故宫太和殿，北京太庙大殿开间为十一间。
6. 承启楼：承启楼位于福建省龙岩市永定区高头乡高北村，据传从明崇祯年间破土奠基，至清康熙年间竣工，历时3代，约半个世纪。承启楼规模巨大，全楼为三圈一中心。外圈高

4层，高16.4米，每层设72个房间；第二圈高2层，每层设40个房间；第三圈为单层，设32个房间，中心为祖堂。全楼共有400个房间，3个大门，2口水井。1986年，中国邮电部发行了一套“中国民居”邮票，其中面值1元的福建民居就是承启楼。

Siheyuan Houses in Beijing[1]

A Siheyuan（四合院）is a historical type of residence that was commonly found throughout China, famous in Beijing. Siheyuan is sometimes referred to as "Chinese quadrangle", which literally means a courtyard surrounded by buildings on four sides. Due to its special layout, it is compared to a box with a garden in the center. Siheyuan dates back as early as the Western Zhou period with a history over 2,000 years. They exhibit outstanding and fundamental characteristics of Chinese traditional architecture throughout Chinese history. The Siheyuan composition was the basic pattern used for all kinds of architecture such as residences, palaces, temples, monasteries, family businesses and government offices etc. (See Fig. 1. 6)

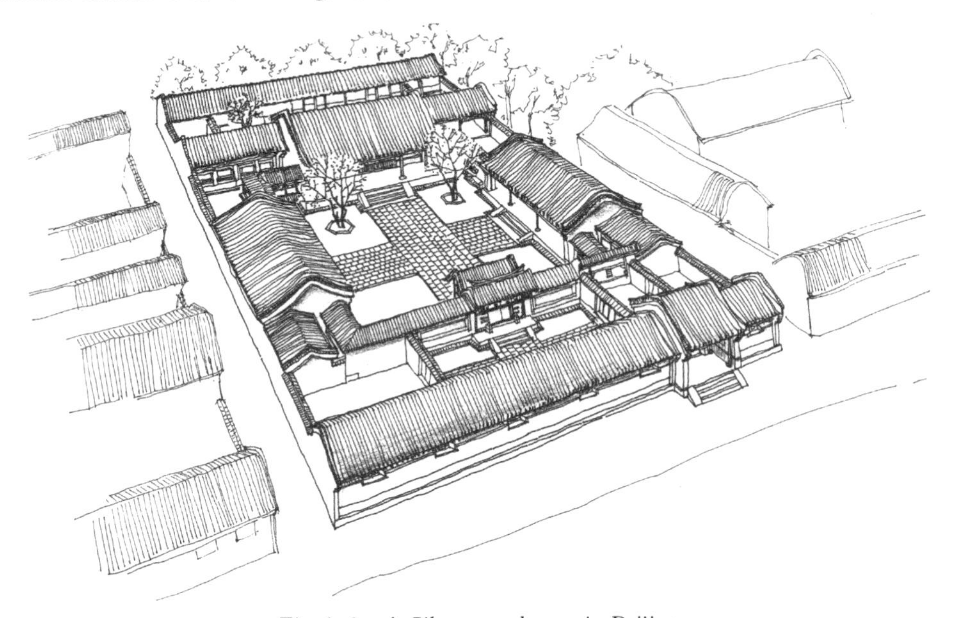

Fig. 1. 6　A Siheyuang house in Beijing

1. The Layout of a Beijing Siheyuan

The fundamental characteristics of a Beijing Siheyuan are being enclosed, symmetrical in middle axis and clear distinction between primary and secondary, outside and inside. The four buildings of a Siheyuan are normally positioned along the north-south and east-west axis respectively. The building positioned to the north and facing the south is considered the main house (正房). The buildings adjoining the main house and facing east and west are called side houses (厢房). The northern, eastern and western buildings are connected by beautifully decorated veranda (廊子). These pathways serve as shelters from the sunshine during the hot daytime and provide a cool place to appreciate the view of the courtyard at night. The building that faces north is known as the opposite house (倒座房). Behind the northern building, there would often be a separate backside building (后罩房), the only place where two-story buildings are allowed to be constructed. (See Fig. 1.7)

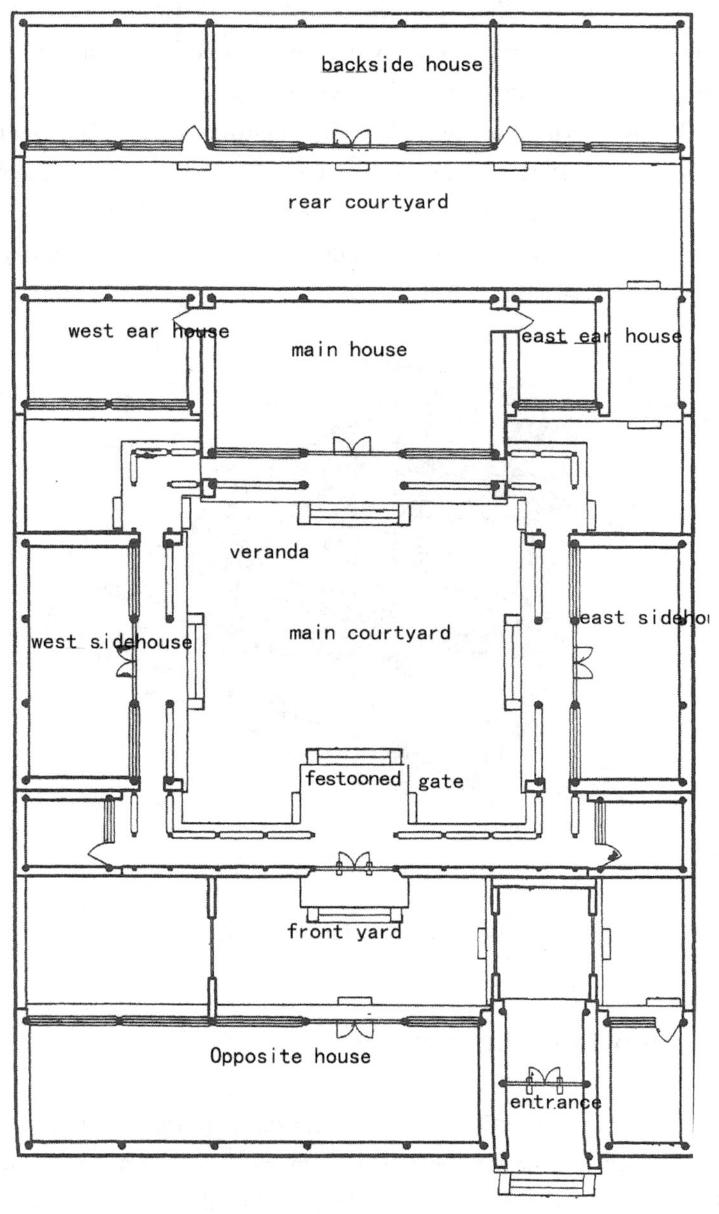

Fig. 1.7 The layout of a Bejing Siheyuan

The entrance gate, usually painted vermilion and with copper door knockers on it, is usually at the southeastern corner. Normally, there is a screen wall (影壁)[2] inside the gate for privacy and superstition. It is said that the screen wall can protect the house from evil spirits. A pair of stone lions are often placed outside the gate on both sides.

Some largesize Siheyuan compounds would have two or more layers of courtyards and even private gardens attached, which is a sign of wealth and status in ancient times.

2. The Significance of Beijing Siheyuan

A spacious Beijing Siheyuan would be occupied by a single, usually large and extended family, signifying wealth and prosperity. Since there is only one gate leading to a hutong[3] (alley), when the gate is closed the courtyard loses touch with the outside world, family members can fully enjoy tranquility and share the happiness of a peaceful family union.

2.1 The Comfortable Design

A Siheyuan is a practically sound, engineered structure, it offers space, comfort, quiet and privacy. From antiquity, the wall of northwestern corner of a Siheyuan is usually higher than the other walls to protect the inside buildings from the harsh winds blowing across northern China in the winter. The building eaves curve downward, so that rainwater will flow along the curve rather than dropping straight down. The rooftop is ridged to provide shade in the summer while retaining warmth in the winter.

With plants, rocks, and flowers, the courtyard is also a garden, and acts like an open-air living room. The veranda divides the courtyard into several big and small spaces that are not very distant from each other. Family members talked with each other here, creating a cordial atmosphere.

Grey is the dominant color for a Siheyuan, the bricks, walls, roofs and the grounds are all in grey shade while the pillars, doors and windows are painted red and green, all of which bring harmony and simple elegance to the courtyard house. The paths in the yard are paved by the bricks with the four corners of the yard left for planting trees. Green trees and red flowers really make good scenery. Some big courtyard houses even have gardens, kiosks, platforms or pavilions.

2.2 The Traditional Hierarchy Design

The layout of Siheyuan was designed according to the traditional concepts of the five elements[4] that were believed to compose the universe, and the eight diagrams of divination[5]. The gate was made at the southeast corner which was the "wind" corner, and the main house was built on the north side which was believed to belong to "water", an element to prevent fire.

The layout of a courtyard also represents the traditional Chinese morality and

Confucian ethics. In Beijing, four buildings enclosed the courtyard receive different amounts of sunlight, and the northern main building receives the most, thus serving as the living room and bedroom of the owner or head of the family. The eastern and western side buildings receive less, and serve as the rooms for children or less important members of the family. The southern building receives the least sunlight, and usually functions as a reception room and the servants' dwelling, or where the family would gather to relax, or take them as dining room or study for children. The backside building is only for unmarried girls and female servants because unmarried women were not allowed the direct exposure to the strange, so they occupied the most secluded building in the Siheyuan.

A more detailed and further stratified Confucian order was followed in Siheyuan house, as we know, the main house in the north was assigned to the eldest member of the family, i. e. the head of the family, usually grandparents. If the main house had enough rooms, a central room would serve as a shrine for ancestral worship. When the head of the household had concubines, the wife would reside in the room to the eastern end of the main house, while the concubines would reside in the room to the western end of the main house. The eldest son of the family and his wife would reside in the eastern side house, while the younger son and his wife would reside in the western side house. If a grandson was fully grown, he would reside in the opposite house in the south side.

3. The Ornaments of Siheyuan Houses

The ornaments of Siheyuan houses are of distinctive features, with some brick carvings or wood carvings dotted on the prominent places such as the screen walls and the lateral walls of the door. The porch, festooned gate (垂花门)[6], drum-shaped bearing stone, as well as doors and windows with wooden partition are also the key points for ornamentation. All these carving decorations and colored drawings are the embodiment of folk customs and traditional culture, which reflect people's pursuit for happiness, goodliness, wealth and auspiciousness.

3.1 The Festooned Gate

A festooned gate, also called hanging lotus gate, stands on the central axis. It is small, with only one bay and four columns. The most special feature of this gate is two short columns with the lotus bud end hanging from the end of each of the overhanging beams above the columns on both sides. The gate is the most exquisite structure in a courtyard.

Most festooned gates have overhanging roofs with round ridges, a more complex form of roof with a overhanging roofridge in front and a round ridge in rear. Two gate leaves fixed onto the front eave columns can be open and closed. Four gate leaves fixed on the back columns is a "screen gate" which remains closed except on important occasions, such as wedding day and funeral day. The gate leaves on the front are painted red while the screen gate is painted green. (See Fig. 1.8)

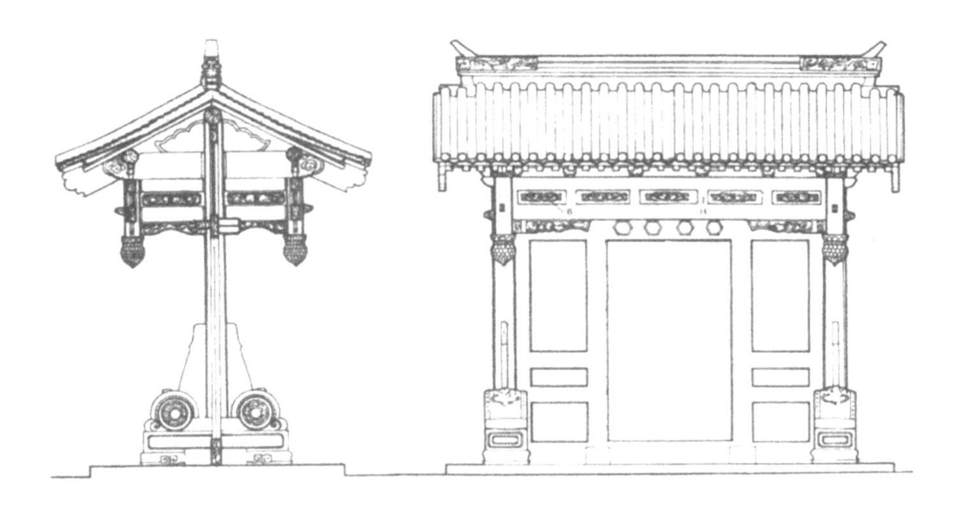

Fig. 1.8 Festooned gate

3.2 The Ornamental Carvings

Carving are highly valued for Beijing Siheyuan house, for there are plenty of exquisite images and designs on components made of wood, brick and stone, showing a superb artistic standard. Wood carvings are mainly found on the hanging lotus columns and flower board of a festooned gate. Continuous designs such as pines, bamboos, flowers and grass are found on suspended lintels under the eaves of side houses.

Brick carvings are often found on roof ridges and the center and top of screen walls, where the common images are pines, bamboos, flowers and birds. The brick carvings at the center of a screen wall resemble a rhombus, for example, the design with the bat the Chinese character "longevity" together means to have both good fortune (福) and longevity (寿); the pattern of a vase with China roses (月季) means being safe and sound for the whole year.

The drum stones on both sides of the front door are the focus of stone carving, there are often lions in various shapes and poses on the front, and there is a single small lion on top of a minority of drum stones. Besides lions, there can be kylin(麒麟), pine, crane flower and grass on the drum head.

Notes:

1. Siheyuan：四合院，也称四合房，是中国汉族的一种传统合院式建筑，通常为大家庭所居住，其格局为一个院子四面建有房屋，从四面将庭院合围在中间。四合院提供了对外界比较隐密的庭院空间，其建筑和格局体现了中国传统的尊卑等级思想以及阴阳五行学说。呈"口"字形的称为一进院落；"日"字形的称为二进院落；"目"字形的称为三进院落。在大宅院中，第一进为门屋，第二进是厅堂，第三进或后进为私室或闺房，是妇女或眷属的活动空间，一般人不得随意进入，因此古人有诗云："庭院深深深几许"。四合院在中国各地有多种类型，其中以北京四合院为典型。
2. Screen wall：影壁，也称照壁，古称萧墙，是汉族传统建筑中用于遮挡视线的墙壁，即使大门敞开，外人也看不到宅内。旧时人们认为自己的住宅中，不断有鬼来访。如果孤魂野鬼溜进宅子，就要给自己带来灾祸。影壁还可以烘托气氛，增加住宅气势。四合院常见的影壁有三种：第一种位于大门内侧，呈一字形，叫做一字影壁。第二种是位于大门外面的影壁，这种影壁座落在胡同对面，正对宅门，主要用于遮挡对面房屋和不甚整齐的房角檐头。还有一种影壁，位于大门的东西两侧，平面呈八字形，称做"反八字影壁"。四合院宅门的影壁，绝大部分为砖料砌成。
3. Hutong：胡同，也叫"里弄""巷"，是指城镇或乡村里主要街道之间的、比较小的街道，一直通向居民区的内部，是沟通当地交通不可或缺的一部分。北京胡同最早起源于元代，最多时有6000多条。元朝的《析津志辑佚》中记载："三百八十四火巷，二十九胡同。"也就是说，共有街巷胡同413条，其中29条直接称为胡同，384条称为火巷，是广义上的胡同。胡同与胡同之间的距离大致相同。南北走向的一般为街，相对较宽，因过去以走马车为主，所以也叫马路。东西走向的一般为胡同，相对较窄，宽度一般超不过九米，以走人为主。胡同两边都是四合院。
4. five elements：五行，是中国古代的一种系统观，广泛地用于中医学、堪舆、命理、相术和占卜等方面。五行是指阴阳演变过程的五种基本动态。古人把宇宙万物划分为五种性质的事物。《尚书·洪范》记载："五行：一曰水，二曰火，三曰木，四曰金，五曰土。水曰润下，火曰炎上，木曰曲直（弯曲，舒张），金曰从革（成分致密，善分割），土爱稼穑（意指播种收获）。润下作咸，炎上作苦，曲直作酸，从革作辛，稼穑作甘。"古人将宇宙万物进行了分类，而且对每类的性质与特征都做了界定，用来说明世界万物的形成及其相互关系。
5. eight diagrams of divination：八卦，是中国文化的基本哲学概念，其形成源于河图和洛书。所谓八卦就是八个不同的卦相，表示事物自身变化的阴阳系统，用"—"代表阳，用"- -"代表阴，用这两种符号，按照大自然的阴阳变化平行组合，组成八种不同形式。八卦与阴阳五行一样，是用来推演世界空间时间各类事物关系的工具。每一卦形代表一定的事物。乾代表天，坤代表地，巽(xùn)代表风，震代表雷，坎代表水，离代表火，艮(gèn)代表山，兑代表泽。八卦互相搭配又变成六十四卦，用来象征各种自然现象和人事现象。
6. festooned gate：垂花门，是汉族传统民居建筑院落内部的门，是四合院中一道很讲究的门，它是内宅与外宅（前院）的分界线和唯一通道。因其檐柱不落地，垂吊在屋檐下，称为垂柱，其下有一垂珠，通常彩绘为花瓣的形式，故被称为垂花门。旧时人们常说的“大门不出，二门不迈”中的“二门”，即指此垂花门。垂花门与院中十字甬路、正房一样，同在一条南北走向的主轴线上。进内宅后的抄手游廊、十字甬路均以垂花门为中轴左右分开。垂花门一般建在三层或五层的青石台阶上，两侧为磨砖对缝精致的砖墙。垂花门从外边看像一座华丽的砖木结构门楼，从院内看则似一座类似亭榭的方形小屋，顶部多为卷棚式，门外部分的顶部为清水脊，两顶勾联搭在一起的交汇处形成天沟。垂花门是装饰性极强的建筑，向外一侧的梁头常雕成云头形状，称为麻叶梁头。在麻叶梁头之下，有一对倒悬的短柱，柱头向下，头部雕饰出莲瓣、串珠、花萼云或石榴头等形状，酷似一对含苞待放的花蕾，称为垂莲柱。

Terraced House[1] in Great Britain

A terraced house is a type of town house in Great Britain, originally referred to the city residence of someone whose main or largest residence was a country house. Historically, a town house was the city residence of a noble or wealthy family, who would own one or more country houses in which they lived for much of the year. From the 18th century, landowners and their servants would move to a townhouse during the social season when major balls took place. It was one of the successes of Georgian architecture[2] to persuade the rich to buy terraced houses, even aristocrats whose country houses had grounds of hundreds or thousands of acres often lived in terraced houses in town.

1. The Development of Terraced House in Great Britain

Terraced houses became introduced to London from Italy in the 1630s. Terraces first became popular in England when London began rebuilding after the Great Fire[3] in 1666. The terrace was designed to hold family and servants together in one place, as opposed to separate servant quarters, and came to be regarded as a “higher form of life”. They became a trademark of Georgian architecture in Britain, including Grosvenor Square, London, in 1727 and Queen's Square, Bath, in 1729. The parlor became the largest room in the house where the aristocracy would entertain and impress their guests. (See Fig. 1. 9)

In 17th and 18th-century the building materials were supplied locally, using stone where possible, otherwise firing brick from clay. The house was divided into small rooms partly for structural reasons, and partly because it was more economical to supply timber in shorter lengths. The London Building Act of 1774 made it a legal requirement for all terraced houses there to have a minimum wall thickness and specified other basic building requirements.

Fig. 1. 9 A terraced house neighboring ①

Terraced houses were still considered desirable architecture at the start of the 19th century. The architect John Nash[4] included terraced houses when designing Regent's Park in 1811, as it would allow individual tenants to feel as if they owned their own mansion.

The terraced house reached mass popularity in the mid-19th century as a result of increased migration to urban areas. Between 1841 and 1851, towns in England grew over 25% in size, at which point over half the population lived in urban areas; this increased further to nearly 80% by 1911. Terraced houses became an economical solution to fit large numbers of people into a relatively constricted area. Many terraced houses were built in the South Wales Valleys in the mid to late 19th century owing to the large-scale expansion of coal mining in the area.

Nationwide legislation for terraced housing began to be introduced during the Victorian era[5]. The 1858 Local Government Act stated that a street containing terraced houses had to be at least 36 feet (11m) wide with houses having a minimum open area at the rear of 150 square feet ($14m^2$), and specified the distance between properties should not be less than the height of each. Other building codes inherited from various local councils defined a minimum set of requirements for drainage, lighting and ventilation. The various legislations led to a uniform design of terraced houses that was replicated in streets throughout the country.

Though many working-class people lived in terraces, they were also popular with

① D. R. Bally, Clip ground. com.

middle classes in some areas, particularly the North of England. Houses were generally allowed to be used for commercial purposes, with many front rooms being converted to a shop front and giving rise to the corner shop. By the 1890s, larger terraces designed for lower-middle-class families were being built. These houses contained eight or nine rooms each and included upstairs bathrooms and indoor toilets.

2. The Architectural Features of Terraced Houses

The Terraced houses for all social classes remained resolutely tall and narrow, each dwelling occupying the whole height of the building. This contrasted with well-off continental dwellings, which had already begun to be formed of wide apartments occupying only one or two floors of a building. A curving crescent, often looking out at gardens or a park, was popular for terraces where space allowed. In early and central schemes of development, plots were sold and built on individually, though there was often an attempt to enforce some uniformity, but as development reached further out schemes were increasingly built as a uniform scheme and then sold.

The terraced houses are generally two-to three-storey structures that share a wall with a neighboring unit. As opposed to an apartment building, terraced houses do not have neighboring units above or below them. They are usually divided into smaller groupings of homes. The first and last of these houses is called an end terrace, and is often a different layout from the houses in the middle, sometimes called mid-terrace. (See Fig. 1. 10)

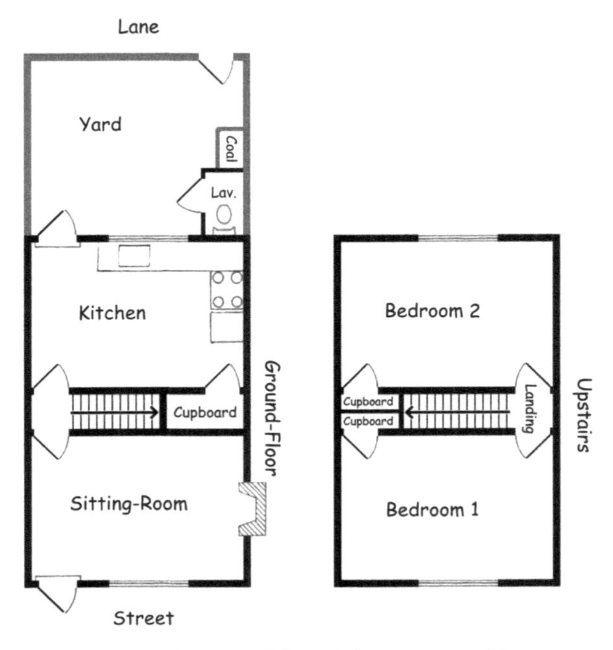

Fig. 1. 10 Layout of 2-up 2-down terraced house

A house may be several storeys high, one or two rooms deep, and optionally contain a basement and attic. In this configuration, a terraced house may be known as a "two-up two-down", having a ground and first floor with two rooms on each. Most terraced houses have a duo pitch[6] gable roof.

For a typical two-up two-down house, the front room has historically been the parlor, or reception room, where guests would be entertained, while the rear would act as a living room and private area. Many terraced houses are extended by a back projection, which may or may not be the same height as the main building.

A terraced house has windows at both the front and the back of the house; if a house connects directly to a property at the rear, it is a back to back house. The 19th century terraced houses, especially those designed for working-class families, did not typically have a bathroom or toilet with a modern drainage system; instead these would have a privy using ash to deodorize human waste.

The terraced houses, which typically opened straight onto the street, are often with a few steps up to the door. There was often an open space, protected by iron railings, dropping down to the basement level, with a discreet entrance down steps off the street for servants and deliveries; this is known as the "area". This meant that the ground floor front was now removed and protected from the street and encouraged the main reception rooms to move there from the floor above. Where, as often, a new street or set of streets was developed, the road and pavements were raised up, and the gardens or yards behind the houses at a lower level, usually representing the original one.

3. The Royal Crescent[7] in Bath

The Royal Crescent is a row of 30 terraced houses laid out in a sweeping crescent in the city of Bath, England. Designed by the architect John Wood the Younger and built between 1767 and 1774, it is among the greatest examples of Georgian architecture to be found in the United Kingdom and is a Grade I listed building[8]. Although some changes have been made to the various interiors over the years, the Georgian stone facade remains much as it was when it was first built.

The Royal Crescent now includes a hotel and a Georgian house museum, while some of the houses have been converted into flats and offices.

John Wood designed the great curved facade with Ionic columns on a rusticated ground floor. The 114 columns are 76cm in diameter reaching 14.3m, each with an entablature 1.5m deep. The central house, now the Royal Crescent Hotel boasts two sets of coupled columns. (See Fig. 1.11)

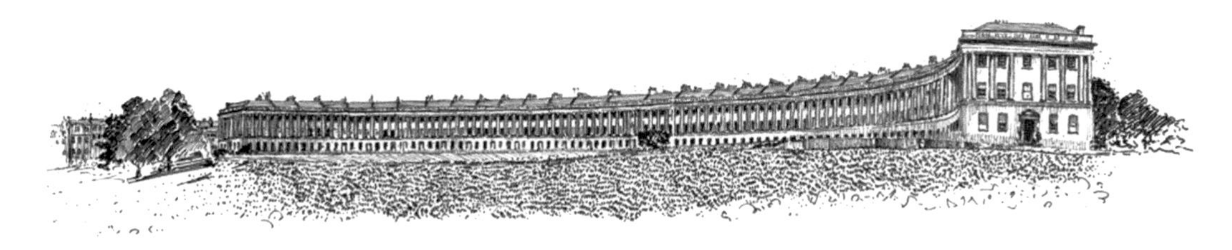

Fig. 1. 11 Bath royal crescent ①

The original purchaser bought a length of the facade, and then employed their own architect to build a house behind the facade to their own specifications; hence what can appear to be two houses is occasionally just one. This system of town planning is betrayed at the rear and can be seen from the road behind the Crescent: while the front is uniform and symmetrical, the rear is a mixture of differing roof heights, juxtapositions and fenestration. This architecture, described as "*Queen Anne fronts and Mary-Anne backs*", occurs repeatedly in Bath.

In front of the Royal Crescent is a Ha-ha[9], a ditch on which the inner side is vertical and faced with stone, with the outer face sloped and turfed, making an effective but invisible partition between the lower and upper lawns. The Ha-ha is designed so as not to interrupt the view from Royal Victoria Park, and to be invisible until seen from close by. It is not known whether it was contemporary with the building of the Royal Crescent, however it is known that when it was first created it was deeper than it is at present.

During the 20th century many of the houses which had formerly been the residences of single families with maids or other staff were divided into flats and offices. However, the tradition of distinguished gentle folk retiring to the crescent continued.

During the Bath Blitz of World War II, some bomb damage occurred. After World War II, during a period of redevelopment which is described as the *Sack of Bath*, the City Council considered plans that would have seen the Crescent transformed into Council offices. These were unsuccessful.

In the 1970s the resident of No. 22, Miss Wellesley-Colley, painted her front door yellow instead of the traditional white. Bath City Council issued a notice insisting it should be repainted. A court case ensued which resulted in the Secretary of State for the Environment declaring that the door could remain yellow. Other proposals for alteration and development including floodlighting and a swimming pool have been defeated.

For many years the residents had to endure numerous tour buses passing their homes every few minutes. However, the road has now been closed to coaches and buses and is a more peaceful environment for tourists and locals alike.

---

① Summerson, J. Architecture in Britain, 1530—1830. New Haven: Yale University Press, 1993.

Notes：

1. Terranced House：联排住宅，在历史上最早建于古罗马，是欧美国家的一种城市住宅形式，曾经在17世纪伦敦大火之后担负起伦敦重建的重任。英国的联排一般是3层左右，建筑面积一般是每户150至250平方米。每户独门独院，拥有一个“有天有地”的院落，大都为50平方米左右，有的达到近100平方米。每户设有1至2个车位，还有地下室。
2. Georgian architecture：乔治王时代建筑，是英语国家对1720—1840年间的古典建筑风格，以乔治命名的英国君主（乔治一世至乔治四世）而定义的称谓。乔治王时代风格继承了克里斯托弗·莱恩（Christopher Wren）爵士等几位设计师的英国巴洛克风格。乔治王时代建筑的特点是比例和平衡感，运用简单的数学比例来决定建筑结构，例如窗户的高度与宽度相关，房间的形状是一个double cube。沿着一条街道的房屋正面的整齐规则性是乔治王时代城镇规划追求的目标。乔治王时代设计通常遵守古典建筑秩序，采用源自于古罗马和古希腊的装饰词汇。最常采用的建筑材料是砖和石头，一般采用红色、棕褐色、白色。乔治王朝建筑风格的主要特点包括：①帕拉迪奥的古典比例；②门廊要素，门头的扇形窗，六嵌板的标准门形式，屋檐上有齿饰；③窗户的六对六标准分割，简化了窗棂线脚的处理，推拉窗的普及；④家装传统的三段式墙面的风格方式：墙裙、墙面、檐壁；⑤壁炉成为装饰重点。
3. Great Fire：伦敦大火，1666年9月2日星期日凌晨1点左右，伦敦普丁巷（Pudding Lane）有一间面包铺失火。一阵大风将火焰很快吹过几条全是木屋的狭窄街道，然后又进入了泰晤士河北岸的一些仓库里。大火延烧了整个城市，连续烧了4天，包括87间教堂、44家公司以及13000间民房尽被焚毁，欧洲最大城市——伦敦大约六分之一的建筑被烧毁。伦敦重建工作由克里斯托弗·雷恩主导，54间教堂中有51间是他重新设计的，包括著名的圣保罗大教堂。
4. John Nash：约翰·纳什（1752—1835），生活在国王乔治四世时代，为英国著名建筑师，曾设计伦敦的Regent's Park（瑞金公园）和Regent Street（瑞金街）。他被国王任命修建伦敦的著名建筑，其中包括如今白金汉宫的一部分。
5. Victorian era：维多利亚时代，1837—1901年，维多利亚女王（Alexandrina Victoria）的统治时期，被认为是英国工业革命和大英帝国的峰端。领土达到了3600万平方公里，英国经济文化也达到全盛时期，英国的经济总量占到了全球的70%。维多利亚时代的文艺运动流派包括古典主义、新古典主义、浪漫主义、印象派艺术及后印象派等。
6. duo pitch：垛间距。砌筑普通墙时墙垛与墙垛的距离一般是4至5米。
7. Royal Crescent：皇家新月楼是英国乔治王朝时代的代表性建筑，由建筑师小约翰·伍德（John Wood）设计，建于1767年至1774年间，堪称18世纪最宏大的城市建筑之一，并且是巴斯最具代表性的帕拉迪奥式建筑。
8. listed building：登录建筑物，是英国有特殊意义和价值的历史建筑。登录建筑不经过当地规划机构的特别批准不得拆除和改建。
9. Ha-ha：英国园林设计的元素之一，是为保护景观而设计的一道矮墙。Ha-ha的意思是走近之后才看见的垂直墙的产生的瞬间感叹。Ha-ha 的设计包括陡然下降的草坡和前面的石墙,以防止牲畜和人进入花园,但又不阻挡观赏花园的视线。

Insights: Fengshui

The term Fengshui as "wind-water" in English is a cultural shorthand taken from the passage of the now-lost Classic of Burial(葬书) recorded in Guo Pu's(郭璞) commentary: Fengshui is one of the Five Arts of Chinese Metaphysics, classified as physiognomy. The Fengshui practice discusses architecture in metaphoric terms of "invisible forces" that bind the universe, earth, and humanity together, known as Qi(气).

Qi(气) is a movable positive or negative life force which plays an essential role in Fengshui. A traditional explanation of Qi would include the orientation of a structure, its age, and its interaction with the surrounding environment, including the local microclimates, the slope of the land, vegetation, and soil quality.

The Book of Burial(葬书) says that burial takes advantage of "Vital Qi"(生气). Historically, Fengshui was widely used to orient buildings—often spiritually significant structures such as tombs, but also dwellings and other structures—in an auspicious manner. Depending on the particular style of Fengshui being used, an auspicious site could be determined by reference to local features such as bodies of water, stars, or a compass.

The magnetic compass(罗盘), which can reflect local geomagnetism is often used to detect the flow of Qi,as weather changes over time, the quality of Qi rises and falls over time, the compass might be considered a form of divination that assesses the quality of the local environment.

Chapter Two

Temple

A temple is a structure reserved for religious or spiritual rituals and activities such as prayer and sacrifice. It is also referred to the dwelling places of a god or gods. Despite the specific set of meanings associated with the Ancient Roman religion, the word has now become quite widely used to describe a house of worship for any number of religions.

Temple is typically used for such buildings belonging to all faiths where a more specific term such as church, mosque or synagogue is not generally used in English. These include Hinduism, Buddhism, and Jainism among religions with many modern followers, as well as other ancient religions such as Ancient Egyptian religion.

Temples typically have a main building and a larger precinct which may contain many other buildings. The form and function of temples is thus very variable, though they are often considered by believers to be in some sense the "house" of one or more deities. Typically offerings of some sort are made to the deity, and other rituals enacted, and a special group of clergy maintain and operate the temple. The degree to which the whole population of believers can access the building varies significantly; often parts or even the whole main building can only be accessed by the clergy.

The Parthenon[1]

The Parthenon is a former temple on the Athenian Acropolis[2], Greece, dedicated to the goddess Athena, whom the people of Athens considered their patron. It is the most important surviving building of Classical Greece, generally considered the zenith of the Doric order[3]. Construction began in 447 B. C. when the Athenian citystate was at the peak of its power and it was completed in 438 B. C. although decoration of the building continued until 432 B. C. Its decorative sculptures are considered some of the high points of Greek art. The Parthenon is regarded as an enduring symbol of Ancient Greece, Athenian democracy and western civilization, and one of the world's greatest cultural monuments. (See Fig. 2. 1)

Fig. 2.1 The Parthenon in Athenian Acropolis ①

1. History of Construction

The Parthenon itself replaced an older temple of Athena, which was destroyed in the Persian invasion of 480 B. C., it was actually used primarily as a treasury of the Delian League[4], which later became the Athenian Empire. In the final decade of the sixth century AD, the Parthenon was converted into a Christian church dedicated to the Virgin Mary[5]. After the Ottoman conquest[6], it was turned into a mosque in the early 1460s. On 26 September 1687, an Ottoman ammunition dump inside the building was ignited by Venetian bombardment. The resulting explosion severely damaged the Parthenon and its sculptures. From 1800 to 1803, Thomas Bruce, 7th Earl of Elgin removed some of the surviving sculptures with the alleged permission of the Ottoman Empire. These sculptures were sold in 1816 to the British Museum in London, where they are now displayed.

The first endeavor to build a sanctuary for Athena was begun shortly after the Battle of Marathon[7] (490 B. C.—488 B. C.) upon a solid limestone foundation that extended and leveled the southern part of the Acropolis summit. This building replaced a "hundred-footer" and would have stood beside the archaic temple dedicated to Athena Polias. The Older or Pre-Parthenon, as it is frequently referred to, was still under construction when the Persians sacked the city in 480 B. C. and razed the Acropolis. The existence of both the proto-Parthenon and its destruction were known from Herodotus[8], and the drums of its columns were plainly visible built into the curtain wall north of the Erechtheon[9].

In the mid-5th century B. C., when the Athenian Acropolis became the seat of the Delian League and Athens was the greatest cultural centre of its time, Pericles[10] initiated an ambitious building project that lasted the entire second half of the century. The most important buildings visible on the Acropolis today—the Parthenon, the Propylaia, the

① Darfrek. Deviant art. com.

Erechtheion and the temple of Athena Nike—were erected during this period. The Parthenon was built under the general supervision of the artist Phidias, who also had charge of the sculptural decoration. The architects Ictinos and Callicrates began their work in 447 B. C. , and the building was substantially completed by 432, but work on the decorations continued until at least 431. Some of the financial accounts for the Parthenon survive and show that the largest single expense was transporting the stone from Mount Pentelicus, about 16 kilometers from Athens to the Acropolis. The funds were in part stolen by Pericles from the treasury of the Delian League, which was moved from the Panhellenic sanctuary at Delos to the Acropolis in 454 B. C.

2. The Description of the Parthenon

The Parthenon is a peripteral octastyle Doric temple with Ionic architectural features. It stands on a platform or stylobate of three steps. In common with other Greek temples, it is of post and lintel construction and is surrounded by columns carrying an entablature. There are eight columns at either end and seventeen on the sides. There is a double row of columns at either end. The colonnade surrounds an inner masonry structure, the cella which is divided into two compartments. At either end of the building the gable is finished with a triangular pediment originally filled with sculpture. The columns are of the Doric order, with simple capitals, fluted shafts and no bases. Above the architrave of the entablature is a frieze of carved pictorial panels, separated by formal architectural triglyphs, typical of the Doric order. Around the cella and across the lintels of the inner columns runs a continuous sculptured frieze in low relief. This element of the architecture is Ionic in style rather than Doric.

The dimensions of the base of the Parthenon are 69. 5 by 30. 9 meters . The cella was 29. 8 meters long by 19. 2 meters wide, with internal colonnades in two tiers, structurally necessary to support the roof. On the exterior, the Doric columns measure 1. 9 meters in diameter and are 10. 4 meters high. The corner columns are slightly larger in diameter. The Parthenon had 46 outer columns and 23 inner columns in total, each column containing 20 flutes. The stylobate has an upward curvature towards its centre of 60 millimeters on the east and west ends, and of 110 millimeters on the sides. The roof was covered with large overlapping marble tiles known as imbrices and tegulae. (See Fig. 2. 2)

The Parthenon is regarded as the finest example of Greek architecture. The temple, wrote John Julius Cooper, "enjoys the reputation of being the most perfect Doric temple ever built. Even in antiquity, its architectural refinements were legendary, especially the subtle correspondence between the curvature of the stylobate, the taper of the naos walls and the entasis of the columns." Entasis refers to the slight swelling, of 1/8 inch, in the centre of the columns to counteract the appearance of columns having a waist, as the swelling makes them look straight from a distance. The stylobate is the platform on which

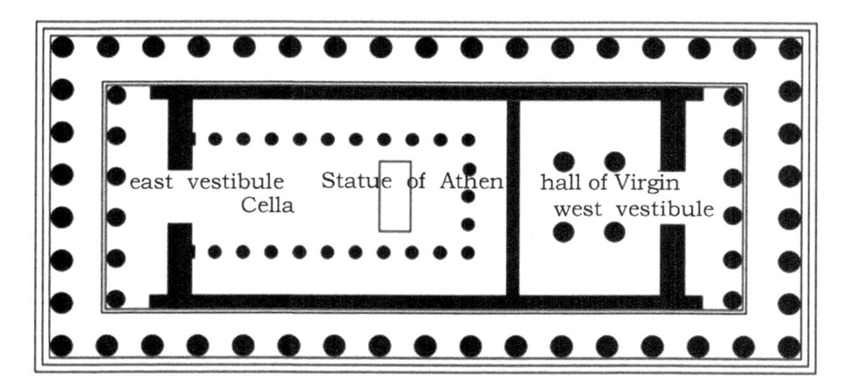

Fig. 2. 2 Floor plan of the Parthenon

the columns stand. As in many other classical Greek temples, it has a slight parabolic upward curvature intended to shed rainwater and reinforce the building against earthquakes. The columns might therefore be supposed to lean outwards, but they actually lean slightly inwards so that if they carried on, they would meet almost exactly a mile above the centre of the Parthenon; since they are all the same height, the curvature of the outer stylobate edge is transmitted to the architrave and roof above.

It is not universally agreed what the intended effect of these "optical refinements" was; they may serve as a sort of "reverse optical illusion". As the Greeks may have been aware, two parallel lines appear to bow, or curve outward, when intersected by converging lines. In this case, the ceiling and floor of the temple may seem to bow in the presence of the surrounding angles of the building. Striving for perfection, the designers may have added these curves, compensating for the illusion by creating their own curves, thus negating this effect and allowing the temple to be seen as they intended. It is also suggested that it was to enliven what might have appeared an inert mass in the case of a building without curves.

3. The Decoration of Partheon

The cella of the Parthenon housed the chryselephantine statue of Athena Parthenos sculpted by Phidias and dedicated in 439 B. C. or 438 B. C. The appearance of this is known from other images. The decorative stonework was originally highly colored. By the year 438 B. C., the sculptural decoration of the Doric metopes on the frieze above the exterior colonnade, and of the Ionic frieze around the upper portion of the walls of the cella, had been completed. The richness of the Parthenon's frieze and metope decoration is in agreement with the function of the temple as a treasury. In the back room of the cella were stored the monetary contributions of the Delian League, of which Athens was the leading member.

The frieze of the Parthenon's entablature contained ninety-two metopes, fourteen each on the east and west sides, thirty-two each on the north and south sides. They were carved in high relief, a practice employed until then only in treasuries. According to the building records, the metope sculptures date to the years 446 B. C. —440 B. C. The metopes present examples of the Severe Style in the anatomy of the figures' heads, in the limitation of the corporal movements to the contours and not to the muscles. Several of the metopes still remain on the building, but, with the exception of those on the northern side, they are severely damaged. Some of them are located at the Acropolis Museum, others are in the British Museum, and one is at the Louvre museum.

The most characteristic feature in the architecture and decoration of the temple is the Ionic frieze running around the exterior walls of the cella, which is the inside structure of the Parthenon. The bas-relief frieze was carved in situ; it is dated to 442 B. C. —438 B. C.

One interpretation is that it depicts an idealized version of the Panathenaic procession[11] from the Dipylon Gate in the Kerameikos to the Acropolis. In this procession held every year, with a special procession taking place every four years, Athenians and foreigners were participating to honor the goddess Athena, offering sacrifices and a new peplos.

The traveler Pausanias, when he visited the Acropolis at the end of the 2nd century A. D. , only mentioned briefly the sculptures of the pediments of the temple, reserving the majority of his description for the gold and ivory statue of the goddess inside.

The east pediment narrates the birth of Athena from the head of her father, Zeus[12]. According to Greek mythology, Zeus gave birth to Athena after a terrible headache prompted him to summon Hephaestus for assistance. To alleviate the pain, he ordered Hephaestus to strike him with his forging hammer, and when he did, Zeus's head split open and out popped the goddess Athena in full amour. The sculptural arrangement depicts the moment of Athena's birth.

The west pediment faced the Propylaia and depicted the contest between Athena and Poseidon[13] during their competition for the honor of becoming the city's patron. Athena and Poseidon appear at the center of the composition, diverging from one another in strong diagonal forms, with the goddess holding the olive tree and the god of the sea raising his trident to strike the earth. At their flanks, they are framed by two active groups of horses pulling chariots, while a crowd of legendary personalities from Athenian mythology fills the space out to the acute corners of the pediment.

4. Restoration of Parthenon

In 1975, the Greek government began a concerted effort to restore the Parthenon and other Acropolis structures. The project attracted funding and technical assistance from the European Union. An archaeological committee thoroughly documented every artifact remaining on the site, and architects assisted with computer models to determine their original locations. Particularly important and fragile sculptures were transferred to the Acropolis Museum. A crane was installed for moving marble blocks; the crane was designed to fold away beneath the roofline when not in use. In some cases, prior reconstruction was found to be incorrect. These were dismantled, and a careful process of restoration began. Originally, various blocks were held together by elongated iron H pins that were completely coated in lead, which protected the iron from corrosion. Stabilizing pins added in the 19th century were not so coated, and corroded. Since the corrosion product is expansive, the expansion caused further damage by cracking the marble.

Notes:

1. Partheon：帕特农神庙，是雅典卫城最重要的主体建筑，是卫城的最高点，代表了古希腊建筑艺术的最高水平。帕特农神庙是为了歌颂雅典战胜波斯侵略者而建的，设计者为建筑师伊克梯诺(Ictinus)和卡里克利特(Callicrates)。帕特农神庙里供奉被认为是雅典守护神的雅典娜女神，传说为一尊黄金象牙镶嵌的全希腊最高大的雅典娜女神像。帕特农神庙公元前447年开始兴建，9年后封顶，6年之后各项雕刻才告完成。1687年，威尼斯人与土耳其人作战时，神庙遭到破坏。19世纪下半叶，曾对神庙进行过部分修复。
2. Athenian Acropolis：雅典卫城，位于雅典市中心的卫城山丘上，原意为“高处的城市”或“高丘上的城邦”，始建于公元前580年，最初是用于防范外敌入侵的要塞，山顶四周筑有围墙，古城遗址在卫城山丘南侧。卫城中最早的建筑是雅典娜神庙和其他宗教建筑。根据古希腊神话传说，雅典娜将纺织、裁缝、雕刻、制作陶器和油漆工艺传授给人类，是战争、智慧、文明和工艺女神，后来成为城市保护神，雅典城市因故得名。雅典卫城面积约有4平方千米，坚固的城墙筑在四周。高地东面、南面和北面都是悬崖绝壁，地形十分险峻，人们只能从西侧登上卫城。雅典卫城内前门、山门、帕特农神庙、阿尔忒弥斯神庙等建筑，现都仅存残垣。
3. Doric order：多立克柱式，是希腊古典建筑的三种柱式中出现最早的一种，又被称为男性柱，约在公元前7世纪出现。另外两种柱式爱奥尼柱式和科林斯柱式，也都源于古希腊。在古希腊，多立克柱式一般都建在阶座之上，特点是粗大雄壮，柱头是倒圆锥台，没有柱础。柱身有时雕成20条槽纹，有时是平滑的，柱头没有装饰。柱下部约占全柱三分之一的地位槽纹很浅，往上越来越深。多立克柱建造比例通常是：柱下径与柱高的比例是1∶5.5；柱高与柱直径的比例是4∶1或6∶1，柱与柱之间的距离是柱直径的两倍到两倍半。
4. Delian League：提洛同盟。公元前478年希波战争期间，以雅典为首的一些希腊城邦结成军事同盟，因盟址及金库曾设在提洛岛，故称“提洛同盟”。同盟初期的宗旨是以集体力量解放遭受波斯奴役的希腊城邦和防御波斯再次入侵。最初入盟的主要是小亚细亚和爱琴海诸岛的希腊城邦，后来增至约200个。入盟各邦保持原有的政体，同盟事务由在提洛岛召开的同盟会议决定，按入盟城邦实力大小各出一定数量的舰船、兵员和盟捐。
5. Virgin Mary：圣母玛利亚，耶稣基督之母，《圣经》新约和《古兰经》里耶稣的生母。基督教徒和穆斯林认为，玛利亚还是处女时受圣灵感应而怀孕，玛利亚的丈夫名叫约瑟。在天主教会中，人们对圣母玛利亚的敬礼特别隆重。教会认为，玛利亚在基督的救世事业中所起的独一无二的配合作用，使圣母能够成为恩宠的中保。也就是说，天主因耶稣的救世之功而赐给人们的恩宠或直接或间接地要通过玛利亚而分施给人们。

6. Ottoman conquest：奥斯曼帝国统治。希腊自1460年就处于奥斯曼帝国（土耳其）的统治之下，1821年3月希腊本土爆发起义，并迅速发展到整个伯罗奔尼撒半岛、克里特岛、爱琴海诸岛屿、卢麦里以及马其顿等地。1822年1月1日第一届国民大会宣布希腊独立。
7. Battle of Marathon：马拉松战役，是希波战争中的一次重要战役。公元前491年，波斯皇帝大流士派遣使者到希腊各城邦要求希腊对波斯表示屈服，遭到了雅典和斯巴达的拒绝。公元前490年，大流士亲率波斯军队再次入侵希腊，在雅典城东北60千米的马拉松平原登陆，妄图一举消灭雅典。雅典军队与斯巴达援军1000人组成希腊联军，在马拉松平原与波斯大军展开激战，把波斯军队追赶到海边，波斯军队慌忙登船而逃。为了把胜利喜讯迅速告诉雅典人，米太亚得将军派士兵斐力庇第斯以最快速度从马拉松跑到雅典中央广场，对着盼望的人们说了一声"大家欢乐吧，我们胜利了"之后，就倒在地上牺牲了。为了纪念马拉松战役的胜利和表彰斐力庇第斯的功绩，1896年在雅典举行的第一届奥林匹克运动会上，增加了马拉松赛跑项目，以此来纪念为马拉松战果传递的英雄。
8. Herodotus：希罗多德（公元前484—425），伟大的古希腊历史学家，史学名著《历史》一书的作者。希罗多德诞生在小亚细亚西南海滨的一座古老的城市，其父是一个奴隶主。大约从30岁开始，希罗多德开始了一次范围广泛的旅游，向北走到黑海北岸，向南到达埃及最南端，向东至两河流域下游一带，向西抵达意大利半岛和西西里岛。每到一地，希罗多德就到历史古迹名胜处浏览凭吊，考察地理环境，了解风土人情，听当地人讲述民间传说和历史故事。所著《历史》一书，共9卷，亦作《希腊波斯战争史》。
9. Erechtheon：伊瑞克提翁神庙，是雅典卫城的著名建筑之一，传说是雅典娜女神和海神波塞东为争做雅典保护神而斗智的地方。伊瑞克提翁神庙位于埃雷赫修神庙的南面的一块凹凸不平的高地上，建于公元前421年—公元前405年之间，是雅典卫城建筑中爱奥尼亚样式的典型代表。伊瑞克提翁神庙有三个神殿，分别供奉希腊的主神宙斯、海神波塞冬和铁匠之神赫菲斯托斯，由两条别具特色的柱廊把它们连接起来。伊瑞克提翁神庙南端用6根大理石雕刻而成的少女像柱代替石柱顶起屋顶，充分体现了建筑师的智慧。她们长裙束胸，轻盈飘忽，头顶千斤，亭亭玉立。由于石顶的份量很重，6位少女为了顶起沉重的石顶，颈部必须设计得足够粗，建筑师给每位少女颈后保留了一缕浓厚的秀发，再在头顶加上花篮，成功地解决了建筑美学上的难题。
10. Pericles：伯里克利（约公元前495—公元前429），古希腊奴隶主民主政治的杰出代表和政治家。从公元前443年到公元前429年，伯里克利连选连任雅典最重要的官职——首席将军，完全掌握国家政权。在伯里克利的领导下，雅典的奴隶制经济、民主政治、海上霸权和古典文化臻于极盛。
11. Panathenaic procession：泛雅典人巡游，是古代雅典人祭祀城邦女神雅典娜的宗教节日泛雅典人节中的主要活动。每四年举办一次，雅典人为此专门修建了一座供队伍整装待发的建筑——列队厅（Pompeion），地点在城市的西北郊。巡游通过的道路被称为"泛雅典人节大道"，是雅典城最重要的道路之一，起点是位于雅典城西北的迪庇隆门（Dipylon Gate），经过陶工区，然后从西北角进入市政广场（Agora），之后从广场的东南角延伸向卫城陡峭的北坡，最后向东直至卫城脚下的山门(Propylaia)。这条道路全长1千多米，有10到20米宽，沿途经过雅典几乎所有的重要建筑。参加巡游的主体是雅典公民，一些外邦人和获释的奴隶也可参与，但进入卫城献祭的只能是雅典人。

12. Zeus：宙斯，是古希腊神话中众神之王，奥林匹斯十二神之首，统治宇宙的至高无上的主神，与罗马神话的朱庇特(Jupiter)是同一神祇。

13. Poseidon：波塞冬，是希腊神话中的海神，宙斯的哥哥，其象征物为三叉戟，他的坐骑是由白马驾驶的黄金战车。当他愤怒时，海底就会出现怪物，他挥动三叉戟不但能轻易掀起滔天巨浪引起风暴和海啸，使大陆沉没、天地崩裂，还能将万物打得粉碎，甚至引发震撼整个世界的强大地震。波塞冬的三叉戟也被用来击碎岩石，从裂缝中流出的清泉浇灌大地，使农民五谷丰登，所以波塞冬又被称为丰收神。据说当他的战车在大海上奔驰时，波浪会变得平静，并且周围有海豚跟随。

St. Peter's Basilica[1]

St. Peter's Basilica, located within the Vatican City[2], is famous as a place of pilgrimage, for its liturgical functions and for its historical associations. Tradition and some historical evidence hold that Saint Peter's[3] tomb is directly below the altar of the basilica. For this reason, many Popes have been interred at St. Peter's since the Early Christian period. There has been a church on this site since the 4th century. Construction of the present basilica, over the old Constantinian basilica, began on 18 April 1506 and was completed on 18 November 1626. St. Peter's Basilica has the largest interior of any Christian church in the world and has been described as "holding a unique position in the Christian world" and as "the greatest of all churches of Christendom". (See Fig. 2.3)

Fig. 2.3 St. Peter's Basilica

1. The Description of the Building

St. Peter's Basilica is a huge church in the Renaissance style[4] located in west of the River Tiber and near the Janiculum Hill and Emperor Hadrian's[5] Mausoleum. Its central dome dominates the skyline of Rome. The basilica is approached via St. Peter's Square, a forecourt in two sections, both surrounded by tall colonnades. The first space is oval and the second trapezoid. The facade of the basilica, with a giant order of columns, stretches across the end of the square and is approached by steps on which stand two 5. 55 meters statues of the 1st century apostles to Rome, Saints Peter and Paul[6]. The entrance is through a narthex, or entrance hall, which stretches across the building. One of the decorated bronze doors leading from the narthex is the Holy Door, only opened during Jubilee year[7].

1. 1 The Layout of St. Peter's Basilica

The St. Peter's Basilica is cruciform in shape, with an elongated nave in the Latin cross[8] form but the early designs were for a centrally planned structure and this is still in evidence in the architecture. Pope Julius' scheme for the grandest building in Christendom was the subject of a competition. It was the design of Donato Bramante[9] that was selected, and for which the foundation stone was laid in 1506. This plan was in the form of an enormous Greek Cross[10] with a dome inspired by that of the huge circular Roman temple, the Pantheon. When Pope Julius died in 1513, Bramante died the previous year. On 1 January 1547 in the reign of Pope Paul III, Michelangelo, then in his seventies, succeeded as the superintendent of the building program at St. Peter's. He is to be regarded as the principal designer of a large part of the building as it stands today. St. Peter's has been extended with a nave by Carlo Maderno. It is the chancel end with its huge centrally placed dome that is the work of Michelangelo. Because of its location within the Vatican State and because the projection of the nave screens the dome from sight when the building is approached from the square in front of it, the work of Michelangelo is best appreciated from a distance. (See Fig. 2. 4)

1. 2 The Dome

The central space of the interior space of St. Peter's Basilica is dominated both externally and internally by one of the largest domes in the world. The dome is constructed of a single shell, surrounded at its base by a continuous colonnade and surmounted by a temple-like lantern with a ball and cross on top. The dome of St. Peter's rises to a total height of 136. 57 meters from the floor of the basilica to the top of the external cross. It is the tallest dome in the world. Its internal diameter is 41. 47 meters, slightly smaller than two of the three other huge domes that preceded it, those of the Pantheon[11] of Ancient Rome, 43. 3 meters, and Florence Cathedral of the Early Renaissance, 44 meters. It has a

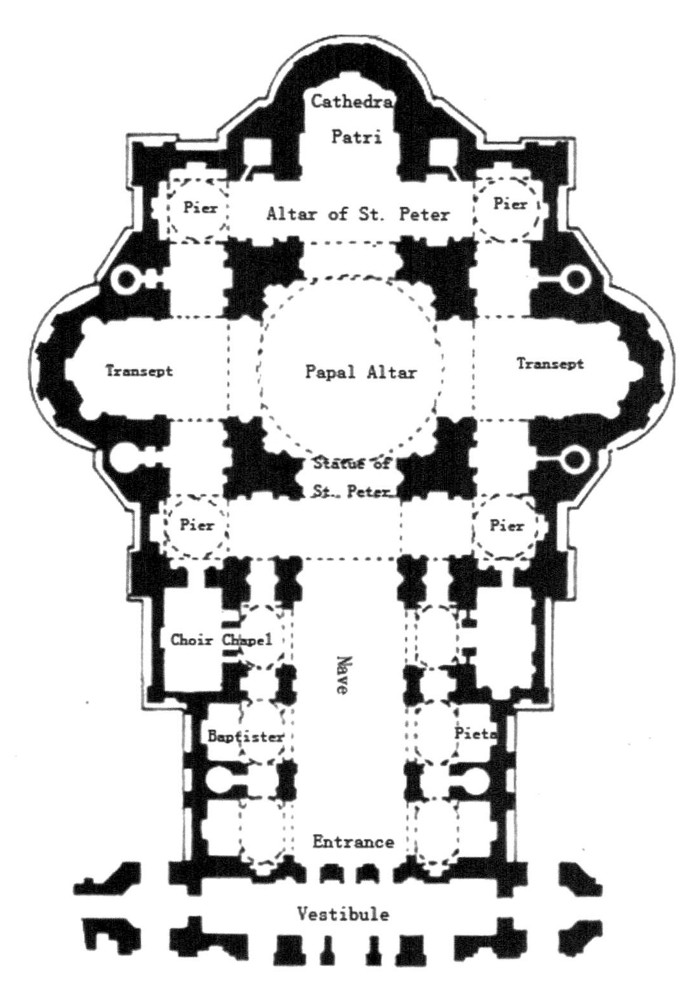

Fig. 2.4 Floor plan of St. Peter's Basilica

greater diameter by approximately 9.1 m than Constantinople's Hagia Sophia[12] church, completed in 537. It was to the domes of the Pantheon and Florence duomo that the architects of St. Peter's looked for solutions as to how to go about building what was conceived, from the outset, as the greatest dome of Christendom. The dome is decorated at the top with a band of script. Around its base are windows through which the light streams. The decoration is divided by many vertical ribs which are ornamented with golden stars.

1.3 The facade of St. Peter's Basilica

The facade designed by Maderno, 114.69 meters wide and 45.55 meters high, is built of travertine stone, with a giant order of Corinthian columns and a central pediment rising in front of a tall attic surmounted by thirteen statues: Christ flanked by eleven of the Apostles and John the Baptist. The inscription below the cornice on the 1 meter tall frieze reads: In honor of the Prince of Apostles, Paul V Borghese, a Roman, Supreme Pontiff, in the year 1612, the seventh of his pontificate, born in Rome but of a Sienese family, liked to emphasize his "Romanness." The facade is often cited as the least satisfactory part of the design of St. Peter's. The reasons are that it was not given enough consideration by the Pope and committee because of the desire to get the building completed quickly, coupled with the fact that Maderno was hesitant to deviate from the pattern set by Michelangelo at the other end of the building. Lees-Milne describes the problems of the facade as being too broad for its height, too cramped in its details and too heavy in the attic story. The breadth is caused by modifying the plan to have towers on either side. These towers were never executed above the line of the facade because it was discovered that the ground was not sufficiently stable to bear the weight. One effect of the facade and lengthened nave is to screen the view of the dome, so that the building, from the front, has no vertical feature, except from a distance.

1.4 The decoration of interior space of St. Peter's Basilica

The entire interior of St Peter's is lavishly decorated with marble, reliefs, architectural sculpture and gilding. The basilica contains a large number of tombs of popes and other notable people, many of which are considered outstanding artworks. There are also a number of sculptures in niches and chapels, including Michelangelo's *Pieta*[13]. The central feature is a baldachin, or canopy over the Papal Altar, designed by Gianlorenzo Bernini. The sanctuary culminates in a sculptural ensemble, also by Bernini, and containing the symbolic *Chair of St. Peter*.

Bernini's first work at St. Peter's was to design the baldacchino, a pavilion-like structure 30 meters tall and claimed to be the largest piece of bronze in the world, which stands beneath the dome and above the altar. Its design is based on the ciborium, of which there are many in the churches of Rome, serving to create a sort of holy space above and around the table on which the Sacrament[14] is laid for the Eucharist[15] and emphasizing the significance of this ritual. These ciborium are generally of white marble, with inlaid colored stone. Bernini's concept was for something very different. He took his inspiration in part from the baldachin or canopy carried above the head of the pope in processions, and in part from eight ancient columns that had formed part of a screen in the old basilica. Their twisted barley-sugar shape had a special significance as the column to which Jesus was bound before his crucifixion was believed to be of that shape. Based on these columns, Bernini created four huge columns of bronze, twisted and decorated with olive leaves and bees, which were the emblem of Pope Urban.

The baldacchino is surmounted not with an architectural pediment, like most baldacchini, but with curved Baroque brackets supporting a draped canopy, like the brocade canopies carried in processions above precious iconic images. In this case, the draped canopy is of bronze, and all the details, including the olive leaves, bees, and the portrait heads of Urban's niece in childbirth and her newborn son, are picked out in gold leaf.

2. St. Peter's Square

To the east of the basilica is the St. Peter's Square. The present arrangement, constructed between 1656 and 1667, is the Baroque[16] inspiration of Bernini who inherited a location already occupied by an Egyptian obelisk which was centrally placed . The obelisk, known as "The Witness", at 25.5 meters and a total height, including base and the cross on top, of 40 meters, is the second largest standing obelisk, and the only one to remain standing since its removal from Egypt and re-erection at the Circus of Nero in 37 A.D., where it is thought to have stood witness to the crucifixion of St. Peter.

The other object in the old square with which Bernini had to contend was a large fountain designed by Maderno in 1613 and set to one side of the obelisk, making a line parallel with the facade. Bernini's plan uses this horizontal axis as a major feature of his unique, spatially dynamic and highly symbolic design. Bernini's ingenious solution was to create a piazza in two sections. That part which is nearest the basilica is trapezoid, but rather than fanning out from the facade, it narrows. This gives the effect of countering the visual perspective. It means that from the second part of the piazza, the building looks nearer than it is, the breadth of the facade is minimized and its height appears greater in proportion to its width. The second section of the piazza is a huge elliptical circus which gently slopes downwards to the obelisk at its center. The two distinct areas are framed by a colonnade formed by doubled pairs of columns supporting an entablature of the simple Tuscan Order. The part of the colonnade that is around the ellipse does not entirely encircle it, but reaches out in two arcs, symbolic of the arms of "the Catholic Church reaching out to welcome its communicants".

Notes:

1. St. Peter's Basilica：圣彼得大教堂，是位于梵蒂冈的天主教宗座圣殿，建于1506年至1626年，为天主教会重要的象征之一。根据天主教会圣传，大教堂是宗徒之长圣彼得的安葬地点，历任教宗也大都安葬于此。作为最杰出的文艺复兴建筑和世界上最大的教堂，圣彼得大教堂占地23000平方米，可容纳超过六万人。教堂中央是直径42米的穹窿，顶高约138米，堂内保存有欧洲文艺复兴时期许多艺术家如米开朗基罗、拉斐尔等的壁画与雕刻。意大利文艺复兴时期的多位建筑师与艺术家如伯拉孟特、拉斐尔、米开朗基罗和小安东尼奥·达·桑加罗等都曾参与圣彼得大教堂设计。
2. Vatican City：梵蒂冈城，是世界天主教的中心、罗马教廷所在地，位于意大利首都罗马西北角的梵蒂冈高地上，北、西、南3面有高墙与罗马市隔开，东面的圣彼得广场同罗马市畅行无阻。梵蒂冈面积仅0.44平方千米，人口约1400人，常住人口仅500人。修筑在使徒圣彼得墓上的大教堂是全城的中心，也是世界上最大的宗教建筑。

3. Saint Peter：圣彼得，是耶稣十二门徒之一，也是耶稣的第一个门徒，曾经是来自加利利的一个渔夫，在耶稣的信徒中占据领导地位，对基督教会的成立影响重大。圣彼得名字在拉丁语中是"磐石"的意思。他在公元64年罗马皇帝尼禄统治时期到罗马传教，被头朝下钉在十字架上，在此地殉难。
4. Renaissance style：文艺复兴风格，是欧洲建筑史上继哥特式建筑之后出现的一种建筑风格，有时也包括巴洛克建筑和新古典主义建筑，十五世纪起源于意大利佛罗伦萨，后传播到欧洲其他地区。该风格在理论上以文艺复兴思潮为基础；在造型上排斥象征神权至上的哥特建筑风格，提倡复兴古罗马时期的建筑形式，特别是古典柱式比例、半圆形拱券、以穹隆为中心的建筑形体等。
5. Emperor Hadrian：哈德良皇帝(Publius Aelius Traianus Hadrianus，公元76年—138年)，罗马帝国第三位皇帝，图拉真皇帝的表侄，是一位博学多才的皇帝。哈德良统治时期，帝国基本上是平静的，没有战争。和图拉真一样，哈德良在罗马大兴土木，重建了奥古斯都时期兴建的万神殿，并修建了维纳斯女神庙。
6. Saints Paul：圣保罗，也是耶稣十二门徒之一，是神所拣选，外邦人的使徒，也被历史学家公认是对于早期教会发展贡献最大的使徒。他一生中至少进行了三次漫长的宣教之旅，足迹遍至小亚细亚、希腊、意大利各地，在外邦人中建立了许多教会，影响深远。
7. Jubilee year：圣年。根据圣经规定，每50年为一个圣年，也称为朱比利年，该年到来时，要大赦囚犯，免除债务和显示神的怜悯。天主教传统中，圣年还包括到宗教圣地，特别是罗马的朝圣活动。
8. Latin cross：拉丁十字，为不等臂十字，是基督教会崇尚的建筑形制。之所以采取这样的布局可能和基督教对十字架的崇拜有关系。教堂中央的大厅仍然以一个穹窿为主体，但大厅往四周各伸出一个矮矮的"翼廊"。
9. Donato Bramante：伯拉孟特(1444—1514年)，是意大利文艺复兴时期著名的画家、建筑师，在建筑方面，他与同时期的米开朗基罗和拉斐尔各领风骚。其代表作品坦比哀多礼拜堂被认为是文艺复兴盛期建筑的代表作，对后世有很大的影响。
10. Greek Cross：希腊十字，原为基督教十字架的一种形式，其十字架的四臂相等，后指集中式教堂形制的一种，这种教堂的中央穹顶和它四面筒形拱成等臂的十字，故称希腊十字，主要见于拜占庭式教堂建筑。
11. Pantheon：万神殿，至今完整保存的唯一一座罗马帝国时期建筑，始建于公元前27—公元前25年，由罗马帝国首任皇帝屋大维的女婿阿格里帕建造，用以供奉奥林匹亚山上诸神，是奥古斯都时期的经典建筑。公元80年的火灾使万神殿的大部分被毁，仅余一长方形的柱廊，有12.5米高的花岗岩石柱16根，这一部分被作为后来重建的万神殿的门廊。现存的万神殿是一个宽度与高度相等的巨大圆柱体，上面覆盖着半圆形的屋顶。拉斐尔等许多著名艺术家就葬在这里。
12. Hagia Sophia：圣索菲亚大教堂，是位于现今土耳其伊斯坦布尔的拜占庭式建筑典范。公元532年拜占庭皇帝查士丁尼一世下令建造教堂，由物理学家米利都的伊西多尔及数学家特拉勒斯的安提莫斯设计，公元537年完成。圣索非亚大教堂的平面采用希腊十字，在空间上创造了巨型的圆顶，室内没有用柱子来支撑。圣索非亚大教堂的大圆顶离地55米高,在17世纪圣彼得大教堂完成前,一直是世界上最大的教堂。

13. *Pieta*:圣殇,也称"哀悼基督",是米开朗基罗为圣彼得大教堂所作的大理石雕塑,是他早期最著名的代表作。作品的题材取自圣经故事中基督耶稣被钉死在十字架上之后,圣母玛丽亚抱着基督的身体痛哭的情景。米开朗基罗创作这幅雕塑时年仅24岁,这也是他唯一签名的作品。
14. Sacrament:圣礼,是初期教会所承认与遵守的仪节,是由耶稣基督亲自设立的。圣礼有两种,即洗礼和圣餐,是通过可以看见的物质形式来表明看不见的属灵恩典。
15. Eucharist:圣餐,是基督教各主要派别共有的重要圣事。圣餐的设立源于耶稣与门徒共进最后晚餐,掰饼分酒给门徒时所说"这是我的身体"、"这是我的血"。基督教认为饼和酒是耶稣为救赎人类被钉于十字架的象征。基督教各派的圣餐礼均由神职人员主持。
16. Baroque:巴洛克,是一种代表欧洲文化的典型艺术风格。这个词最早来源于葡萄牙语(BARROCO),意为"不圆的珍珠",指"缺乏古典主义均衡性的作品",原本是18世纪崇尚古典艺术的人们对17世纪不同于文艺复兴风格的艺术一个带贬义的称呼。巴洛克艺术有如下特点:一是既有宗教的特色,又有享乐主义的色彩;二是具有浓郁的浪漫主义色彩,非常强调艺术家的丰富想象力;三是强调运动与变化;四是关注作品的空间感和立体感。

Foguang Monastery[1]

Foguang Monastery(佛光寺) is a Buddhist building complex located five kilometers from Doucun, Wutai County, Shanxi Province of China. The major hall of the temple is the Great East Hall(东大殿) built in 857 A.D. during the Tang Dynasty (618—907). According to architectural records, it is the third earliest preserved timber structure in China. It was rediscovered by the famous architectural historian Liang Sicheng (梁思成, 1901—1972)[2] in 1937, while an older hall at Nanchan Monastery(南禅寺)[3] was discovered by the same team a year later. The monastery also contains another significant hall dating from 1137 called the Manjusri Hall(文殊殿). In addition, the second oldest existing pagoda in China, after the Songyue Pagoda(嵩岳塔) in Henan Province, dating from the 6th century, is also located within the monastery. (See Fig. 2. 5)

Fig. 2. 5 East Hall of Foguang Monastery

1. History of the Monastery Construction

The Foguang Monastery was established in the fifth century during the Northern Wei dynasty. From the years of 785 to 820, the monastery underwent an active building period when a three storied, 32 meters tall pavilion was built. In 845, Emperor Wuzong banned Buddhism in China[4], as part of the persecution, Foguang Monastery was burned to the ground, with only the Zushi pagoda(祖师塔) survived. Twelve years later in 857 the monastery was rebuilt, with the Great East Hall being built on the former site of a three storey pavilion. The construction was led by a monk named Yuancheng(愿诚)and a woman benefactor named Ning Gongyu(宁公遇) provided most of the funds needed to construct the hall, In the 10th century, a depiction of Foguang Monastery was painted in cave 61 of the Mogao Grottoes(莫高窟)[5], which indicates that Foguang Monastery was an important stop for Buddhist pilgrims. However, it is likely the painters had never seen the temple because the main hall in the painting is a two-storied white building with a green-glaze roof, which is very different from the red and white of the Great East Hall.

In 1137 of the Jin Dynasty, the Manjusri Hall was constructed on the north side of monastery, along with another hall dedicated to Samantabhadra(普贤)[6], which was burnt down in the Qing Dynasty (1644—1912).

In 1930, the Society for Research in Chinese Architecture(营造学社)[7] began a research in China for ancient buildings. In 1937, an architectural team led by Liang Sicheng discovered that Foguang Monastery was a relic of the Tang Dynasty. Liang was able to date the building after his wife found an inscription on one of the rafters. The date's accuracy was confirmed by Liang's study of the building which matched with known information about Tang buildings.

2. The Description of the Great East Hall

The Foguang monastery consists of two main side halls, the northern hall called The Hall of Manjusri(文殊殿), which was constructed in 1147 during the Jin Dynasty(金代) and south hall called The Samantabhadra Hall, once existed on the south side of the monastery but is no longer extant. The Great East Hall(东大殿), which is the largest hall, was constructed in 857 during the Tang Dynasty. Unlike most of Chinese temples which are oriented in a south-north position, the East Hall is oriented in an east-west position due to there being mountains located on the east, north and south. According to Chinese tradition, the mountains behind a building is believed to improve its Fengshui(风水). (See Fig. 2. 6)

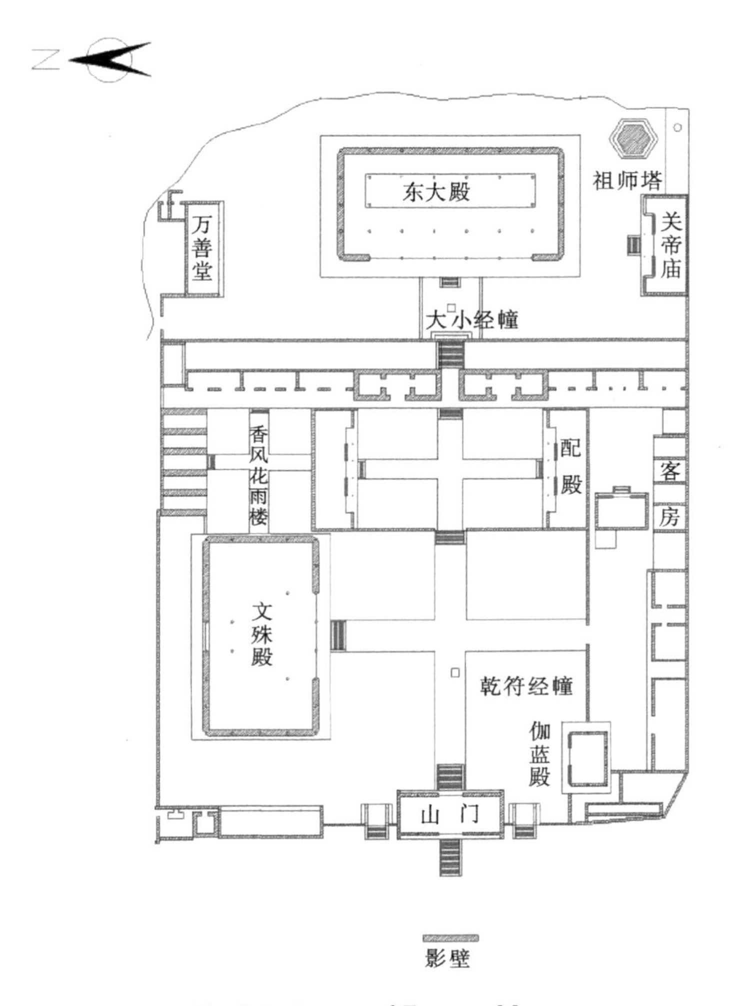

Fig. 2. 6 Layout of Foguang Monastery

The Great East Hall is the third oldest wooden building in China after the main hall of the Nanchan Monastery dated to 782, and the main hall of the Five Dragons monastery(五龙寺)dated to 831. The hall is located on the far east side of the monastery, atop a large stone platform. It is a single storey structure measuring seven bays by four or 34 by 17.7 meters, and is supported by inner and outer sets of columns. On top of each column is a complicated set of brackets containing seven different bracket types that are one-third as high as the column itself. Supporting the roof of the hall, each of the bracket sets are connected by crescent shaped crossbeams, which create an inner ring above the inner set of columns and an outer ring above the outer columns. The hall has a lattice ceiling that conceals much of the roof frame from view. The hipped-roof and the extremely complex bracket sets are testament to the Great East Hall's importance as a structure during the Tang Dynasty. According the 11th century architectural treatise, *Yingzao Fashi*(营造法

式), the Great East Hall closely corresponds to a seventh rank building in a system of eight ranks. The high rank of the Great East Hall indicates that even in the Tang Dynasty it was an important building, since no other buildings with such a high rank from the period survive until now.

Inside the hall are thirty-six sculptures, as well as murals on each wall that date from the Tang Dynasty and later periods. Unfortunately the statues lost much artistic value when they were repainted in the 1930s. The centre of the hall has a platform with three large statues of Sakyamuni(释迦牟尼佛), Amitabha(阿弥陀佛) and Maitreya(弥勒佛) sitting on lotus shaped seats. Each of the three statues is flanked by four assistants on the side and two bodhisattvas(菩萨) in front. Next to the platform, there are statues of Manjusri riding a lion as well as Samantabhadra on an elephant and two heavenly kings standing on either side of the dais. A statue representing the hall's benefactor, Ning Gongyu and one of the monk who helped build the hall Yuancheng, are present in the back of the hall. There is one large mural that in the hall shows events that took place in the *Jataka*[8], which chronicles Buddha's past life. Smaller murals in the temple show Manjusri and Samantabhadra gathering donors to help support the upkeep of the temple.

3. Manjusri Hall

On the north side of the main courtyard is the Manjusri Hall. It was constructed in 1137 during the Jin Dynasty and is roughly the same size as the East Hall, also measuring seven bays by four. Located on a high platform, the hall has three front doors and one central back door. In spite of 300 years' difference in age, the facades of the east hall and Manjusri Hall are rendered in very much the same style. Experts wonder if the Jin building deliberately imitated the Tang hall, or if perhaps this is an artifact of the latest restoration.

Manjusri Hall features a single-eave hip gable roof, unlike most examples of major temple buildings, it is a fold-over gable roof that looks like an inverted "V" in cross-section. Inside, the roof is supported by trusses. The trussed frame that supports the hall's roof is an unusual design for China. Its purpose was to reduce the number of columns needed inside to support the frame. This approach turned out to be very successful in providing a large, uninterrupted, interior space. The interior of the hall has only four support pillars. In order to support the large roof, diagonal beams are used. For some reason, though, it was not followed through in later structures. On each of the four walls are murals of arhats[9] painted in 1429 during the Ming Dynasty.

The statue of bodhisattva enshrined in this hall is riding his lion and surrounded by attendants, in hands of bodhisattva presents what appears to be a restored lotus vine. According to the temple signage, the statues are sculptured in Jin Dynasty and their paint, has been renewed in modern times.

4. Zushi Pagoda

The Zushi Pagoda is a small funerary pagoda located to the south of the Great East Hall. While it is unclear as to the exact date of its construction, it was either built during the Northern Wei Dynasty (386—534) or Northern Qi Dynasty (550—577) and possibly contains the tomb of the founder of the Foguang Monastery.

It is a white, hexagonal shaped 6meters tall pagoda made of the brick. The pagoda is divided into two storeys, but there is only one chamber inside on the ground floor; the upper level is purely ornamental. A door on the front opens to a hexagonal chamber originally intended to enshrine statues of the two founders of a sect of Buddhism. Above the flat arched doorway is an ornamental headpiece formed by lotus petals. The other five sides on the pagoda's first storey are plain. The walls, however, incline slightly inward. The protruding eaves are composed of three layers of carved lotus petals and six layers of bricks[10], each layer reaching beyond the one below. Under the eaves are brick brackets, nine on each side. The pent roofs were built by stacking bricks, level upon level, in an inward direction, creating a solid image for the pagoda. (See Fig. 2. 7)

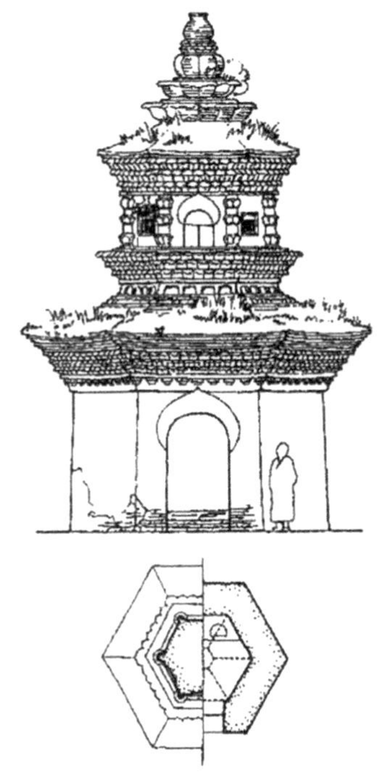

Fig. 2. 7 Facade and plan of Zushi Pagoda by Liang Sicheng[①]

① 梁思成

The second level of the pagoda is a hexagonal pavilion-style structure. Between the first and second levels there is a sub-base in the form of a Sumeru pedestal; bricks form square patterns in the lower part and lotus petals, nine on each side, on the upper part. Vase-shaped columns at the corners of the pedestal support the second storey, which differs from most other pagoda structures. First, the two leaves on the flame-shaped false door were installed at different angles, as if the door were half open. Second, the corner columns are decorated at the top, in the middle and at the foot with carved lotus petals full of Indian influence. Third, the surface of the second storey is painted with red clay, a decoration for wooden structures; on the inside walls of the arched doorway are remnants of painted patterns. On the small window on the northwestern side are ornamental paintings showing two tiers of beams with five short columns in between and filled with inverted V patterns. Such patterns are usually found in sculptures and murals in grottoes cut during the Southern and Northern Dynasties (南北朝), but rarely seen on real buildings.

The steeple of the pagoda is also made of brick, but the style is quite peculiar. The lower part has carvings of lotus petals to support a precious bottle in the shape of a six-petal flower. There are two more layers of lotus petals on top of the bottle and a bead at the highest point.

Notes:

1. Foguang Monastery：佛光寺，位于山西五台县的佛光新村，距县城 30 千米，是全国重点文物保护单位。佛光寺建在半山坡上。东、南、北三面环山，西面地势低下开阔。寺庙坐东朝西，全寺有院落三重，分建在梯田式的寺基上，现有殿、堂、楼、阁等 120 余间。其中，东大殿七间，为唐代建筑；文殊殿七间，为金代建筑；其余的均为明、清时期的建筑。正殿东大殿建于公元 857 年，仅次于建于唐建中三年（公元 782 年）的五台县南禅寺正殿，在全国现存的木结构建筑中居第二。1937 年 6 月，著名建筑学家梁思成和夫人林徽因赴山西五台县对佛光寺进行了考察、测绘。
2. Liang Sicheng (1901—1972)：梁思成，建筑历史学家、建筑教育家和建筑师，毕生致力于中国古代建筑的研究和保护，参与了人民英雄纪念碑、中华人民共和国国徽等作品的设计。梁思成在 1932 年主持了故宫文渊阁的修复工程，同年，著成《清式营造则例》手稿。从 1937 年起，他和林徽因等人先后踏遍中国 15 省 200 多个县，测绘和拍摄 2000 多件唐、宋、辽、金、元、明、清各代保留下来的古建筑遗物，包括天津蓟县辽代建筑独乐寺观音阁、宝坻辽代建筑广济寺、河北正定辽代建筑隆兴寺、山西辽代应县木塔、大同辽代寺庙群华严寺和善化寺、河北赵州隋朝建造的安济桥等。
3. Nanchan Temple：南禅寺，位于山西省忻州市五台县西南的东冶镇李家庄。寺院坐北向南，占地面积 3078 平方米。寺内主要建筑有山门（观音殿）、东西配殿（菩萨殿和龙王殿）

和大殿，组成一个四合院。南禅寺大殿为中国现存最古老的一座唐代木结构建筑，也是中国现存最早的木构建筑，距今1200多年。

4. Emperor Wuzong banned Buddhism in China：唐武宗灭佛。唐代后期，由于佛教寺院土地不输课税，僧侣免除赋役，佛教寺院经济过分扩张，损害了国库收入，与普通地主也存在着矛盾。唐武宗崇信道教，深恶佛教，会昌年间财政急需，在道士赵归真的鼓动和李德裕的支持下，从会昌二年(842年)开始渐进地进行毁佛，在会昌五年(845年)达到高潮，于会昌六年(846年)武宗死后终止。佛教徒称这次毁佛运动为“会昌法难”。
5. Mogao Grottoes：莫高窟，俗称千佛洞，坐落在河西走廊西端的敦煌。莫高窟始建于十六国的前秦时期，历经十六国、北朝、隋、唐、五代、西夏、元等历代的兴建，形成巨大的规模，有洞窟735个，壁画4.5万平方米、泥质彩塑2415尊，是世界上现存规模最大、内容最丰富的佛教艺术地。北宋、西夏和元代，莫高窟渐趋衰落，仅以重修前朝窟室为主，新建极少。
6. Samantabhadra：普贤菩萨，中国佛教四大菩萨之一，象征理德、行德，同文殊菩萨的智德、正德相对应，是娑婆世界释迦牟尼佛的右、左胁侍，被称为“华严三圣”。日本真言宗认为卫护佛门的金刚萨埵是普贤菩萨化身。
7. The Society for Research in Chinese Architecture：中国营造学社，于1930年2月在北平正式创立，朱启钤任社长，梁思成、刘敦桢分别担任法式组、文献组的主任。学社从事古代建筑实例的调查、研究和测绘，以及文献资料搜集、整理和研究，编辑出版《中国营造学社汇刊》，1946年停止活动。中国营造学社为中国古代建筑史研究作出重大贡献。
8. *Jataka*：《本生经》，讲述释迦牟尼在成佛之前，经过了无数次轮回转生的故事。每一次转生，便有一个行善立德的故事，现存有547个收集在巴利文经藏《小尼迦耶》中。
9. arhats：罗汉，阿罗汉的简称，最早从印度传入中国。一说可以帮人除去生活中一切烦恼；二说可以接受天地间人天供养；三说可以帮人不再受轮回之苦。即杀贼、应供、无生，是佛陀得道弟子修证最高的果位。
10. stacking bricks：叠涩，一种古代砖石结构建筑的砌法，用砖、石，有时也用木材通过一层层堆叠向外挑出或收进，向外挑出时要承担上层的重量。叠涩法主要用于早期的叠涩拱、砖塔出檐、须弥座的束腰和墀头墙的拔檐，常见于砖塔、石塔、砖墓室等建筑物。

Süleymaniye Mosque[1]

The Süleymaniye Mosque is an Ottoman imperial mosque located on the Third Hill of Istanbul, Turkey, it is the largest mosque in the city, and one of the best-known sights of Istanbul. This vast religious complex blended Islamic and Byzantine architectural elements. It combines tall, slender minarets with large domed buildings supported by half domes in the style of the Byzantine church Hagia Sophia[2]. As with other imperial mosques in Istanbul, the Süleymaniye Mosque was designed as a külliye[3], or complex with adjacent structures to serve both religious and cultural needs. (See Fig. 2. 8)

Fig. 2. 8 Süleymaniye Mosque, Istanbul

1. History of Construction of the Mosque

The Süleymaniye Mosque is a group of buildings, which were built between the years 1550—1557. The construction was commissioned to great Architect Sinan[4] by the Sultan of the time Suleyman the Magnificent (1520—1566)[5]. The buildings represent the unification of art and political power, symbolize the Ottoman classical era with the perfection in their plan schemes as well as their economical and cultural functions.

The Süleymaniye Mosque, built on the order of Sultan Süleyman, the construction work began in 1550 . It is very probable that the ceremony for laying the foundation took place at the date of 13th of June, 1550. At that day, Sultan Suleyman came to the palace on horseback and the governmental officials, the Muslim theologians and the religious persons were all gathered there. When the selected fortunate time came and by the order of the Sultan, the first stone on the foundation of the altar to start the construction was placed.

The Süleymaniye was ravaged by a fire in 1660 and was restored by Sultan Mehmed IV. Part of the dome collapsed again during the earthquake of 1766. Subsequent repairs damaged what was left of the original decoration of Sinan. During World War I the courtyard was used as a weapons depot, and when some of the ammunition ignited, the mosque suffered another fire. Not until 1956 was it fully restored again.

The original complex of Süleymaniye consisted of the mosque itself, a hospital, primary school, public baths, a caravanserai, four Qur'an schools, a specialized school for the learning of *Hadith*[6], a medical college, and a public kitchen which served food to the poor. Many of these structures are still in existence, and the former imaret is now a noted restaurant. The former hospital is now a printing factory owned by the Turkish Army.

2. Description of the Mosque

The Süleymaniye complex is situated within approximately 18 acres of land. It consists of a mosque in the centre and the mausoleums of Sultan Suleyman and Hurrem Sultan in the immediate garden of the mosque. The buildings which serves the public; hospital, school, inns, imaret, fountains and water collection units, were built as a part of the Suleymaniye Foundation. As a medieval traveler and writer records, "this place with the buildings surrounding it is covered with a thousand domes and 3 thousand people are serving here". But a complex of this size were never seen again after the 16th century.

2.1 The Layout of the Süleymaniye Complex

The Süleymaniye Complex has adopted an approach that blends in according to the conditions of the land. During the planning of the second intellectual centre of the time, the buildings that surround the mosque protect the natural settings and the buildings and the streets are serving an approach set up on alive and correct principles.

The front facades of the buildings face Mecca[7], the appearance upon completion of all the buildings from various parts of the city was determined by Sinan and were drawn on paper. The slightly rising hill on the land was transformed into a pyramidal mass which seemed to be the continuation of the foothill.

The Suleymaniye Mosque at the center has a plan close to a square, and again an inner courtyard with a square-like shape and inside a large rectangular shaped area that contains the large courtyard and the adjacent buildings. Due to its huge dimensions, the Suleymaniye Mosque and the building complex is one of the largest construction works in the Ottoman Empire but probably because of the devotion and personal preferences of the Sultan Suleyman, it is generally a plain construction.

In the garden behind the main mosque there are two mausoleums including the tombs of Sultan Suleiman I, his wife Hürrem Sultan and their daughter Mihrimah Sultan. The sultans Suleiman II, Ahmed II and also Saliha Dilasub Sultan and Safiye Sultan, the daughter of Mustafa II, are buried here. Just outside the mosque walls, to the north is the tomb of architect Sinan. (See Fig. 2.9)

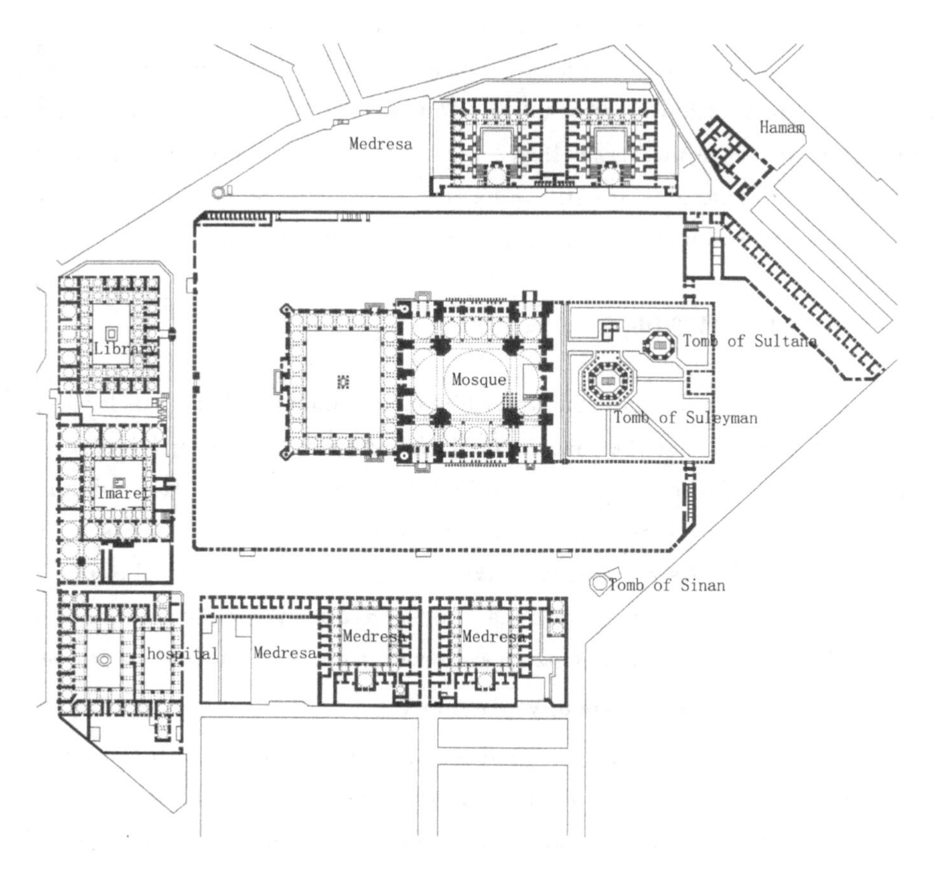

Fig. 2. 9 Layout of the Süleymaniye Mosque

2. 2 The Dome

The dome which makes the external architecture influential also solves the sound and spacing problems of the building. The weight of dome is transferred to ground through four supports, whereas its horizontal thrust is transferred to sides through the two semi-domes and buttresses. The arches used for bearing the main dome of the mosque with 3163m^2 internal area, are placed over four piers. These piers are placed on the foundations with dimensions of 6. 2 x 5. 1 meters. These huge piers which each has a footprint equal to a small mosque were constructed elegantly and altar figures were carved on front and rear surfaces.

At the interior area, the dimensions of the dome are 48. 23m height, 26. 30m diameter and 74cm width. The load of the dome, which is approximately 1000 tons, is transferred to the foundation via the two semi-sphere domes and four piers. It is assumed that each pier with 9. 02 meter height, 1. 4 meter diameter and approximately 30 tons weight, each are transferring the load of 8000 tons to the base.

The main dome is supported with 2 large and 5 small semispherical domes, and surrounded by 30 small domes. There are 2-3 meter spacing between internal and external domes. These spaces are for balancing the temperature in the mosque through seasonal changes and also used for giving way to the main dome. They may also play a role for acoustics.

In the accountancy records of the construction of the mosque, it is specified that 225 pots were used in the coverings of the dome and a hammered irontie beam was placed at the bottom of the dome. 134 scales of linen were used for the grout of the dome. At the middle of the dome a legend of verse was written in flooding Thuluth style[8] and 26 muralists worked for the hand-carved decorations of the domes. The crescents of the minaret, chandeliers and other gilding works were done by using 2137 foreign gold coins. It was recorded that 1010 of these coins were used for the crescent of the main dome.

2.3 The Interiors of the Mosque

The interior of the mosque is almost a square, 59 meters in length and 58 meters in width, forming a single vast space. The dome is flanked by semi-domes, and to the north and south arches with tympana-filled windows, supported by enormous porphyry monoliths. Sinan decided to make a radical architectural innovation to mask the huge north-south buttresses needed to support these central piers. He incorporated the buttresses into the walls of the building, with half projecting inside and half projecting outside, and then hid the projections by building colonnaded galleries. There is a single gallery inside the structure, and a two-story gallery outside.

The interior decoration is subtle, with very restrained use of Iznik tiles[9]. The white marble mihrab and minbar are also simple in design, and woodwork is restrained, with simple designs in ivory and mother of pearl.

The mihrab which raises between chine decorations was made as a single piece having two columns made with raised grooves on top of stalactite base covered with golden gilding. It has a plain structure and was made by using best marbles and completed at the end of April 1554. Additionally the two windows on both sides of the mihrab have brass fencings.

The minbar which was constructed according to the esthetics of the mosque was made up of marble. It was also informed that the crescent of the minbar was gilded with golden coins and kiosk of the minbar is a hanged ball covered with golden leafs.

The mosque receives light through 138 windows. Sinan arranged the sunlight illumination so well and found the sunbeam received by the mosque analogous to the wings of angel Gabriel[10]. On the accounting records of the mosque, the existence of 5 sets of windows on the altar wall, 2 sets of windows on each side wall and 9 ornamental windows with different sizes on the qibla wall were mentioned.

3. The courtyard

The courtyard of the mosque has an area of 19 acres, which separates the three sides from the neighboring buildings and gives a change for viewing the mosque from different angles. There are two artesian wells and three cisterns within the courtyard. The artesian well on the front end has a depth of 65 meters and was opened 35 years ago, the artesian well on the Golden Horn side was opened during the arrangements of the courtyard. This well has a depth of 165 meters with huge amount of water source.

The inner courtyard with area of 1377m^2 has a ground covered with marble and the sides are covered with a total of 28 domes, 7 on each side and 9 on the western sides. Like the other imperial mosques in Istanbul, the Süleymaniye mosque itself is preceded by a monumental courtyard on its west side. The courtyard is of exceptional grandeur with a colonnaded peristyle with columns of marble, granite and porphyry. At the four corners of the courtyard are the four minarets, a number only allowable to mosques endowed by a sultan. The minarets have a total of 10 galleries, which by tradition indicates that Suleiman I was the 10th Ottoman sultan.

Entrance to the inner courtyard is through 3 separate doorways. At this area where the main entrance doorway to the mosque is placed, a water scale which looks like Kaaba[11] is existing at the center of the area.

Notes:

1. Süleymaniye Mosque：苏莱曼清真寺，是土耳其著名清真寺，坐落在伊斯坦布尔金角湾西岸，被称为伊斯坦布尔最美的清真寺。苏莱曼清真寺为奥斯曼帝国第十代素丹苏莱曼一世(1495—1566)于1550—1557年敕建，由奥斯曼帝国鼎盛时期最著名的建筑师锡南(Sinan)设计和督建。清真寺周围有医院、学校、浴场等大小20余处公共设施。清真寺由方形的大厅和前面的庭院组成，规模宏大。大厅的中央由4座柱墩支撑着中央穹顶，其侧推力通过周围一系列较小的穹顶和半穹顶及拱来平衡，延续了拜占庭时代的结构体系。中央大穹顶内径26m、内顶高52m，内部装饰庄重华丽，是奥斯曼土耳其帝国鼎盛时期的见证物。
2. Hagia Sophia：圣索菲亚大教堂，是位于土耳其伊斯坦布尔的宗教建筑，有近1500年的漫长历史，因其巨大的圆顶而闻名于世，是一幢“改变了建筑史”的拜占庭式建筑典范。圣索菲亚大教堂由君士坦丁大帝为供奉智慧之神索菲亚而建，始建于公元325年，后受损于战乱，公元537年查士丁尼皇帝为标榜自己的文治武功进行重建。它作为基督教教堂，持续了9个世纪。公元1453年，奥斯曼土耳其将君士坦丁堡改名为伊斯坦布尔，并将索菲亚大教堂改成清真寺。苏丹穆罕默德二世下令将教堂内所有拜占庭的壁画全部用灰浆遮盖住，所有基督教雕像也被搬出，还在周围修建了4个高大的尖塔。

3. külliye：库里耶。在奥斯曼时期有一种名为“库里耶”的社会管理制度，围绕着一座清真寺，周边建有医院、学校、图书馆、店铺和公共浴池，统一由一个机构进行管理。旁边有一所传授伊斯兰教和科学课程的伊斯兰学院、一所神学学校、一排店铺以及一处墓地。
4. Sinan：锡南(1489—1588)，为安纳托利亚人，出身于信奉基督教的建筑工匠家庭，奥斯曼帝国著名建筑师。锡南青年时承父业为建筑木石工匠，后在巴格达寄居，考察研究当地伊斯兰清真寺、陵墓等建筑，并掌握了伊斯兰文化。1538 年，他被奥斯曼宫廷聘用，被素丹苏莱曼一世任命为宫廷建筑总监。此后 40 多年间，一直在宫廷主持全国的建筑工程。锡南一生设计建造了 79 座清真寺、34 座宫殿、55 所学校、19 座陵墓、33 所公共浴室、16 幢住宅、7 所伊斯兰教经学院、12 家商队客栈和 18 座殡仪馆，此外还建有谷仓、军械库、桥梁、喷泉、医院和大型渠道等。他的设计构思将罗马建筑、波斯建筑和阿拉伯伊斯兰风格融为一体，形成了土耳其建筑的基本格调。他建造的大清真寺多覆盖以宏伟的圆顶，四周耸立着尖塔，构架高大雄伟，布局合理，选料精细，精工细作，装饰雕刻华丽，多呈几何图案，外观色调和谐，庄严肃穆，被称为“伊斯兰建筑大师”。
5. Suleyman the Magnificent：苏莱曼一世(Suleyman I,1494 年 11 月 06 日—1566 年 9 月 7 日)是奥斯曼帝国第 10 位、也是在位时间最长的苏丹。由于苏莱曼一世的文治武功，他在西方被普遍誉为苏莱曼大帝，在伊斯兰发展史上与阿巴斯大帝、阿克巴大帝齐名。他在位时完成了对奥斯曼帝国法律体系的改造，在他的统治下，奥斯曼帝国在政治、经济、军事和文化等诸多方面都进入极盛时期。
6. *Hadith*：《圣训》，是穆斯林对伊斯兰先知穆罕默德说过的话、做过的事的尊称。《圣训》是伊斯兰教重要文化遗产之一，经辑录定本的《圣训集》被视为仅次于《古兰》启示的基本经典，是对《古兰》基本思想的具体阐释，是对伊斯兰教义、教律、教制、礼仪和道德的全面回答和论述，是后世各派法学家立法、执法的第二位渊源和依据，也是历代教职人员、学者进行宣教、立论、立说的依据，故受到穆斯林高度尊崇。
7. Mecca：麦加，意为“荣誉的麦加”，是伊斯兰教的圣地，非穆斯林不得进入。麦加座落在沙特阿拉伯西部赛拉特山区一条狭窄的山谷里，面积不到 760 平方千米，为伊斯兰教创始人穆罕默德诞生地。穆罕默德在麦加创立和传播伊斯兰教。公元 630 年，穆罕默德率兵攻占麦加，宣布麦加是伊斯兰教最神圣的地方，让该地成为穆斯林朝觐的中心。
8. Thuluth style：苏鲁斯体。“苏鲁斯”系阿拉伯语音译，意为“三分之一”，故又译为“三一体”，属于阿拉伯文大楷，是阿拉伯文书法体之一。该体具有端庄、高雅、华丽、壮观等特点，其字体高大而豪放。书写时词与词、字母与字母可以互相交叉、重叠，或串联、盘缠在一起，其中有一半以上的字母有 2～3 种写法，且能以花草、树木、水果、人物、鸟兽、建筑或自然风景等形象编织成变种字体，故书写技术难度较大。阿拉伯人称苏鲁斯体为“书法之母”。苏鲁斯体的应用范围比较广泛，诸如书写《古兰经》的章名、经文警句、机关单位的牌匾、书刊的题目以及各种商业、文艺、体育广告等。穆斯林家庭中的装饰如中堂、匾额、壁挂、镜框等也常使用苏鲁斯体书写。
9. Iznik tiles：伊兹尼克瓷砖，是一种装饰性的陶制品，生产始于 1475 年左右，一直持续到 17 世纪末。伊兹尼克瓷砖融合了奥斯曼风格和中国、巴尔干半岛诸国甚至欧洲等国家和地区的风格，是奥斯曼制陶艺术发展的顶峰。伊兹尼克瓷砖最开始的颜色主要为蓝绿色和暗钴蓝色，之后灰绿色和淡紫色等较为清淡的颜色也出现在陶瓷之上。

10. angel Gabriel：天使加百列，本为炽天使，身份显赫而高贵。加百列为大天使长，位列天堂重要的警卫长职位，担任整个天界的警戒工作，又成为炽天使的最佳后备人选，传信为其主要职能。传说末日审判的号角就是由他吹响的。他被认为象征智慧。
11. Kaaba："克尔白"，是阿拉伯语音译，意为立方体，是麦加禁寺内的一立方体殿宇。据伊斯兰教传说，克尔白由人祖阿丹(Adam)依天上原型而建造。公元623年，穆罕默德奉真主"启示"改定克尔白为礼拜朝向；公元628年，宣布了朝觐克尔白是伊斯兰教的"天命"。公元630年，穆罕默德光复麦加后，清除了克尔白石殿内外360尊偶像，克尔白天房遂成为穆斯林朝拜的中心。

Insights: Classical Order

The Architectural Orders are the ancient styles of classical architecture, each distinguished by its proportions and characteristic profiles and details, and most readily recognizable by the type of column employed. Three ancient orders of architecture—the Doric, Ionic, and Corinthian originated in Greece. (See Fig. 2.10)

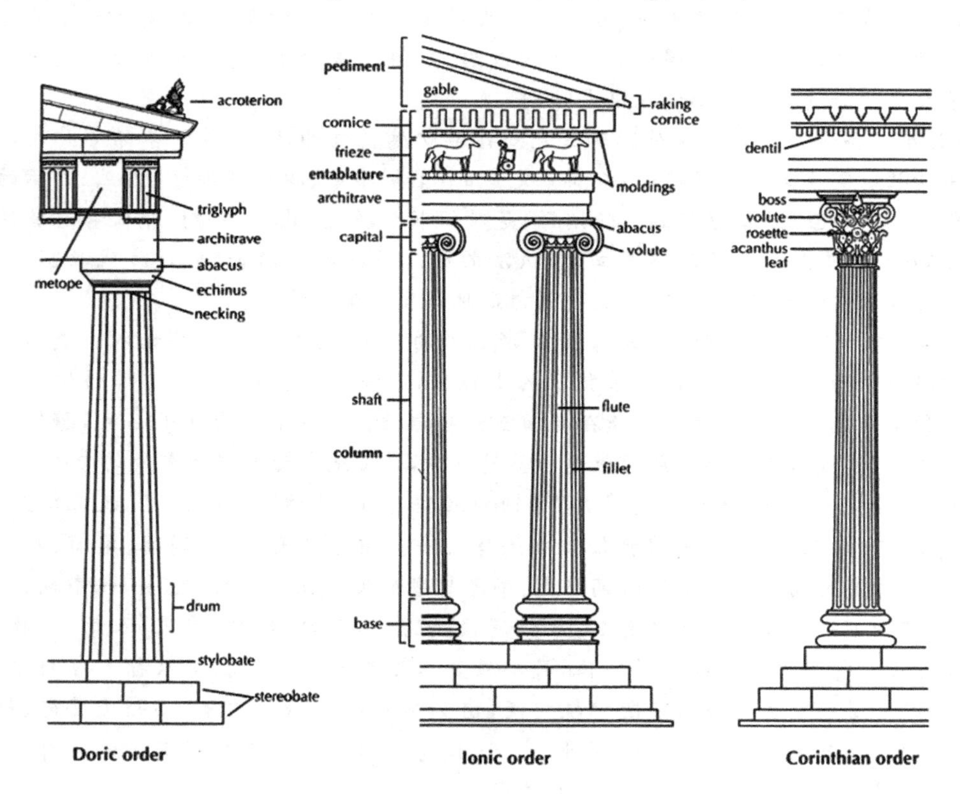

Fig. 2.10 Greek classical order

Doric order is considered the earliest and simplest of the orders, characterized by short, faceted, heavy columns with plain, round capitals and no base. With a height that is only four to eight times its diameter, the most squat of all orders. The shaft of the Doric order is channeled with 20 flutes. The capital consists of a necking which is of a simple form. The echinus is convex and the abacus is square.

The Ionic order is distinguished by slender, fluted pillars with a large base and two opposed volutes, also called scrolls in the echinus of the capital. The echinus itself is decorated with an egg-and-dart motif. The Ionic shaft comes with four more flutes than the Doric counterpart. The Ionic base has two convex moldings which are separated by a scotia. The Ionic order is also marked by an entasis, a curved tapering in the column shaft.

The Corinthian order is the most ornate of the Greek orders, characterized by a slender fluted column having an ornate capital decorated with two rows of acanthus leaves and four scrolls. The shaft of the Corinthian order has 24 flutes. The column is commonly ten diameters high.

The Romans adapted all the Greek orders and developed two orders of their own, basically modifications of Greek orders. It was not until the Renaissance that these were named and formalized as the Tuscan and Composite respectively.

The Tuscan order has a very plain design, with a plain shaft, and a simple capital, base, and frieze. The Tuscan order is characterized by an unfluted shaft and a capital that only consists of an echinus and an abacus. The column is normally seven diameters high.

The Composite order is a mixed order, combining the volutes of the Ionic with the leaves of the Corinthian order. It was considered as a late Roman form of the Corinthian order.

Chapter Three

Palace

A palace is a grand residence, especially a royal residence or the home of astate head or some other high-ranking dignitary, such as a bishop or archbishop, in many parts of Europe, the term is also applied to ambitious private mansions of the aristocracy. The word is also sometimes used to describe a lavishly ornate building for public entertainment or exhibitions.

The earliest known palaces were the royal residences of the Egyptian Pharaohs at Thebes, featuring an outer wall enclosing labyrinthine buildings and courtyards. Other ancient palaces include the Assyrian palaces at Nimrud and Nineveh, the Minoan palace at Knossos, and the Persian palaces at Persepolis and Susa. Palaces in East Asia, such as the imperial palaces of Korea, Thailand, Vietnam, Japan, large stone and wooden structures in the Philippines and China's Forbidden City, consist of many low pavilions surrounded by vast, walled gardens, in contrast to the single building palaces of Medieval Western Europe.

In modern times, the term has been applied by archaeologists and historians to large structures that housed combined ruler, court and bureaucracy in "palace cultures". Many historic palaces are now put to other uses such as parliaments, museums, hotels or office buildings.

The Alhambra[1]

The Alhambra is a palace and fortress complex located in Granada, Andalusia, Spain (See Fig. 3. 1). It was originally constructed as a small fortress in 889 A. D. on the remains of Roman fortifications and then largely ignored until its ruins were renovated and rebuilt in the mid-13th century by the Moorish emir[2] Mohammed ben Al-Ahmar of the Emirate of Granada. Moorish poets described it as "a pearl set in emeralds".

Fig. 3. 1 The Alhambra ①

1. The History of the Alhambra

There is no reference to the Alhambra as being a residence of kings until the 13th century, even though the fortress had existed since the 9th century. The Nasrites were probably the emirs who built the Alhambra, starting in 1238. After reclaiming Andalucía, the conquering Catholic Monarchs restored the Palacio Nazaries, and eventually, the Alhambra's mosque was replaced by a church. After the conclusion of the Christian Reconquista[3] in 1492, the site became the Royal Court of Ferdinand and Isabella and the palaces were partially altered to Renaissance tastes[4]. In 1526 Charles I & V commissioned a new Renaissance palace better befitting of the Holy Roman Emperor[5] in the revolutionary Mannerist style[6] influenced by Humanist philosophy in direct juxtaposition with the Nasrid Andalusian architecture, but which was ultimately never completed due to Morisco rebellions in Granada.

The Alhambra Palace entered a dark age of sorts when it was abandoned to thieves and vagrants during the 18th century. During the time Napoleon controlled Spain, the Alhambra was

① city illustration. com.

used as a barracks for French soldiers, and was almost blown up. After it was declared a national monument in 1870, travelers and romantic artists of all countries had railed against those who scorned the most beautiful of their monuments. Washington Irving, author of Tales of the Alhambra, brought the fortress to the world's attention and prompted the restoration of the palace and the Alhambra gardens.

2. The Layout of Alhambra

Despite long neglect, willful vandalism and some ill-judged restoration, the Alhambra endures as a typical example of Muslim art in its final European stages, relatively uninfluenced by the direct Byzantine influences. The majority of the palace buildings are quadrangular in plan, with all the rooms opening on to a central court, and the whole reached its present size simply by the gradual addition of new quadrangles designed on the same principle though varying in dimensions, and connected with each other by smaller rooms and passages.

Alhambra was extended by the different Muslim rulers who lived in the complex, however, each new section that was added followed the consistent theme of "paradise on earth". Column arcades, fountains with running water, and reflecting pools were used to add to the aesthetic and functional complexity. In every case, the exterior was left plain and austere. Sun and wind were freely admitted. Blue, red, and a golden yellow, all somewhat faded through lapse of time and exposure, are the colors chiefly employed. (See Fig. 3. 2)

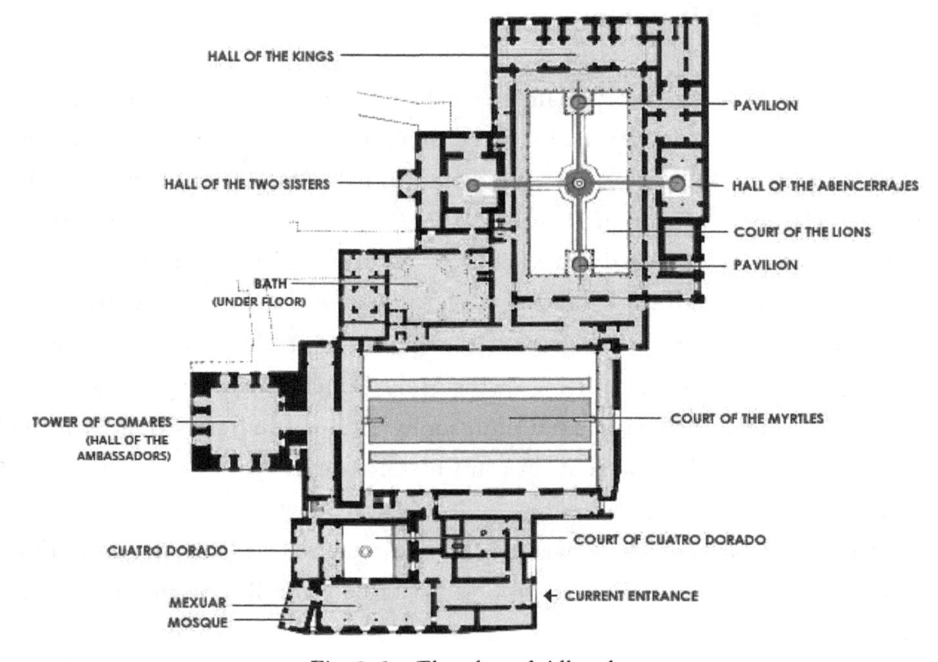

Fig. 3. 2 The plan of Alhambra

3. The Main Courts of Alhambra

3.1 Court of the Myrtles

The present entrance to the Moorish palace is by a small door from which a corridor connects to the Court of the Myrtles, also called the Court of the Blessing or Court of the Pond. Because water was usually in short supply, the technology required to keep these pools full was expensive and difficult. This court is 42 meters long by 22 meters broad, and in the centre there is a large pond set in the marble pavement full of goldfish, and with myrtles growing along its sides. There are galleries on the north and south sides; the southern gallery is 7m high and supported by a marble colonnade. Underneath it to the right was the principal entrance, and over it are three windows with arches and miniature pillars.

3.2 The Court of the Lions

The Court of the Lions is an oblong courtyard, 35 meters in length by 20 meters in width surrounded by a low gallery supported on 124 white marble columns. A pavilion projects into the court at each extremity, with filigree walls and a light domed roof. The square is paved with colored tiles and the colonnade with white marble, while the walls are covered 1.5 meters up from the ground with blue and yellow tiles, with a border above and below of enameled blue and gold. The columns supporting the roof and gallery are irregularly placed. They are adorned by varieties of foliage; about each arch there is a large square of stucco arabesques; and over the pillars is another stucco square of filigree work. (See Fig. 3.3)

Fig. 3.3 The Court of the Lions

In the centre of the court is the Fountain of Lions, an alabaster basin supported by the figures of twelve lions in white marble, not designed with sculptural accuracy but as symbols of strength, power, and sovereignty. Each hour one lion would produce water from its mouth. At the edge of the great fountain there is a poem written by Ibn Zamrak. It praises the beauty of the fountain and the power of the lions, and it also describes their ingenious hydraulic systems and how they actually worked.

Notes:

1. The Alhambra：阿尔罕布拉宫，是西班牙的著名故宫，为中世纪摩尔人在西班牙建立的格拉纳达埃米尔国的王宫，为摩尔人留存在西班牙所有古迹中的精华，有“宫殿之城”和“世界奇迹”之称。阿尔罕布拉宫始建于13世纪阿赫马尔王及其继承人统治期间。宫殿由众多的院落组成，其中以爱神木之院为最大，也最漂亮，其院中水池倒映出的北廊的倒影十分有名。宫殿的布局其实是很严格的，被划分为行政场所、活动仪式场所和王族居所等三大部分，它们既独立又巧妙地相连。在阿尔罕布拉宫中，最著名的是大使厅、两姊妹厅、狮子中庭及大浴场等地方。宫殿中的“桃金娘中庭(Patio de los Arrayanes)”是一处引人注目的大庭院，也是阿尔罕布拉宫最为重要的群体空间，是外交和政治活动的中心。1492年，摩尔人被逐出西班牙后，建筑物开始荒废。1828年，在斐迪南七世资助下，经建筑师何塞·孔特雷拉斯与其子、孙三代进行长期的修缮与复建，才恢复原有风貌。
2. Moorish emir：摩尔人酋长。摩尔人多指在11—17世纪创造了阿拉伯安达卢西亚文化、随后在北非作为难民定居下来的西班牙穆斯林居民或阿拉伯人，是西班牙人及柏柏尔人的混血后代。此词由罗马人最先使用，指罗马毛里塔尼亚省的居民，偶尔也指一般的穆斯林。
3. Christian Reconquista：收复失地运动，是公元718至1492年间，位于西欧伊比利亚半岛北部的西班牙、葡萄牙各国逐渐战胜南部摩尔人政权的运动。公元710年，阿拉伯帝国的穆斯林跨过直布罗陀海峡，在伊比利亚半岛登陆，把伊比利亚南部，西班牙境内的安达卢西亚地区完全占据。
4. Renaissance tastes：文艺复兴风格。文艺复兴风格建筑15—19世纪流行于欧洲，有时也包括巴洛克建筑和古典主义建筑，起源于意大利佛罗伦萨。该风格在造型上排斥象征神权至上的哥特建筑风格，讲究秩序和比例，拥有严谨的立面和平面构图以及从古典建筑中继承下来的柱式系统；提倡复兴古罗马时期的建筑形式，特别是古典柱式比例、半圆形拱券、以穹隆为中心的建筑形体等。
5. Holy Roman Emperor：神圣罗马皇帝，是欧洲中世纪时的一个君主头衔，是历史学家称呼中世纪时获教宗赐予“罗马皇帝”头衔的东法兰克国王和罗马人民的国王，以及1356年后统治神圣罗马帝国的选举君主。首个获教宗加冕为“罗马皇帝”的君主是加洛林王朝的查理曼。最后一个神圣帝国皇帝当选人是弗朗茨二世。
6. Mannerist style：手法主义，是16世纪晚期欧洲的一种艺术风格。其主要特点是追求怪异和不寻常的效果，如以变形和不协调的方式表现空间，以夸张的细长比例表现人物等。建筑史中手法主义指1530—1600年间意大利某些建筑师的作品中体现前期巴洛克风格的倾向。

The Potala Palace[1]

The Potala Palace is located in Lhasa named after Mount Potalaka, the mythical abode of Avalokitesvara[2] (See Fig. 3. 4). The Potala Palace was the chief residence of the Dalai Lama through the history of Tibet. Lozang Gyatso, the Great Fifth Dalai Lama[3], started the construction of the Potala Palace in 1645 after one of his spiritual advisers pointed out that the site which was situated between Drepung and Sera monasteries[4] and the old city of Lhasa was ideal as a seat for government.

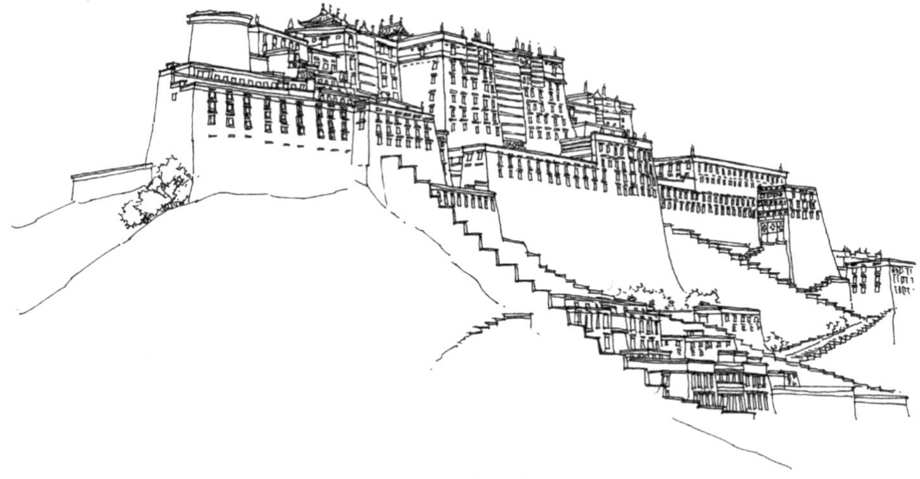

Fig. 3. 4 The Potala Palace

1. The History of Construction of Potala Palace

The site on which the Potala Palace is built used to have a palace erected by Songtsen Gampo[5] on the Red Hill. The site was used as a meditation retreat by King Songtsen Gampo, where he in 637 built the first palace there in order to greet his bride, Princess Wencheng of China's in Tang Dynasty. The Potala contains two chapels on its the northwest corner that conserve parts of the original palace building. One is Phakpa Lhakhang, the other is the Chogyel Drupuk, a recessed cavern which is identified as Songtsen Gampo's meditation cave. These chapels are considered the oldest surviving structures on the hill and the most sacred ones since the Potala's most venerated statue, the Arya Lokeshvara(观世音菩萨)[6] is housed inside the Phapka Lhakhang and it draws thousands of Tibetan pilgrims each day.

Lozang Gyatso, the Great Fifth Dalai Lama, started construction of the Potala Palace in 1645 after one of his spiritual advisers Konchog Chophel pointed out thatthe Red Hill sitting between Drepung(哲蚌寺) and Sera monasteries(色拉寺) and the old city of Lhasa was ideal as a seat of government. Dalai Lama and his government moved into the Potrang Karpo ("White Palace") in 1649. The Potala was then used as a winter palace of Dalai Lama from that time. The Potrang Marpo ("Red Palace") was added between 1690 and 1694 and got its name from a sacred hill in India. It is said that a rocky point refered to the God of Mercy whom the Indians call Avalokitesvara. The Construction lasted until 1694, some twelve years after death of Fifth Dalai Lama. The external structure was built within 3 years while the interior, together with its furnishings took 45 years to complete.

2. Description of Potala Palace

Built at an altitude of 3,700 meters on the side of Marpo Ri in the center of Lhasa Valley, the Potala Palace is not unlike a fortress in appearance with its vast inward-sloping walls broken only in the upper parts by straight rows of many windows and its flat roofs at various levels. At the south base of the rock is a large space enclosed by walls and gates, with great porticos on the inner side. A series of tolerably easy staircases, broken by intervals of gentle ascent, leads to the summit of the rock which is occupied by the palace.

The building complex measures 400 meters east-west and 350 meters north-south, with sloping stone walls averaging 3 meters thick, and 5 meters thick at the base, where copper liquid was poured into the foundations to help against earthquakes. The Palace buildings stand 13 stories high and contain over 1,000 rooms, 10,000 shrines and 200,000 statues. The palaces are 117 meters high above Marpo Ri, and more than 300 meters above the valley floor.

The Potala was the residence of the Dalai Lama and his large staff, the interior space of Potala is in excess of 130,000 square meters, fulfilling numerous functions, it was the seat of Tibetan government and housed a school for training monks and it was one of Tibet's major pilgrimage destinations because of the tombs of past Dalai Lamas.

2.1 The White Palace

The White Palace or Potrang Karpo makes up the living quarters of the Dalai Lama. The first White Palace was built during the lifetime of the Fifth Dalai Lama, he and his government moved into it in 1649. It then was extended to its size today by the thirteenth Dalai Lama in the early 20th century. The palace was for secular uses and contained the living quarters, offices, the seminary and the printing house. A central, yellow-painted courtyard known as a Deyangshar separates the living quarters of Dalai Lama and his monks with the Red Palace, the other side of the sacred Potala, which is completely devoted to religious study and prayer.

The White Palace of Potala is a majestic 7-storey palace with various halls and galleries, the inside of which is painted with colorful resplendent Buddha murals by masters. The white outside walls of White Palace suggests source of its name and the symbol of mercy of Buddhism. The kalsomine on the walls is mainly made from lime and milk, which are consecrated by a great number of pious Buddhism believers every year. (See Fig. 3. 5)

Fig. 3. 5 The facade of the White Palace

The zig-zag porch outside the White Palace is the access to the entrance; the courtyard on the eastern mountainside was Dalai Lamas' exclusive theater and the site for outdoor Buddhist activity. On both sides of the courtyard is the official training school.

The White Palace and the Red Palace are connected by the Zhasha building where more than 25,000 monks used to live together there. Zhasha is actually located at the west of the Red Palace but usually seen as a part of the White Palace because of its white walls.

Cuoqinsha(措钦夏) on the fourth floor is the largest hall of the White Palace. In the hall a throne for Dalai Lama is set up under a plaque inscribed by Emperor Tongzhi of Qing Dynasty (1644—1911). It was on this floor that important Buddhist ceremonies such as the enthronement of Dalai Lama took place.

The fifth and sixth floors are living and working areas where Dalai Lamas enjoyed the privilege of living in the Sunlight Hall on the garret floor of the palace, the name of which comes from the sunshine shooting from some part of the roof. The Sunlight Hall comprises two parts: the western hall was built first and the eastern one modeled the western afterward where the thirteenth and fourteenth Dalai Lamas lived and worked respectively and only senior monk and layman officials were allowed to enter.

2.2 Red Palace

Covering an overall floorage of some 16,000 square meters with seven actual floors and six supporting foundation floors, the Red Palace of Potala contains some marvelous structures such as the Dalai Lamas' stupas, Lha-khang(Buddha hall), chanting sutra hall. Distinguishing from the kalsomine of the White Palace, the paint of Red Palace is extracted from a local plant signifying wisdom and strength. The Red Palace was once the supreme religionary ruling center of Tibet where important Buddhist ceremonies were held.

Floors from the first to the seventh of the Red Palace were used as storeroom, the seventh to the ninth as Buddha halls and sutra halls, and the tenth to the thirteenth are set up with windows and inner skylights. Around the skylights are the stupas of the Dalai Lamas and some halls. (See Fig. 3. 6)

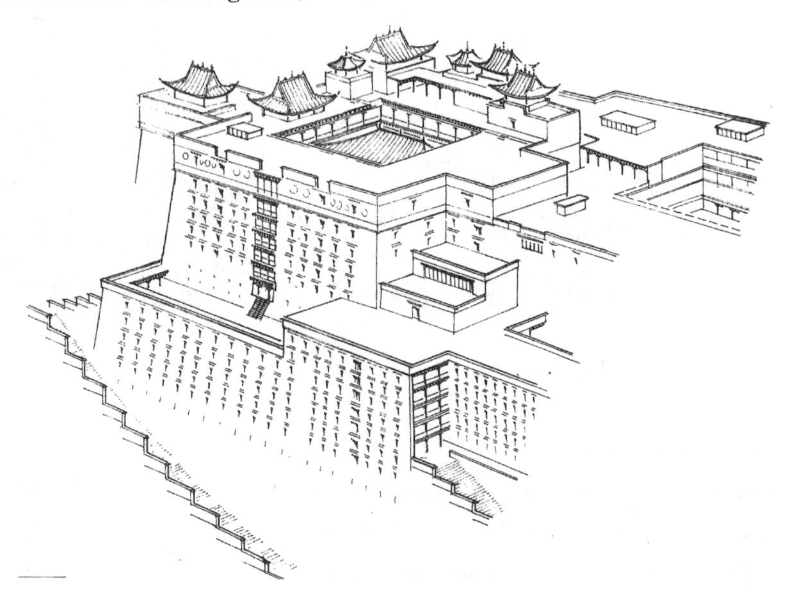

Fig. 3. 6 The bird view of Red Palace

Being the core of Potala, Red Palace is famous as a house of great treasure of Buddhist arts, such as the West Hall on the fifth floor, holding over 1250 square meters is largest and the enthronement of a Dalai Lama was held here; The 14-meter-high stupa of the fifth Dalai Lam built with 3700 kilograms of gold and about 15,000 diamonds is great treasure without any doubt. The Dhammassami Hall and the Saint Avalokitesvara Hall are said the only two existing structures constructed during the reign of Songtzen Gampo.

The central hall of the Red Palace the West Hall consists of four great chapels that proclaim the glory and power of the builder of the Potala, the Fifth Dalai Lama. Special cloth from Bhutan wraps the Hall's numerous columns and pillars. The hall is noted for its fine murals reminiscent of Persian miniatures[7], depicting events in the Fifth Dalai Lama's life, including the famous scene of his visit to Emperor Shun Zhi in Beijing located on the east wall outside the entrance. On the north side of the central hall is the holiest shrine of the Potala the Saint's Chapel which can be dated to the seventh century. The North Chapel centers on a crowned Sakyamuni Buddha[8] on the left and the Fifth Dalai Lama on the right seated on magnificent gold thrones. On the far left of the chapel is the gold stupa tomb of the Eleventh Dalai Lama who died as a child with rows of benign Medicine Buddhas who were the heavenly healers. On the right of the chapel contains a small ancient jewel encrusted statue of Avalokiteshvara, his two attendants and his incarnations including Songsten Gampo and the first four Dalai Lamas.

The South Chapel centers on Padmasambhava[9], the 8th century Indian magician and saint. His consort Yeshe Tsogyal, a gift from the King is by his left knee and his other wife from his native land of Swat by his right. On his left hand eight of his holy manifestations meditate with an inturned gaze. On his right hand eight wrathful manifestations wield instruments of magic powers to subdue the demons of the Bön faith[10].

The East chapel is dedicated to Tsong Khapa[11], founder of the Gelug tradition. His figure is surrounded by lamas from Sakya Monastery who had briefly ruled Tibet and formed their own tradition until converted by Tsong Khapa.

West Chapel contains the five golden stupas. The enormous central stupa, 14.85 meters high, contains the mummified body of the Fifth Dalai Lama. This stupa is built of sandalwood and is remarkably coated in 3,727 kg of solid gold and studded with 18,680 pearls and semi-precious jewels. On the left is the funeral stupa for the Twelfth Dalai Lama and on the right that of the Tenth Dalai Lama. On the floor below, a low, dark passage leads into the Dharma Cave where Songsten Gampo is believed to have studied Buddhism. In the holy cave are images of Songsten Gampo, his wives, his chief minister and Sambhota[12] who developed Tibetan writing by borrowing Sanskrit.

Notes:

1. The Potala Palace：布达拉宫，坐落于西藏自治区首府拉萨市西北玛布日山上，是世界上海拔最高，集宫殿、城堡和寺院于一体的宏伟建筑，也是西藏最庞大、最完整的古代宫堡建筑群。布达拉宫最初为吐蕃王朝赞普松赞干布为迎娶尺尊公主和文成公主而兴建。1645年，五世达赖洛桑嘉措重建布达拉宫之后，成为历代达赖喇嘛冬宫居所，以及重大宗教和政治仪式举办地，也是供奉历世达赖喇嘛灵塔之地，旧时西藏政教合一的统治中心。

布达拉宫依山垒砌，主体建筑分为白宫和红宫两部分。白宫是达赖喇嘛的冬宫，也曾是原西藏地方政府的办事机构所在地，高七层。红宫最主要的建筑是历代达赖喇嘛的灵塔殿，共有五座，分别是五世、七世、八世、九世和十三世。宫殿高 200 余米，外观 13 层，内为 9 层。群楼重叠，殿宇嵯峨，气势雄伟，是藏式古建筑的杰出代表。布达拉宫前辟有布达拉宫广场，是世界上海拔最高的城市广场。

2. Avalokitesvara：观世音菩萨，是佛教中慈悲和智慧的象征，无论在大乘佛教还是在民间信仰，都具有极其重要的地位。以观世音菩萨为主导的大慈悲精神，被视为大乘佛教的根本。观世音是西方三圣中的一尊，是继承阿弥陀佛位的菩萨。观世音菩萨具有平等无私的广大悲愿，当众生遇到任何的困难和苦痛时，如能至诚称念观世音菩萨，就会得到菩萨的救护。在佛教的众多菩萨中，观世音菩萨最为民间所熟知和信仰。所谓"家家阿弥陀，户户观世音"。
3. the Great Fifth Dalai Lama：五世达赖喇嘛阿旺罗桑嘉措(1617—1682 年)，出生于西藏山南的琼结家族，是西藏古老而显赫的家族之一。五世达赖喇嘛统治时期结束了西藏分裂局面，建立了噶厦地方政权，巩固了封建农奴制度，发展了与清朝中央王朝的关系。达赖喇嘛在布达拉宫的修建，在藏族文化史、宗教史、建筑史、医学史、艺术史等方面均有高深造诣。
4. Drepung and Sera monasteries：哲蚌寺与色拉寺，位于拉萨郊区，是藏传佛教格鲁派主要寺院，与甘丹寺一起，被称为拉萨三大寺，在历史上对西藏地方政府的政策与统治有重要影响。
5. Songtsen Gampo：松赞干布(617—650 年)，为吐蕃王朝第 33 任赞普，在位期间(629—650 年)，平定吐蕃内乱，确立了吐蕃的政治、军事、经济及法律等制度，将都城从山南琼结迁到逻些(拉萨)。松赞干布娶尼泊尔尺尊公主与唐朝文成公主，将佛教引入吐蕃，修建了西藏第一座佛殿大昭寺，并在红山之上修建了九层高的布达拉宫。
6. Arya Lokeshvara：布达拉宫帕巴拉康殿内正中央供奉的檀香木质自在观音像，为松赞干布所依本尊，属布达拉宫的稀世珍品。殿门上方悬挂清朝同治皇帝御书"福田妙果"匾额。
7. Persian miniatures：细密画，是波斯艺术的重要门类，一种精细刻画的小型绘画，主要作书籍的插图和封面、扉页、徽章、盒子、镜框等物件上和宝石、象牙首饰上的装饰图案，画于羊皮纸、纸或书籍封面的象牙板或木板上。题材多为人物肖像、图案或风景，也有风俗故事。
8. Sakyamuni Buddha：释迦牟尼佛，为佛教的创立者，是古代中印度迦毗罗卫国的释迦族人，存在于公元前 565 年左右。他大概 29 岁时出家，35 岁时得到佛陀的自觉。他的足迹遍布恒河流域，向各阶层说法教化。
9. Padmasambhava：莲花生大士，是印度佛教史上最伟大的大成就者之一。公元八世纪，应藏王赤松德赞迎请入藏弘法，成功创立了西藏第一座佛、法、僧三宝齐全的佛教寺院——桑耶寺。他奠定了西藏佛教的基础。藏传佛教尊称他为古汝仁波切(意为大宝上师)，也称莲花生。

10. Bön faith：苯教，也被称为古象雄佛法，发源于西藏古象雄的冈底斯山和玛旁雍措湖一带。苯教不仅仅涉及到宗教，还涵盖了民风民俗、天文、历算、藏医、哲学、因明学（逻辑）、辩论学、美术、舞蹈、音乐等方面，是西藏及其周边地区人们重要的精神信仰，至今仍对西藏人民的精神文化生活发挥着不可缺少的作用。
11. Tsong Khapa：宗喀巴（1357—1419 年），是藏传佛教格鲁派（黄教）的创立者、佛教理论家。宗喀巴本名罗桑扎巴（善慧称吉祥），青海湟中县人，父亲是蒙古族，母亲是藏族。宗喀巴在他 38 岁时，改戴黄色桃形僧帽，表示他区别于其他教派，不同于那些败坏戒律的修行者，决心继承和遵守印度大师释迦室利所规定的戒律。
12. Sambhota：吞弥·桑布扎，公元 7 世纪人，为藏文创造者，也是一名翻译家，曾经翻译多部书籍。松赞干布统一吐蕃之后，在公元 7 世纪上半叶派遣了 16 名聪颖俊秀青年前往天竺拜师学习梵文。吞弥·桑布扎学成回到吐蕃，以梵文 50 个根本字母为楷模，结合藏语言特点，创制了 4 个母音字及 30 个子音字的藏文。

The Forbidden City[1]

The Forbidden City, located in the centre of Beijing, China, was the Chinese imperial palace from the Ming Dynasty to the end of the Qing Dynasty, it served as the home of emperors and their households as well as the ceremonial and political centre of Chinese government for almost 500 years. The palace complex exemplifies traditional Chinese palatial architecture, and has influenced cultural and architectural developments in East Asia. Since 1925, the Forbidden City has been under the charge of the Palace Museum[2], whose extensive collection of artwork and artifacts were built upon the imperial collections. (See Fig. 3. 7)

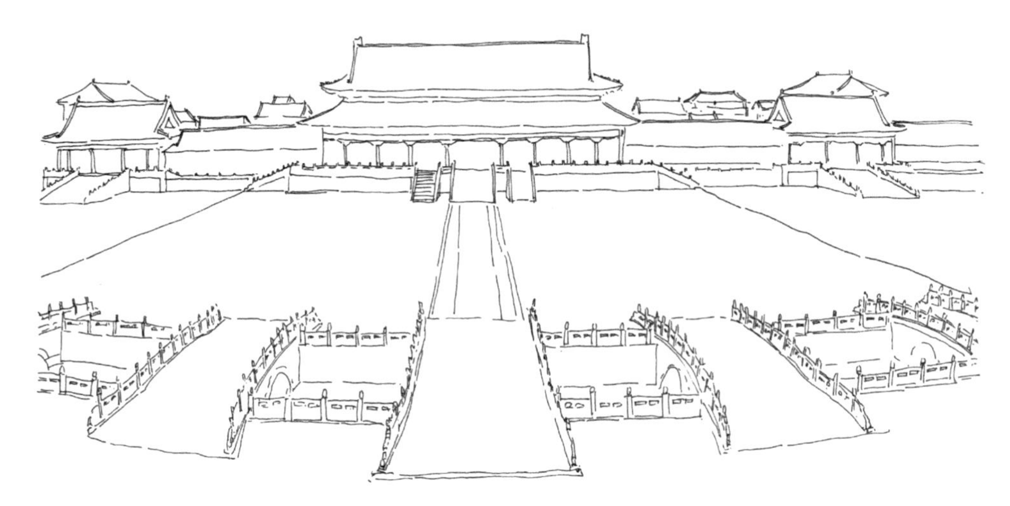

Fig. 3. 7 The Forbidden City

1. The History of the Forbidden City

The History of the Forbidden City or "Purple" Forbidden City, began in the 15th century when it was built as the palace of the Ming emperors of China. The construction began in 1406 and lasted 14 years, more than a million workers were engaged in the work. The building material used include whole logs of precious zhennan wood (楠木) found in the jungles of south-western China, large blocks of marble from quarries near Beijing and the "golden bricks" (金砖)[3] which were specially baked paving bricks from Suzhou(苏州).

When Emperor Yongle moved the capital from Nanjing to Beijing in 1420, the Forbidden City was the seat of the Ming Dynasty. When Manchus had achieved supremacy in northern China, and a ceremony was held at the Forbidden City to proclaim the young Shunzhi Emperor(顺治皇帝) as ruler of all China under the Qing dynasty in 1644. The Qing rulers changed the names on some of the principal buildings to emphasize "Harmony" rather than "Supremacy".

In 1860, during the Second Opium War[4], Anglo-French forces(英法联军) took control of the Forbidden City and occupied it until the end of the war. In 1900 Empress Dowager Cixi(慈禧太后) fled from the Forbidden City during the Boxer Rebellion(义和团), leaving it to be occupied by forces of the treaty powers until the following year.

After being the home of 24 emperors—14 of the Ming Dynasty and 10 of the Qing dynasty—the Forbidden City ceased being the political centre of China in 1912 with the abdication of Puyi(溥仪), the last Emperor of China, he was evicted from the palace in 1924 and a year later, the Palace Museum was established.

2. Description of the Forbidden City Architecture

The Forbidden City, covers 720,000 square meters and remains the central north-south axis of Beijing. This axis extends south through Tian'anmen gate to Tian'anmen Square, the ceremonial centre of the People's Republic of China, and to Yongdingmen(永定门). It extends north through Jingshan Hill(景山) to the Bell and Drum Towers.

The Forbidden City is in rectangle, measuring 961 meters from north to south and 753 meters from east to west. It consists of 980 surviving buildings with 8,886 bays of rooms without including various antechambers. (See Fig. 3.8)

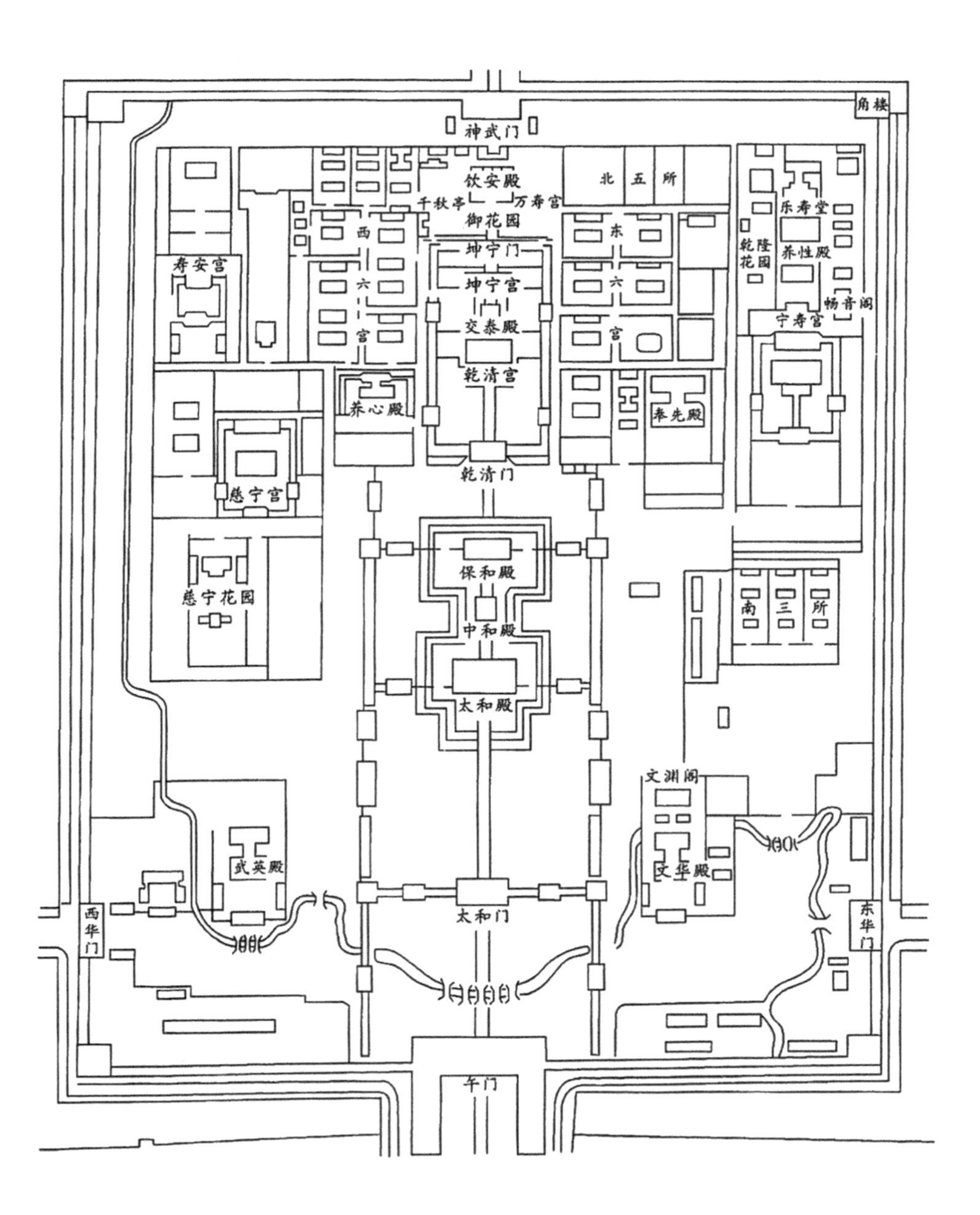

Fig. 3. 8 The layout of the Forbidden City

2.1 Gates

The Forbidden City is surrounded by a 7.9 meters high city wall and a 6 meters deep by 52 meters wide moat. The walls are 8.62 meters wide at the base, tapering to 6.66 meters at the top. These walls served as both defensive walls and retaining walls for the palace. They were constructed with a rammed earth core, and surfaced with three layers of specially baked bricks on both sides, with the interstices filled with mortar. The wall is pierced by a gate on each side. At the southern end is the main Meridian Gate (午门). To the north is the Gate of Divine Might(神武门) which faces Jingshan Park. The east and west gates are called the "East Glorious Gate"(东华门) and "West Glorious Gate"(西华门).

The Meridian Gate has two protruding wings forming a square before it. The gate has five gateways, the central gateway is part of the Imperial Way, a stone flagged path that forms the central axis of the Forbidden City and the ancient city of Beijing leading from the Gate of China(中华门) in the south to Jingshan(景山) in the north. Only the Emperor may walk or ride on the Imperial Way, the Empress can walk on it on the occasion of her wedding, and top students were allowed to walk on it after the Imperial Examination.

At the four corners of the wall sit towerswith intricate roofs boasting 72 ridges, which are the most visible parts of the palace outside the walls.

2.2 Outer Court

The Forbidden City is traditionally divided into two parts : the Inner Court and Outer Court. The Inner Court or Back Palace includes the northern sections of the palace and was the residence of the Emperor and his family, and the Outer or Front Court includes the southern sections and was used for ceremonial purposes and state affairs.

Entering from the Meridian Gate, one encounters a large square, pierced by the meandering Inner Golden Water River which is crossed by five bridges. Beyond the square stands the Gate of Supreme Harmony (太和门), behind the gate is a big square where a three-tiered white marble terrace rising from the center, on which the focus of the palace complex Hall of Supreme Harmony(太和殿), the Hall of Central Harmony(中和殿), and the Hall of Preserving Harmony(保和殿)standing on.

The Hall of Supreme Harmony is the largest and rises some 30 meters above the level of the surrounding square. It is the ceremonial centre of imperial power and the largest surviving wooden structure in China. It is nine bays wide and five bays deep, the numbers 9 and 5 being symbolically connected to the majesty of the Emperor. Set into the ceiling at the centre of the hall is an intricate caisson decorated with a coiled dragon, from the mouth of which issues a chandelier-like set of metal balls, called the "Xuanyuan Mirror"(轩辕镜). In the Ming Dynasty, the Emperor held court here to discuss state affairs. During the Qing Dynasty, the Hall of Supreme Harmony was only used for ceremonial purposes such as coronations, investitures and imperial weddings.

The Hall of Central Peace is a smaller, square hall used by the Emperor to prepare and rest before and during ceremonies. The Hall of Preserving Harmony was used for ceremonies rehearsing and the site of the Imperial examination.

In the southwest and southeast sides of the Outer Court are the halls of Military Eminence(武英殿) and Literary Glory(文华殿). The former one was used for the Emperor to receive ministers and later housed the Palace's printing house. The latter one was used for ceremonial lectures by highly regarded Confucian scholars, and then became the office of the Grand Secretariat where a copy of the Siku Quanshu (四库全书)was stored.

2.3 Inner Court

The Inner Court is separated from the Outer Court by an oblong courtyard, it was the home of the Emperor and his family. In Qing Dynasty, the Emperor lived and worked almost exclusively in the Inner Court.

At the centre of the Inner Court is another set of three halls Palace of Heavenly Court (乾清宫), Hall of Union(交泰殿), and the Palace of Earthly Tranquility(坤宁宫)from south to north. Smaller in size than the Outer Court halls, the three Inner Court halls were the official residences of the Emperor and the Empress. Since the Emperor represented the Heavens, he would occupy the Palace of Heavenly Purity while the Empress represented the Earth, she would occupy the Palace of Earthly Tranquility, in between whom was the Hall of Union where the Heavens and earth mixed to produce harmony.

The Palace of Heavenly Purity is a double-eaved building and set on a single-level white marble platform. It is connected to the Gate of Heavenly Purity to its south by a raised walkway. In the Ming Dynasty, it was the residence of the Emperor, but from the Yongzheng Emperor(雍正皇帝) of the Qing Dynasty, the Emperor lived at the smaller Hall of Mental Cultivation(养心殿), the Palace of Heavenly Purity then became the Emperor's audience hall. Above the throne hangs a tablet reading "Justice and Honour" (正大光明).

The Palace of Earthly Tranquility is a double-eaved building, 9 bays wide and 3 bays deep. In the Ming Dynasty, it was the residence of the Empress. In the Qing Dynasty, large portions of the palace were converted for Shamanist(萨满)[5] worship. From the reign of the Yongzheng Emperor, the Empress moved out of the Palace, but two rooms were retained for use on the Emperor's wedding night.

Between these two palaces is the Hall of Union which is square in shape with a pyramidal roof. Stored here are the 25 Imperial Seals of the Qing Dynasty as well as other ceremonial items.

Behind these three halls lies the Imperial Garden. Relatively small and compact in design, the garden nevertheless contains several elaborate landscaping features.

3. The Symbolism of the Forbidden City

The design of the Forbidden City, from its overall layout to the smallest detail, was meticulously planned to reflect philosophical and religious principles, and above all to symbolize the majesty of Imperial power.

Yellow is the color of the Emperor, thus almost all roofs in the Forbidden City bear yellow glazed tiles except two examples. The library at the Pavilion of Literary Profundity (文渊阁) had black tiles because black was associated with water, and thus fire-prevention. Similarly, the Crown Prince's residences have green tiles because green was associated with wood, and thus growth.

The main halls of the Outer and Inner courts are all arranged in groups of three—the shape of the Qian trigram[6], representing Heaven. The residences of the Inner Court on the other hand are arranged in groups of six-the shape of the Kun trigram[7], representing the Earth.

The sloping ridges of building roofs are decorated with a line of statuettes led by a man riding a phoenix and followed by an imperial dragon. The number of statuettes represents the status of the building—a minor building might have 3 or 5. The Hall of Supreme Harmony has 10, the only building in the country to be permitted this in Imperial times. As a result, its 10th statuette, called a "Hangshi", or "ranked tenth" is also unique in the Forbidden City.

The layout of buildings follows ancient customs laid down in the Classic of Rites. Thus, ancestral temples are in front of the palace. To the south of the Forbidden City were two important shrines—the Imperial Shrine of Family or the Imperial Ancestral Temple (太庙) and the Imperial Shrine of State (太社稷), where the Emperor would venerate the spirits of his ancestors and the spirit of the nation respectively.

Notes:

1. The Forbidden City：故宫，中国明清两代的皇家宫殿，旧称紫禁城，位于北京中轴线的中心，是中国古代宫廷建筑之精华。北京故宫是一座长方形城池，南北长 961 米，东西宽

753 米,四面围有高 10 米的城墙,城外有宽 52 米的护城河。北京故宫于明成祖永乐四年(1406 年)开始建设,以南京故宫为蓝本营建,到永乐十八年(1420 年)建成。故宫是世界上现存规模最大、保存最为完整的木质结构古建筑群。紫禁城建筑分为外朝和内廷两部分。外朝的中心为太和殿、中和殿、保和殿,称三大殿,是国家举行盛大典礼的地方。内廷的中心是乾清宫、交泰殿、坤宁宫,称后三宫,是皇帝和皇后居住的正宫。故宫占地 72 万平方米,建筑面积约 15 万平方米,有大小宫殿 70 多座,房屋 9000 余间。

2. Palace Museum:故宫博物院,位于北京故宫紫禁城内院,建立于 1925 年 10 月 10 日,是在明朝、清朝两代皇宫及其收藏的基础上建立起来的中国综合性博物馆,也是中国最大的古代文化艺术博物馆,其文物收藏主要来源于清代宫中旧藏。
3. golden bricks:金砖,又称御窑金砖,是古时专供宫殿等重要建筑使用的一种高质量的铺地方砖。因其质地坚细,敲之若金属般铿然有声,故名金砖。明清以来金砖受到历代帝王的青睐,成为皇宫建筑的专用产品。明代永乐年间,明成祖朱棣建造紫禁城,苏州陆慕砖窑由于质量优良,博得了永乐皇帝的称赞,赐名窑场为御窑。
4. Opium War:鸦片战争,是 1840 到 1842 年英国对中国发动的一场战争,也是中国近代史的开端。闭关锁国后的清朝逐步落后于世界,但是在外贸中,一直处于贸易顺差地位。为了扭转对华贸易逆差,英国开始向中国走私毒品鸦片,获取暴利。1840 年 6 月,英军舰船 47 艘、陆军 4000 人抵达广东珠江口外,封锁海口,鸦片战争开始。战争以中国失败并赔款割地告终。清政府签订了中国历史上第一个不平等条约《南京条约》。
5. Shamanist:萨满,原词含有智者、晓彻、探究等意,后逐渐演变为萨满教巫师即跳神之人的专称。“萨满”通行于欧亚大陆与北美大陆北方各民族中。萨满被称为神与人之间的中介者,通过萨满的舞蹈、击鼓、歌唱来完成精神世界对神灵的邀请或引诱,使神灵以所谓“附体”的方式附着在萨满体内,并通过萨满的躯体完成与凡人的交流,上述神秘仪式即被称为“跳神”或“跳萨满”。
6. Qian trigram:乾卦,易经学说专用名词,是易经六十四卦的第一卦。“乾”(qián),代表天,天的特性是强健。象曰:天行健,君子以自强不息。乾卦讲的是一个事物从发生到繁荣的过程,即春生夏长。
7. Kun trigram:坤卦,是《易经》六十四卦之第二卦。坤为地,地是静止的,代表当前关系是静止状态,变化较少;地是广大的,地上万物生长,象征当前关系比较宽松而悠闲。

Palace of Versailles[1]

The Palace of Versailles is a royal château, some 20km southwest of the centre of Paris. Versailles was the seat of political power in the Kingdom of France from 1682 when Louis XIV[2] moved the royal court from Paris until the royal family was forced to return to the capital in October 1789. Versailles is famous not only as a building, but as a symbol of the system of absolute monarchy of the Ancien Régime[3]. (See Fig. 3. 9)

Fig. 3. 9 The Palace of Versailes

1. History of Palace of Versailles

Versailles was a small village dating from the 11th century when the château was built, the village centered on a small castle and church and the area was governed by a local lord. Its location on the road from Paris to Dreux and Normandy brought some prosperity to the village but following an outbreak of the Plague and the Hundred Years' War[4], the village was largely destroyed and its population sharply declined.

In 1624, a small château was constructed of stone and red brick, Louis XIV had played and hunted at the site as a boy. He settled on the royal hunting lodge at Versailles and over the following decades had it expanded into one of the largest palaces in the world. Beginning in 1661, the architect Louis Le Vau, landscape architect André Le Nâtre, and painter-decorator Charles Lebrun began a detailed renovation and expansion of the château. This was done to fulfill Louis XIV's desire to establish a new centre for the royal court.

The first phase of the expansion culminated in the addition of three new wings of stone which surrounded Louis XIII's original building on the north, south and west. Charles Le Brun designed and supervised the elaborate interior decoration, and André Le Nâtre landscaped the extensive Gardens of Versailles. During the second phase of expansion, two enormous wings north and south of the wings flanking the Royal Courtyard were added by the architect Jules Hardouin-Mansart. He also replaced Le Vau's large terrace, facing the garden on the west, with what became the most famous room of the palace, the Hall of Mirrors. Work was sufficiently advanced by 1682, that Louis XIV was able to proclaim Versailles his principal residence and the seat of the government of the Kingdom of France.

In 1738, Louis XV remodeled the king's petit apartment on the north side of the Court de Marble, originally the entrance court of the old château, and built a pavilion not far from the Grand Trianon, the Petit Trianon. At the north end of the north wing, a theatre, the Opéra, was completed in 1770, in time for the marriage of Louis-Auguste Dauphin of France to the Austrian Archduchess Marie Antoinette. After his accession to the throne, Louis XVI made few changes to the palace, primarily to the royal private apartments.

During the French Revolution, Versailles fell into disrepair and most of the furniture was sold. Some restoration work was undertaken by Napoleon[5] and Louis XVIII, but the principal effort to restore and maintain Versailles was initiated by Louis-Philippe[6] when he created the Musée de l'Histoire de France dedicated to "all the glories of France".

2. The Description of Versailles Palace

The palace that we recognize today was largely completed in 1715, the eastern facing palace has a U-shaped layout and symmetrical advancing secondary wings terminating with the Dufour Pavilion on the south and the Gabriel Pavilion to the north creating an expansive Royal Court. Flanking the Royal Court are two enormous asymmetrical wings that result in a facade of 1,319 feet in length . The palace has 700 rooms, more than 2,000 windows, 1,250 fireplaces and 67 staircases.

The facade of the original lodge is preserved on the entrance front, built of red brick and cut stone embellishments, the U-shaped layout surrounds a black-and-white marble courtyard. In the center, a 3-storey avant-corps fronted with eight red marble columns supporting a gilded wrought-iron balcony is surmounted with a triangle of lead statuary surrounding a large clock, whose hands were stopped upon the death of Louis XIV. The rest of the facade is completed with columns, painted and gilded wrought-iron balconies and dozens of stone tables decorated with consoles holding marble busts of Roman emperors. Atop the mansard slate roof are elaborate dormer windows and gilt lead roof dressings. (See Fig. 3. 10)

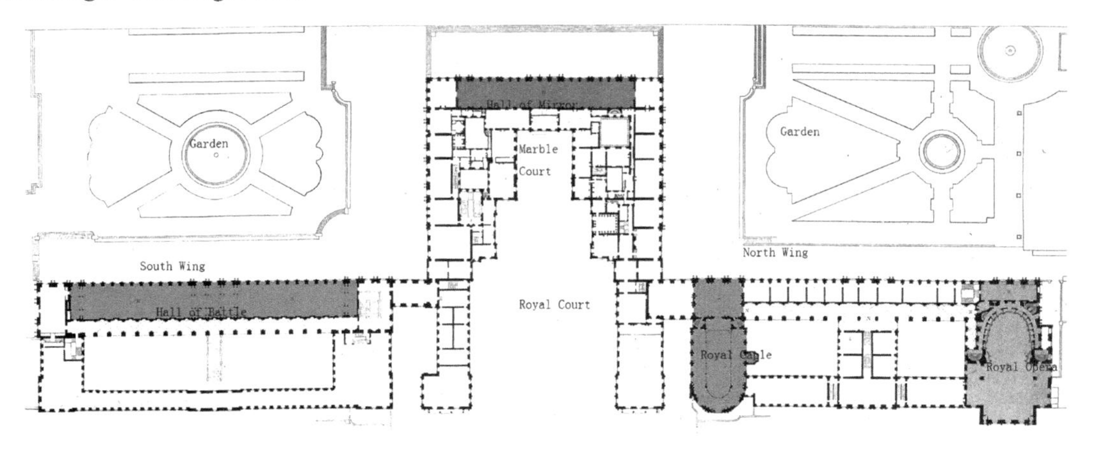

Fig. 3. 10 Layout of Versailles

Inspired by the architecture of baroque Italian villas but executed in the French classical style, the garden front and wings were encased in white cut stone known as the enveloped in 1661—1678. The exterior features an arcaded, rusticated ground floor, supporting a main floor with round-headed windows divided by reliefs and pilasters or columns. The attic storey has square windows and pilasters and crowned by a balustrade bearing sculptured trophies and flame pots dissimulating a flat roof.

2.1 The State Apartments

The State Apartments occupied the main or principal floor of the château neuf. Le Vau's design for the state apartments closely followed Italian models of the day, as evidenced by the placement of the apartments on the next floor up from the ground level—the piano nobile—a convention the architect borrowed from 16th-and 17th-century Italian palace design.

Le Vau's plan called for an enfilade of seven rooms, each dedicated to one of the then-known planets and their associated titular Roman deity. Le Vau's plan was bold as he designed a heliocentric system that centered on the Salon of Apollo. The salon d'Apollon originally was designed as the king's bed chamber, but served as a throne room. The queen's apartment served as the residence of three queens of France.

2.2 The Hall of Mirrors

The Hall of Mirrors is the central gallery of the Palace of Versailles which connect the States apartments. The Hall of Mirrors' dimensions are 73.0m × 10.5m × 12.3m and is flanked by the salon de la guerre (north) and the salon de la paix (south). The principal feature of this hall is the seventeen mirror-clad arches that reflect the seventeen arcaded windows overlooking the gardens. Each arch contains twenty-one mirrors with a total complement of 357 used in the decoration of the galleries des glaces. The arches themselves are fixed between marble pilasters whose capitals depict the symbols of France. These gilded bronze capitals include the fleur-de-lys and the Gallic rooster.

During the 17th century, the Hall of Mirrors was used daily by Louis XIV when he walked from his private apartment to the chapel. At this time, courtiers assembled to watch the king and members of the royal family pass.

3. Royal Opera of Versailles

The Royal Opera of Versailles is the main theatre and opera house of the Palace of Versailles. It was perhaps the most ambitious building project of Louis XV for the château of Versailles. Completed in 1770, the Opéra was inaugurated as part of the wedding festivities of Louis XV's grandson, later Louis XVI, and Marie-Antoinette.

At the time, it represented the finest example in theatre design with 712 seats, and it was the largest theatre in Europe. Today, it remains one of the few 18th century theaters survived to the present day. The Opéra was exceptional for its time since it featured an oval plan. As an economy measure, the floor of the orchestra level can be raised to the level the stage, thus doubling the floor space. The transition from the auditorium to the stage is managed by the introduction of a giant order of engaged Corinthian columns, with a cornice ranging with the whole Ionic entablature. The proscenium is formed by two pairs of columns, coupled in depth, with their entablature.

Notes:

1. Palace of Versailles：凡尔赛宫，位于法国巴黎西南郊外，是巴黎著名宫殿之一。1624 年，法国国王路易十三买下了 117 亩荒地，在这里修建了一座二层的红砖楼房，用作狩猎行宫。凡尔赛宫建于路易十四（1643—1715 年）时代。凡尔赛宫作为法兰西宫庭长达 107 年（1682—1789 年）。全宫占地 111 万平方米，其中建筑面积为 11 万平方米，园林面积 100 万平方米。宫殿建筑布局严密，正宫东西走向，两端与南宫和北宫相衔接，形成对称的几何图案。宫顶建筑采用了平顶形式，显得端正而雄浑。宫殿外壁上端，林立着大理石人物雕像。凡尔赛宫内部陈设和装潢富于艺术魅力。500 多间大殿小厅处处金碧辉煌，豪华非凡。内部装饰以雕刻、巨幅油画及挂毯为主，配有 17、18 世纪造型超绝、工艺精湛的家具。凡尔赛宫拥有 2300 个房间，67 个楼梯和 5210 件家具。1833 年，奥尔良王朝的路易·菲利普国王修复凡尔赛宫，将其改为历史博物馆。
2. Louis XIV：路易十四（1638 年 9 月 5 日—1715 年 9 月 1 日），自号太阳王，是波旁王朝的法国国王，是有确切记录的在欧洲历史中在位唯一最久的独立主权君主。路易十四在法国建立了一个君主专制的中央集权王国，他把大贵族集中在凡尔赛宫居住，将整个法国的官僚机构集中于他的周围，以此强化法王的军事、财政和机构的决策权。他建立起的这一绝对君主制一直持续到法国大革命时期。
3. Ancien Régime：指 1789 年法国大革命之前的旧制度。
4. Hundred Years' War：百年战争，是指英国和法国以及后来加入的勃艮第，于 1337—1453 年间的战争，是世界最长的战争，断断续续进行了长达 116 年。战争胜利使法国完成民族统一，为日后在欧洲大陆扩张打下基础；英格兰几乎丧失所有的法国领地，但也使英格兰的民族主义兴起。
5. Napoleon：拿破仑·波拿巴（1769 年 8 月 15 日—1821 年 5 月 5 日），即拿破仑一世（Napoléon I），19 世纪法国伟大的军事家、政治家，法兰西第一帝国的缔造者。拿破仑于 1804 年 11 月 6 日加冕称帝，把共和国变成帝国。他执政期间多次对外扩张，发动了拿破仑战争，在最辉煌时期，欧洲除英国外，其余各国均向拿破仑臣服或结盟。拿破仑于 1814 年被流放厄尔巴岛，1821 年病逝于圣赫勒拿岛。1840 年，他的灵柩被迎回法国巴黎，隆重安葬在法国塞纳河畔的巴黎荣军院。
6. Louis-Philippe：路易·菲利普（1830—1848 年），法国奥尔良王朝唯一的君主。1789 年法国大革命爆发，他参加支持革命政府的进步贵族团体。1830 年七月革命后，他被资产阶级自由派等拥上王位。1848 年二月革命中，在无产阶级和中产阶级起义的压力下，他于 2 月 24 日逊位，后逃往英国，隐居和老死于英格兰的萨里。

Insights: Dougong

Dougong（斗拱）is a unique structural element of interlocking wooden brackets, one of the most important elements in traditional Chinese architecture. Dougong was developed into a complex set of interlocking parts by its peak in the Tang and Song periods.（See Fig. 3.11）

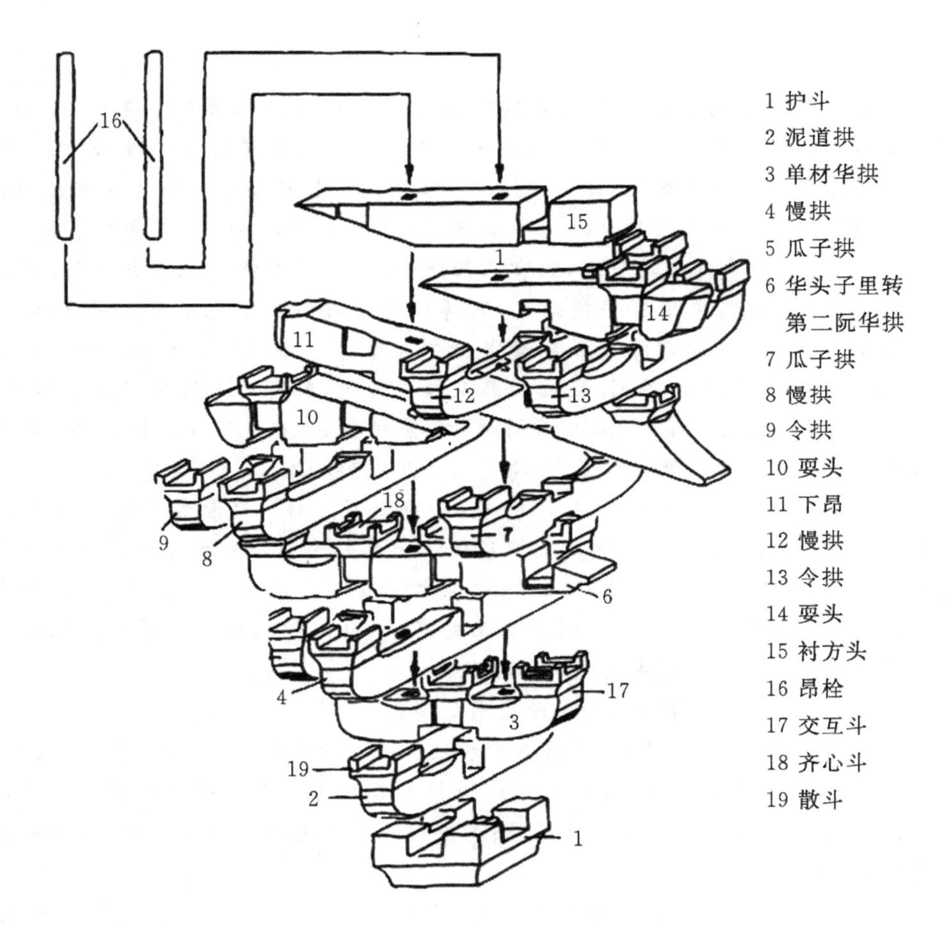

Fig. 3.11 Structure of Dougong

Multiple interlocking bracket sets are formed by placing a large wooden block (dou) on a column to provide a solid base for the bow-shaped brackets (gong) that support the beam or another gong above it. The function of dougong is to provide increased support for the weight of the horizontal beams that span the vertical columns or pillars by transferring the weight on horizontal beams over a larger area to the vertical columns. This process can be repeated many times, and rise many stories. Adding multiple sets of interlocking brackets or dougong reduces the amount of strain on the horizontal beams when transferring their weight to a column. Multiple dougong also allows structures to be elastic and to withstand damage from earthquakes.

After the Song Dynasty, brackets and bracket sets became more ornamental than structural when used in palatial structures and important religious buildings.

During the Ming Dynasty an innovation occurred through the invention of new wooden components that aided dougong in supporting the roof. This allowed dougong to add a decorative element to buildings in the traditional Chinese integration of artistry and function, and bracket sets became smaller and more numerous. Brackets could be hung under eaves, giving the appearance of graceful baskets of flowers while also supporting the roof.

Castle

A castle is a type of fortified structure built in Europe and the Middle East during the Middle Ages by nobility, usually considered to be the private fortified residence of a lord or noble. European castles originated in the 9th and 10th centuries, the fall of the Carolingian Empire resulted in its territory being divided among individual lords and princes. The nobles built castles as offensive and defensive structures to control the area immediately surrounding them; Although their military origins are often emphasized, the structures also served as centers of administration and symbols of power. Urban castles were used to control the local populace and important travel routes, and rural castles were often situated near features that were integral to life in the community, such as mills and fertile land.

Many castles were originally built from earth and timber, but later replaced by stone. Since not all the elements of castle architecture were military in nature, devices such as moats evolved from their original purpose of defense into symbols of power.

Tower of London[1]

The Tower of London is a historic castle located on the north bank of the River Thames in central London. It was founded towards the end of 1066 as part of the Norman Conquest[2] of England. The White Tower, which gives the entire castle its name, was built by William the Conqueror[3] in 1078, and was a resented symbol of oppression. The castle was used as a prison from 1100 until 1952 although that was not its primary purpose. As a royal residence, it was besieged several times, and controlling it has been important to controlling the country (See Fig. 4. 1).

1. The History of the Tower of London

The Tower of London was oriented with its strongest and most impressive defenses overlooking Saxon London, and it would have visually dominated the surrounding area and stood out to traffic on the River Thames. The castle is made up of three "wards", or enclosures: The innermost ward contains the White Tower and is the earliest phase of the

Fig. 4. 1 Tower of London

castle. Encircling it to the north, east, and west is the inner ward built during the reign of Richard the Lion heart[4] (1189—1199). Finally, there is the outer ward which encompasses the castle and was built under Edward I. Although there were several phases of expansion after William the Conqueror founded the Tower of London, the general layout has remained the same since Edward I completed his rebuild in 1285.

The castle encloses an area of almost 4. 9 hectares with a further 2. 4 hectares around the Tower of London constituting the Tower Liberties—land under the direct influence of the castle and cleared for military reasons. The precursor of the Liberties was laid out in the 13th century when Henry III ordered that a strip of land adjacent to the castle be kept clear. Tower Wharf was built on the bank of the Thames under Edward I and was expanded to its current size during the reign of Richard II (1377—1399).

2. The Main Buildings of Tower of London

As a whole, the Tower is a complex of several buildings set within two concentric rings of defensive walls and a moat. (See Fig. 4. 2)

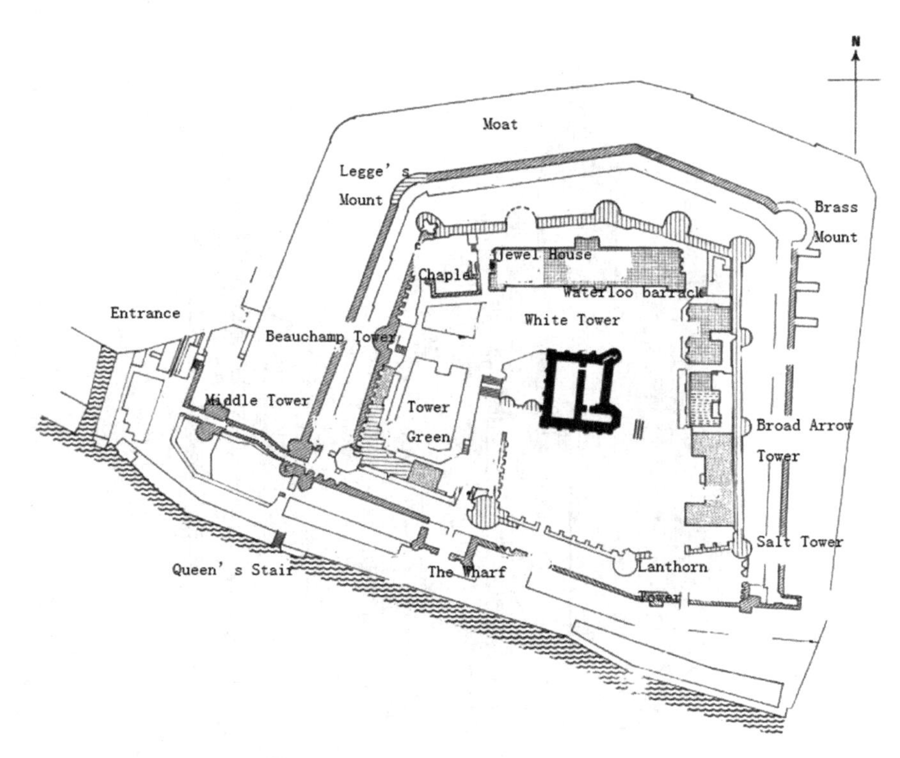

Fig. 4. 2 Layout of Tower of London

2.1 White Tower

The White Tower is a keep, which was often the strongest structure in a medieval castle, and contained lodgings suitable for the lord—in this case the king or his representative. As one of the largest keeps in the Christian world, the White Tower has been described as "the most complete eleventh-century palace in Europe".

The White Tower, not including its projecting corner towers, measures 36 by 32 meters at the base, and is 27 meters high at the southern battlements. The structure was originally three storey's high, comprising a basement floor, an entrance level, and an upper floor. The entrance, as is usual in Norman keeps, was above ground, in this case on the south face, and accessed via a wooden staircase which could be removed in the event of an attack. Each floor was divided into three chambers, the largest in the west, a smaller room in the north-east, and the chapel taking up the entrance and upper floors of the south-east. At the western corners of the building are square towers, while to the north-east a round tower houses a spiral staircase. At the south-east corner there is a larger semi-circular projection which accommodates the apse of the chapel. As the building was intended to be a comfortable residence as well as a stronghold, latrines were built into the walls, and four fireplaces provided warmth.

The tower was terraced into the side of a mound, so the northern side of the basement is partially below ground level. As was typical of most keeps, the bottom floor was an undercroft used for storage. One of the rooms contained a well. Although the layout has remained the same since the tower's construction, the interior of the basement dates mostly from the 18th century when the floor was lowered and the pre-existing timber vaults were replaced with brick counterparts. The basement is lit through small slits.

2.2 Innermost Ward

The innermost ward encloses an area immediately south of the White Tower, stretching to what was once the edge of the River Thames. As was the case at other castles, the innermost ward was probably filled with timber buildings from the Tower's foundation. The lodgings were renovated and elaborated during the 1220s and 1230s, becoming comparable with other palatial residences such as Windsor Castle[5]. Construction of Wakefield and Lanthorn Towers—located at the corners of the innermost ward's wall along the river—began around 1220. They probably served as private residences for the queen and king respectively. The earliest evidence for how the royal chambers were decorated comes from Henry III's reign: the queen's chamber was whitewashed, and painted with flowers and imitation stonework. A great hall existed in the south of the ward, between the two towers. It was similar to, although slightly smaller than, that also built by Henry III at Winchester Castle. Near Wakefield Tower was a postern gate which allowed private access to the king's apartments. The innermost ward was originally surrounded by a protective ditch, which had been filled in by the 1220s. Between 1666 and 1676, the innermost ward was transformed and the palace buildings removed. The area around the White Tower was cleared so that anyone approaching would have to cross open ground.

The entrance floor was probably intended for the use of the Constable of the Tower, Lieutenant of the Tower of London and other important officials. The upper floor contained a grand hall in the west and residential chamber in the east—both originally open to the roof and surrounded by a gallery built into the wall—and St John's Chapel in the south-east. The top floor was added in the 15th century, along with the present roof. St John's Chapel was not part of the White Tower's original design, The chapel's current bare and unadorned appearance is reminiscent of how it would have been in the Norman period.

2.3 The Inner Ward

The inner ward was created during Richard the Lion heart's reign, when a moat was dug to the west of the innermost ward, effectively doubling the castle's size. Henry III created the ward's east and north walls, and the ward's dimensions remain to this day. Between the Wakefield and Lanthorn Towers, the innermost ward's wall also serves as a curtain wall for the inner ward.

The main entrance to the inner ward would have been through a gatehouse, most likely in the west wall on the site of what is now Beauchamp Tower. The inner ward's western curtain wall was rebuilt by Edward I. The 13th-century Beauchamp Tower marks the first large-scale use of brick as a building material in Britain, since the 5th-century departure of the Romans. The Beauchamp Tower is one of 13 towers that stud the curtain wall. While these towers provided positions from which flanking fire could be deployed against a potential enemy, they also contained accommodation. As its name suggests, Bell Tower housed a belfry, its purpose is to raise the alarm in the event of an attack. The royal bow-maker, responsible for making longbows, crossbows, catapults, and other siege and hand weapons, had a workshop in the Bowyer Tower. A turret at the top of Lanthorn Tower was used as a beacon by traffic approaching the Tower at night.

2.4 Outer Ward

A third ward was created during Edward I's extension to the Tower, as the narrow enclosure completely surrounded the castle. At the same time a bastion known as Legge's Mount was built at the castle's north-west corner. Brass Mount, the bastion in the north-east corner, was a later addition. The three rectangular towers along the east wall 15 meters apart were dismantled in 1843. Blocked battlements in the south side of Legge's Mount are the only surviving medieval battlements at the Tower of London. A new 50-metre moat was dug beyond the castle's new limits; it was originally 4.5 meters deeper in the middle than it is today. With the addition of a new curtain wall, the old main entrance to the Tower of London was obscured and made redundant; a new entrance was created in the south-west corner of the external wall circuit. The complex consisted of an inner and an outer gatehouse and a barbican, which became known as the Lion Tower as it was associated with the animals as part of the Royal Menagerie since at least the 1330s. Edward extended the south side of the Tower of London onto land that had previously been submerged by the River Thames. In this wall, he built St Thomas's Tower between 1275 and 1279; later known as Traitors' Gate, it replaced the Bloody Tower as the castle's water-gate. The dock was covered with arrow slits in case of an attack on the castle from the River; there was also a portcullis at the entrance to control who entered. There were luxurious lodgings on the first floor. Edward also moved the Royal Mint into the Tower; its exact location early on is unknown, although it was probably in either the outer ward or the Lion Tower. By 1560, the Mint was located in a building in the outer ward near Salt Tower. Between 1348 and 1355, a second water-gate, Cradle Tower, was added east of St Thomas's Tower for the king's private use.

3. Restoration of Tower of London

The Tower of London has become one of the most popular tourist attractions since at least the Elizabethan period when it was one of the sights of London that foreign visitors wrote about. Over the 18th and 19th centuries, the palatial buildings were slowly adapted for other uses and demolished. Only the Wakefield and St Thomas's Towers survived. The 18th century marked an increasing interest in England's medieval past. One of the effects was the emergence of Gothic Revival architecture[6]. In the Tower's architecture, this was manifest when the New Horse Armoury was built in 1825 against the south face of the White Tower. It featured elements of Gothic Revival architecture such as battlements. Other buildings were remodeled to match the style and the Waterloo Barracks were described as "castellated Gothic of the 15th century".

Public interest was partly fuelled by contemporary writers, William Harrison Ainsworth created a vivid image of underground torture chambers and devices for extracting confessions that stuck in the public imagination. Harrison also suggested that Beauchamp Tower should be opened to the public so they could see the inscriptions of 16th-and 17th-century prisoners. Working on the suggestion, Anthony Salvin refurbished the tower and led a further program for a comprehensive restoration at the behest of Prince Albert[7].

Notes:

1. Tower of London：伦敦塔，是英国伦敦一座标志性的宫殿、要塞，选址在泰晤士河边。已知最早建于此处的要塞是罗马城堡。伦敦塔是由威廉一世为镇压当地人和保卫伦敦城，于 1087 年开始动工兴建的，历时 20 年，堪称英国中世纪的经典城堡。13 世纪时，后人在其外围增建了 13 座塔楼，形成一圈环拱的卫城，使伦敦塔既是一座坚固的兵营城堡，又是富丽堂皇的宫殿，里面还有天文台、监狱、教堂、刑场、动物园、小码头等小建筑。伦敦塔最重要、最古老的建筑是位于要塞中心的诺曼底塔楼，它是整个建筑群的主体，因其是用乳白色的石块建成，史称白塔。伦敦塔在英国王宫中的意义非常重大，英国数代国王都在此居住，国王加冕前住伦敦塔便成了一种惯例。伦敦塔还是一座著名的监狱，英国历史上不少王公贵族和政界名人都曾被关押在这里。此外，古老的伦敦塔在历史上还充任过造币馆、观象台、动物园等。
2. Norman Conquest：诺曼征服战争，是 11 世纪中叶法国诺曼底公爵威廉同英国大封建主哈罗德为争夺英国王位进而征服英国的一场战争。这场战争既是诺曼人对外扩张的继续，又是西欧同英国之间的又一次社会大融合。它以威廉的胜利而告终，对英国历史的发展产生了深远的影响。威廉建立中央政府掌管军事力量，使英国从此未再遭侵略，还不断地在海外进行军事活动。英国文化与诺曼—法国文化交融，大量新词进入英语，使得英国语言得到长足发展。
3. William the Conqueror：威廉一世又称"征服者威廉"(1028 年 9 月 2 日—1087 年 9 月 9 日)，是诺曼底公爵、英格兰国王，中世纪时最重要的国王之一。
4. Richard the Lion heart：狮心王理查(1157 年 9 月 8 日—1199 年 4 月 6 日)，是英格兰金雀花王朝的第二位国王，在位期为 1189—1199 年。理查是那个时代理想的国王，全神贯注于十字军东征，是个伟大的战士，战斗中身先士卒。长期以来，理查被歌颂为十字军的"英雄"。他与萨拉丁的战争改变了地中海东岸的局部政治格局，也留下骑士精神和浪漫传说。

5. Windsor Castle：温莎城堡，位于英国英格兰东南部区域伯克郡温莎·梅登黑德皇家自治市镇温莎，目前是英国王室温莎王朝的家族城堡，也是现今世界上有人居住的城堡中最大的一个。温莎城堡的历史可以回溯到威廉一世时期，城堡的面积大约有45000平方米，与伦敦的白金汉宫、爱丁堡的荷里路德宫一样，是英国君主主要的行政官邸。
6. Gothic Revival architecture：哥特复兴建筑。18世纪40年代，英国诞生了一场以复兴中世纪哥特样式为主题的建筑设计运动，被称为哥特复兴运动。1749年修建的草莓山庄被认为是这一运动的起点，以英国国会大厦威斯敏斯特宫的重建为标志达到运动的高潮。哥特复兴建筑形式被赋予了唤起民族认同感和争取自由民主的政治意义，哥特复兴成为在世界范围内具有一定影响力的建筑思潮，并在19世纪末启发了第一个设计运动——英国工艺美术运动。
7. Prince Albert：阿尔伯特亲王（1819—1861），是英国维多利亚女王的表弟和丈夫。在其20岁时，与表姐维多利亚成婚。阿尔伯特实行教育改革，在全球范围内推行废奴运动，管理宗室事务和女王办公室等，促进了英国君主立宪制的发展。

Himeji Castle[1]

Himeji Castle（姫路城）is a hilltop Japanese castle complex located in Himeji, in Hyōgo Prefecture, Japan which can date to 1333, when Akamatsu Norimura[2] built a fort on top of Himeyama hill. The castle is frequently known as "White Egret Castle" or "White Heron Castle" because of its brilliant white exterior and supposed resemblance to a bird taking flight. The castle, comprising a network of 83 buildings with advanced defensive systems from the feudal period, is regarded as the finest surviving example of prototypical Japanese castle architecture. (See Fig. 4. 3)

Fig. 4. 3 The Himeji Castle ①

① Scotta Moora

1. History of Himeji Castle

Japanese Castles were built to guard important or strategic sites, such as ports, river crossings, or crossroads. Though they were built with more stone than most Japanese buildings, castles were still constructed primarily of wood, and many were destroyed over the years.

Himeji Castle's construction dates to 1333 when a fort was constructed on Himeyama hill by the ruler of the ancient Harima Province. In 1346, his son Sadanori demolished the fort and built Himeyama Castle in its place. In 1545, the Kuroda clan ruler remodeled the castle into Himeji Castle.

In the Meiji Period[3] (1868—1912), many Japanese castles were destroyed. Himeji Castle was abandoned in 1871 and some of the castle corridors and gates were destroyed to make room for Japanese army barracks. The entirety of the castle complex was slated to be demolished by government policy, but it was spared by the efforts of an army colonel.

Himeji was heavily bombed in 1945 at the end of World War II, although most of the surrounding area was burned to the ground, the castle survived intact. One firebomb was dropped on the top floor of the castle but failed to explode. In order to preserve the castle complex, substantial repair work was undertaken in 1956.

2. The Characteristics of Japanese Castle

Japanese castles were always built atop a hill or mound, and an artificial mound would be created for this purpose. This not only aided greatly in the defense of the castle, but also allowed it a greater view over the surrounding land, and made the castle look more impressive and intimidating. In some ways, the use of stone, and the development of the building techniques of the castle was a natural step up from the wooden stockades of earlier centuries. The hills gave Japanese castles sloping walls, which helped to defend them from Japan's frequent earthquakes. The walls' timbers would be left sticking inwards, and planks would simply be placed over them to provide a surface for archers or gunners to stand on. This standing space was often called "stone throwing shelf". Many castles also had trap doors built into their towers, some even suspended logs from ropes to drop on attackers.

The curved walls of Himeji Castle are sometimes said to resemble giant fans, but the principal materials used in the structures are stone and wood. Feudal family crests (家纹) are installed throughout the building, signifying the various lords that inhabited the castle throughout its history.

Japanese castles featured massive stone walls and large moats. Walls were restricted to the castle compound itself; and only very rarely were built along borders. A number of tile-roofed buildings, constructed from plaster over skeletons of wooden beams, lay within the walls, and in later castles, some of these structures would be placed atop smaller stone-covered mounds. Sometimes a small portion of a building would be constructed of stone, providing a space to store and contain gunpowder.

The primary method of defense lay in the arrangement of the baileys, called maru (丸) or kuruwa (曲轮). Maru, meaning 'round' or 'circle' in most contexts, here refers to sections of the castle, separated by courtyards. Some castles were arranged in concentric circles, each maru lying within the last, while others lay their maru in a row.

The castle keep, usually three to five stories tall, is known as the tenshukaku (天守阁), and may be linked to a number of smaller buildings of two or three stories. The least militarily equipped of the castle buildings, the keep was defended by the walls and towers, and its ornamental role was never ignored; few buildings in Japan, least of all castle keeps, were ever built with attention to function purely over artistic and architectural form. Keeps were meant to be impressive not only in their size and in implying military might, but also in their beauty and the implication of a daimyō's(将军) wealth. The intricate gables and windows are a fine example of this.

3. The Description of Himeji Castle

Himeji Castle is the largest castle in Japan, containing many of the defensive and architectural features associated with Japanese castles. The castle complex comprises a network of 83 buildings such as storehouses, gates, corridors, and turrets. Of these 83 buildings, 74 are designated as Important Cultural Assets: 11 corridors, 16 turrets, 15 gates, and 32 earthen walls. Five structures of the castle are designated National Treasures: Main Keep (大天守), northwest small keep (乾小天守), west small keep (西小天守), east small keep (东小天守), and Ni-corridors and kitchen (渡橹附台所). The area within the middle moat of the castle complex is a designated Special Historic Site. (See Fig. 4. 4)

From east to west, the Himeji Castle complex has a length of 950 to 1,600 m, and from north to south, it has a length of 900 to 1,700 m . The castle complex has a circumference of 4,200 m . It covers an area of 233 hectares.

Main Keep at the center of the complex is 46. 4 m high, standing 92 m above sea level. Together with Main Keep, three smaller subsidiary keeps (小天守) form a cluster of towers. Externally, the keep appears to have five floors, because the second and third floors from the top appear to be a single floor; however, it actually has six floors and a basement. The basement of Main Keep has an area of 385m^2, and its interior contains special facilities that are not seen in other castles, including lavatories, a drain board, and a kitchen corridor.

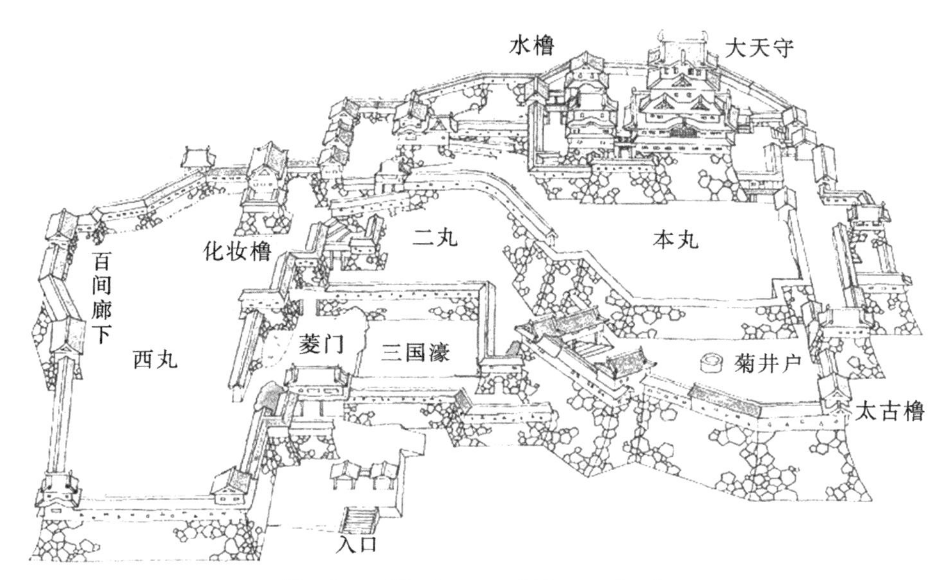

Fig. 4. 4 The layout of Himeji Castle

Main Keep has two pillars, with one standing in the east and one in the west. The east pillar, which has a base diameter of 97 cm, was originally a single fir tree, the base of the west pillar is 85 by 95 cm, and it is made of Japanese cypress. During the Restoration in 1956—1964, a Japanese cypress tree with a length of 26.4 m was brought down from the Kiso Mountains and replaced the old pillar.

The first floor of Main Keep has an area of 554 m^2 and is often called the "thousand-mat room" because it has over 330 Tatami mats. The walls of the first floor have weapon racks (武具掛) for holding matchlocks and spears, and at one point, the castle contained as many as 280 guns and 90 spears. The second floor has an area of roughly 550 m^2. The third floor has an area of 440 m^2 and the fourth floor has an area of 240 m^2. Both the third and fourth floors have platforms situated at the north and south windows called "stone-throwing platforms" (石打棚), where defenders could observe or throw objects at attackers. They also have small enclosed rooms called "warrior hiding places" (武者隐), where defenders could hide themselves and kill attackers by surprise as they entered the keep. The final floor, the sixth floor, has an area of only 115 m^2. The sixth floor windows now have iron bars in place, but in the feudal period the panoramic view from the windows was unobstructed.

4. Defense System of Himeji Castle

Himeji Castle contains advanced defensive systems from the feudal period. Loopholes (狭间) in the shape of circles, triangles, squares, and rectangles are located throughout

Himeji Castle, intended to allow defenders armed with or archers to fire on attackers without exposing themselves. Roughly 1,000 loopholes exist in the castle buildings remaining today. Angled chutes called "stone drop windows"(石落窓) were also set at numerous points in the castle walls, enabling stones or boiling oil to be poured on the heads of attackers passing by underneath, and white plaster was used in the castle's construction for its resistance to fire.

The castle complex included three moats, one of which—the outer moat—is now buried. Parts of the central moat and all of the inner moats survive. The moats have an average width of 20 meters, a maximum width of 34.5 meters, and a depth of about 2.7 meters. The Three Country Moat (三国堀) is a 2,500 m^2 pond existing inside the castle; one of the purposes of this moat was to store water for fire prevention.

The Waist Quarter (腰曲轮) contains numerous warehouses to store rice, salt, and water in case of a siege. A building known as the Salt Turret (塔楼) was used specifically to store salt, and it is estimated that it contained as many as 3,000 bags of salt when the castle complex was in use. The castle complex also contained 33 wells within the inner moat, 13 of which remain; the deepest of these has a depth of 30 meters.

One of the castle's most important defensive elements is the confusing maze of paths leading to the castle's keep. The gates, baileys, and outer walls of the complex are organized so as to confuse an approaching force, causing it to travel in a spiral pattern around the complex on its way to the keep. The castle complex originally contained 84 gates, 15 of which were named according to the Japanese syllabary, 21 gates from the castle complex remain intact, 13 of which are named according to the Japanese syllabary.

In many cases, the castle walkways even turn back on themselves, greatly inhibiting navigation. For example, the straight distance from the Hishi Gate (菱门) to Main Keep (大天守) is only 130 meters, but the path itself is a much longer 325 meters. The passages are also steep and narrow, further inhibiting entry. This system allowed the intruders to be watched and fired upon from the keep during their lengthy approach, but Himeji Castle was never attacked in this manner so the system remains untested.

Notes:

1. Himeji Castle:姬路城,是位于日本兵库县姬山的古城堡,其白色的外墙和蜿蜒屋檐造型犹如展翅欲飞的白鹭,也被称为白鹭城,是世界文化遗产。由于其保存度高,姬路城被称为"日本第一名城"。姬路城最早建成于1346年(正平元年),现存建筑大多建于17世纪早期。城堡由83座建筑物组成,包括主要城堡主楼的8座建筑被视为国宝,其余74座建筑被确认为国家的重要文化财产。姬路城拥有高度发达的防御系统和精巧的防护装置,是17世纪早期日本防御建筑技术的顶峰。
2. Akamatsu Norimura:赤松泽村,在元弘之乱的时候,奉了护良亲王的旨意在苔绳城举兵,登上了历史舞台。赤松氏兴起于佐用郡赤松村，经代代繁衍后分出了许多支族，多数支族以其分配的领地为名。早期分出的家族中，赤松円心最早使用赤松氏的氏名，在元弘之乱和南北朝时期为宗家做出了很大的贡献。

3. Meiji Period：明治时期。19世纪中期，日本处于德川幕府时代，德川幕府对外实行"锁国政策"，具有资产阶级色彩的藩地诸侯、武士，和要求进行制度改革的商人们组成政治性联盟，与反对幕府的基层农民共同形成"倒幕派"。新政府积极引入欧美各种制度及废藩置县等，这些各项改革被称为明治维新。

Red Fort[1] in Delhi

The Red Fort, located in the centre of Delhi, was the residence of the Mughal[2] emperor for nearly 200 years until 1857. Constructed in 1648 by the fifth Mughal Emperor Shah Jahan[3] as the palace of his fortified capital Shahjahanabad, the Red Fort is named for its massive enclosing walls of red sandstone. In addition to accommodating the emperors and their households, it was the ceremonial and political centre of Mughal government and the setting for events critically impacting the region. Every year on India's Independence Day, the Prime Minister of India hoists the national flag at the Red Fort and delivers a nationally-broadcast speech from its ramparts. (See Fig. 4. 5)

Fig. 4. 5 Delhi gate of Red Fort ①

① imagarcade. com

1. History of Red Fort

Emperor Shah Jahan commissioned construction of the Red Fort in 1638, when he decided to shift his capital from Agra to Delhi. Originally red and white, the Shah's favorite colors, its design is credited to architect Ustad Ahmad Lahauri who also constructed the Taj Mahal. The fort lies along the Yamuna River[4], which fed the moats surrounding most of the walls. Supervised by Shah Jahan, it was completed in 1648. Unlike other Mughal forts, the Red Fort's boundary walls are asymmetrical to contain the older Salimgarh Fort. The fortress-palace was a focal point of the medieval city of Shahjahanabad which is present-day Old Delhi. Its planning and aesthetics represent the zenith of Mughal creativity prevailing during Shah Jahan's reign. His successor Aurangzeb[5] added the Pearl Mosque to the emperor's private quarters, constructed barbicans in front of the two main gates to make the entrance to the palace more circuitous.

The administrative and fiscal structure of the Mughal dynasty declined after Aurangzeb, and the 18th century saw a degeneration of the palace. To raise money, the silver ceiling of the Rang Mahal was replaced by copper during this period. Muhammad Shah took over the Red Fort in 1719. In 1739, Persian emperor Nadir Shah easily defeated the Mughal army, plundering the Red Fort including the Peacock Throne.

With the end of Mughal reign, the British sanctioned the systematic plunder of valuables from the fort's palaces. All furniture was removed or destroyed; the harem apartments, servants' quarters and gardens were destroyed, and a line of stone barracks built. Only the marble buildings on the east side at the imperial enclosure escaped complete destruction, but were looted and damaged. While the defensive walls and towers were relatively unharmed, more than two-thirds of the inner structures were destroyed by the British. Lord Curzon, Viceroy of India from 1899-1905, ordered to repair the fort including reconstruction of the walls and the restoration of the gardens complete with a watering system.

1911 saw the visit of the British king and queen for the Delhi Durbar. In preparation of the visit, some buildings were restored. The Red Fort Archaeological Museum was also moved from the drum house to the Mumtaz Mahal.

After Indian Independence the site experienced few changes, and the Red Fort continued to be used as a military cantonment. A significant part of the fort remained under Indian Army control until 2003, when it was given to the Archaeological Survey of India for restoration.

2. Description of Red Fort

The fort complex is considered to represent the zenith of Mughal creativity under Shah

Jahan although the palace was planned according to Islamic prototypes, each pavilion contains architectural elements typical of Mughal buildings that reflect a fusion of Timurid[6] and Persian traditions. The Red Fort's innovative architectural style, including its garden design, influenced later buildings and gardens in Delhi, Rajasthan, Punjab, Kashmir and elsewhere.

The Red Fort has an area of 254.67 acres enclosed by 2.41 kilometers of defensive walls, punctuated by turrets and bastions and varying in height from 18 meters on the river side to 33 meters on the city side. The fort is octagonal, with the north-south axis longer than the east-west axis. The imperial apartments consist of a row of pavilions, connected by a water channel known as the Stream of Paradise. (See Fig. 4.6)

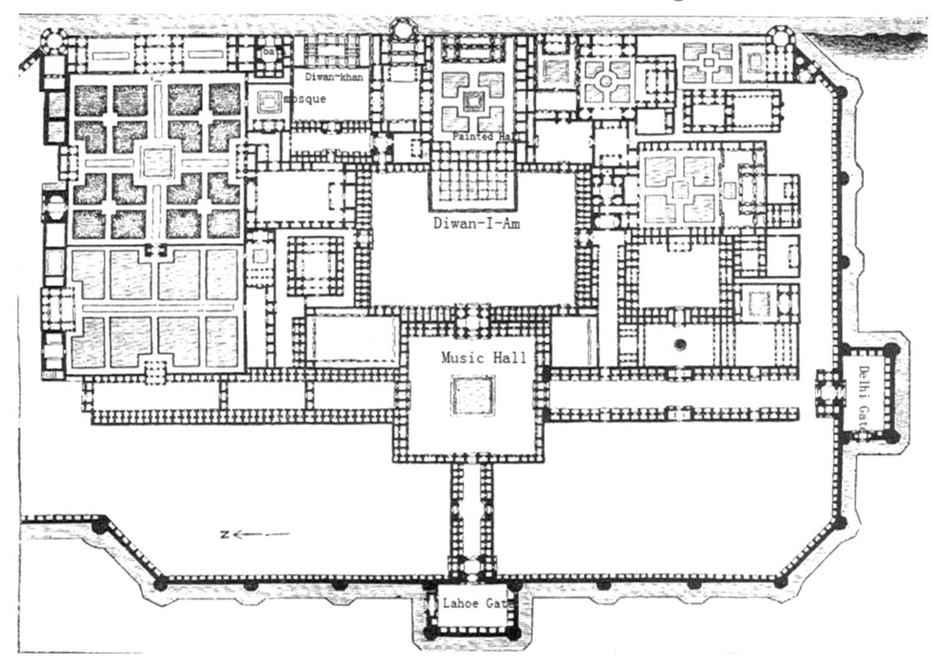

Fig. 4.6 The layout of Red Fort in 1901

The marble, floral decorations and double domes in the fort's buildings exemplify later Mughal architecture. It show cases a high level of ornamentation, and the Kohinoor diamond was reportedly part of the furnishings. The fort's artwork synthesizes Persian, European and Indian art, resulting in a unique Shahjahani style rich in form, expression and color.

2.1 Lahori Gate

The Lahori Gate is the main entrance to the Red Fort. The fort is approached through a covered street flanked by arcaded apartments called the covered bazaar. Situated on the western wall of the fort, the gate received its name because it led to the city of Lahore, in

Punjab, nowadays Pakistan.

The gateway consists of three stories, each decorated with square, rectangular and cusped arched panels. These are flanked by semi-octagonal towers crowned by two open octagonal pavilions. The whole gate is clad in red sandstone, except the roofs of the pavilions, where white stone is used. Between the two pavilions is a screen of miniature chhatris[7] having seven miniature marble domes. Continuing around the whole wall are flame-shaped battlements.

The gate was provided with a 10. 5 meter high barbican by Aurangzeb, with its entrance to the north. It is said that Shah Jahan, while under house arrest, wrote to Aurangzeb and criticized his decision: "You have made a fort a bride, and set a veil on it."

Every year since Indian Independence Day in 1947, the national flag has been raised and the Prime Minister has made a speech from the ramparts at the gate. In the 1980s, the security of the area was increased by blocking the tower windows as a security measure against sniper attacks.

2.2 The Naubat Khana

The Naubat Khana is the drum house that stands at the entrance between the outer and inner court, which was connected to the side arcades. Many Indian royal palaces have a drum house at the entrance. Musicians from the Naubat Khana would announce the arrival of the emperor and other dignitaries at the court of public audience. Music was also played five times a day at chosen hours.

The ground plan is a rectangular structure consisting of three large stories. The construction material is red sandstone, the surface covered in white chunam plaster. The richly carved floral designs on its red sandstone walls appear to have been originally painted with gold. The interior was colorfully painted. Several layers of these paintings can be found at the entrance chamber.

2.3 The Diwan-i-Am

The Diwan-i-Am, or Hall of Audience was where the Mughal emperor Shah Jahan and his successors received members of the general public and heard their grievances. The Diwan-i-Am consists of a front hall, open on three sides and backed by a set of rooms faced in red sandstone. The hall is 100 ft x 60 ft and divided into 27 square bays on a system of columns which support the arches. The roof is spanned by sandstone beams. The proportions of this hall, of its columns, and of the engraved arches show high aesthetics and fine craftsmanship. With an impressive facade of nine engraved arch openings, the hall was ornamented with gilded and white shell lime chunam plaster work. Its ceiling and columns were painted with gold.

In the centre of the eastern wall stands a marble canopy covered by a "Bengal" roof. A marble dais below the throne, inlaid with semi-precious stones, was used by the prime minister to receive petitions. The emperor was separated from the courtiers by a gold-plated railing, while a silver railing ran around the remaining three sides of the hall. The audience ceremony is known as Jharokha Darshan.

Behind the canopy, the wall is decorated with panels inlaid with multi-colored stones. They represent flowers and birds and are reputedly carved by Austin de Bordeaux, a Florentine jeweler. The hall was restored by Lord Curzon, while the inlay work of the throne recess and the plaques of the arch to the west side of the throne were restored by the Florentine artist.

3. The Hayat Baksh Bagh

The Hayat Baksh Bagh, which means "Life-bestowing garden", is the largest of the gardens in the Red Fort. It was laid out by Shah Jahan . It had the size of around 200 square feet. The garden was largely destroyed by the British colonial forces following the failed 1857 rebellion.

The garden is divided into four squares, with causeways, water channels and a star-shaped parterre framed with red sandstone. Originally, flowers in blue, white and purple were planted throughout. The Sawan and Bhadon pavilions are two almost identical structures facing on opposite ends of the canal. The names Sawan and Bhadon are the two rainy months in the Hindu calendar during the monsoon, they are carved out of white marble. A feature is a section of a wall with niches. Originally small oil lamps would be lit and placed in these niches at night, or vases with golden flowers be placed during the day. The water from the channel would cascade over it, creating the impression of a golden curtain.

In the middle between the two pavilions lies the Zafar Mahal. This structure was constructed during the reign of Bahadur Shah II in 1842 and named after him. This pavilion stands in the middle of a pre-existing water tank. It is made out of red sandstone, which was cheaper than white marble. Originally a red sandstone bridge led into the pavilion, which was probably lost after the Indian Rebellion of 1857.

4. The Chhatta Chowk

The Chhatta Chowk, meaning "covered bazaar", is a unique example of Mughal architecture in which bazaars were typically open-air. It was inspired by another market place which Shah Jahan had seen in Peshawar in 1646. It is a long passage way that contains a bazaar located behind the Lahori Gate and is set within an arched passage. It is lined with two-story flats that contain 32 arched bays serving as shops. It is 230 foot in length and 13 foot in width, with octagonal court in the middle for sunlight and natural ventilation, known as Chattar Manzil. This divides the market into two sections, eastern and western, which have vaulted roofs supported on a series of broad arches given at regular intervals. The width of this courtyard is 30 feet. The roof of the chatta is made of inlay work in which there are various types of waves and curves. It appears that the whole of the market, in the interior and on the exterior, was originally stuccoed, painted and gilded to give a gorgeous effect.

On the right and left sides of this courtyard are small doors which used to open to the most populated places in Mughal days. Their edges, supported by stone, and the intermediary space bears stalactite in stucco, which has been universally used in Islamic art, structurally as well as ornamentally. During Shah Jahan's reign, the Chhatta Chowk was very exclusive, specializing in trading goods such as silk, brocades, velvet, gold, silverware, jewellery, gems and precious stones, catering to the luxurious tastes of imperial households.

Notes:

1. Red Fort:德里红堡,是莫卧儿帝国时期的皇宫,位于德里东部老城区,亚穆纳河西岸,因整个建筑主体呈红褐色而得名红堡,是典型的莫卧儿风格伊斯兰建筑。德里红堡由莫卧儿王朝第五代皇帝沙·贾汗(Shah Jahan)所建,1639 开始建造,耗费近 10 年时间完成。自沙·贾汗皇帝时代开始,莫卧儿首都自阿格拉迁址于此。红堡有护城河环绕,四面环以厚重的围墙。围墙为石质,总长度约 2500 米,高度临亚穆纳河一侧稍低,临德里主城区偏高。德里红堡里面建有许多功能性的宫殿,如 Rang Mahal 是皇帝会见国内知名学士学者的地方,Khas Mahal 是皇帝会见各国使节和王朝高级官员的地方,Diwant Khas 则是类似于议事厅的地方等。
2. Mughal:莫卧儿帝国(1526—1857 年),是突厥化的蒙古人帖木儿的后裔巴布尔在印度建立的封建专制王朝。在帝国的全盛时期,领土几乎囊括整个南亚次大陆以及阿富汗等地。莫卧儿帝国上层建筑是穆斯林,而基础则是印度教,波斯语是宫廷、公众事务、外交、文学和上流社会的语言。帝国在第三代皇帝阿克巴时期进入全盛时期。莫卧儿帝国衰落后的 1858 年,英国的维多利亚女王被授予印度女皇称号,成立英属印度,莫卧儿王朝灭亡。
3. Shah Jahan:沙·贾汗(1592 年—1666 年 1 月 22 日),是印度莫卧儿帝国的皇帝。“沙·贾汗”在波斯语中的意思是“世界的统治者”。沙·贾汗在位期间,为他的第二个妻子 Mumtaz Mahal 修筑了举世闻名的泰姬陵。
4. Yamuna River:亚穆纳河,发源于喜马拉雅山脉,向南流经喜马拉雅山麓丘陵,沿着北方邦和哈里亚纳邦边界,进入印度北部平原。亚穆纳河全长约 1376 千米,在安拉阿巴德附近与恒河汇合,汇流处为印度教圣地。
5. Aurangzeb:奥朗则布,印度莫卧儿帝国第六任君主,建造著名泰姬陵的沙·贾汗的第三子。奥朗则布学识渊博,通晓经训和伊斯兰教法,足智多谋,尤精于武略,被赞誉为“帝位之荣缀”。他采用招抚为主、武力为辅的手段，将帝国版图扩大到除最南端外的整个南亚次大陆和阿富汗，后被胜利冲昏了头脑，舍弃了阿克巴大帝以来的宗教宽容国策，对国内的非穆斯林征收人头税，并把他们从官僚机构中驱逐出去，从而激化了国内矛盾。帝国在他死后很快就分崩离析。

6. Timurid：帖木儿帝国(1370—1507年)，是中亚河中地区的蒙古族贵族帖木儿于1370年开创的帝国，首都为撒马尔罕，后迁都赫拉特。鼎盛时期，其疆域以中亚乌兹别克斯坦为核心，从今格鲁吉亚一直到印度，囊括中、西亚各一部分和南亚一小部分，1507年亡于突厥的乌兹别克人。帖木儿帝国时期，中亚突厥人文化特性与根基得到了进一步的延伸与发展，为现代乌兹别克族的形成奠定了重要基础，帖木儿帝国因此也是现代乌兹别克族定型的一个时期。在帖木尔帝国的建立过程中，当时周围所有强大的帝国无一能够迎其锋芒，经三十多年的征服战争，建立了一个领土从德里到大马士革，从咸海到波斯湾的大帝国。帖木儿帝国末代大汗，帖木儿五世孙巴布尔兵败逃至今天的印度，并在那里开创了莫卧儿王朝。

7. chhatris：圆顶风亭，是印度建筑基本元素，流行于印度拉贾斯坦一带，多见于宫殿、陵墓与重要建筑屋顶。

New Swanstone Castle[1]

New Swanstone Castle(Neuschwanstein Castle) is a nineteenth-century Romanesque Revival[2] palace on a rugged hill above the village of Hohenschwangau near Füssen in southwest Bavaria, Germany. The palace was commissioned by Ludwig II of Bavaria[3] as a retreat and as a homage to Richard Wagner[4]. The palace was intended as a personal refuge for the reclusive king, but it was opened to public immediately after his death in 1886. (See Fig. 4.7)

Fig. 4.7 New Swanston Castle

1. History of New Swanstone Castle

The New Swanstone Castle lies at an elevation of 800 meters at the south west border of the German state of Bavaria, its surroundings are characterized by the transition between the Alpine foothills in the south and a hilly landscape in the north. In the Middle Ages, there were three castles overlooking the villages. In 1832, King Maximilian II of Bavaria, Ludwig's father bought the ruins of the castles to replace them with the comfortable neo-Gothic[5] palace known as Hohenschwangau Castle (High Swan Castle). Finished in 1837, the palace became his family's summer residence, and Ludwig spent a large part of his childhood here.

Ludwig the crown prince came to power in 1864, and the construction of a new palace in place of the two ruined castles became the first in his series of palace building projects. In the nineteenth century, many castles were constructed or reconstructed, often with significant changes to make them more picturesque. The building design was drafted by the stage designer Christian Jank and realized by the architect Eduard Riedel. The king insisted on a detailed plan and on personal approval of every draft. Ludwig's control went so far that the palace has been regarded as his own creation, rather than that of the architects involved.

The palace can be regarded as typical for nineteenth-century architecture. The shapes of Romanesque, Gothic and Byzantine architecture and art were mingled in an eclectic fashion and supplemented with 19th-century technical achievements. Characteristic of New Swanstone Castle's design are theatre themes: Christian Jank drew on coulisse drafts from his timeas a scenic painter.

The basic style was originally planned to be neo-Gothic but the palace was primarily built in Romanesque style in the end.

In 1868, the ruins of the medieval twin castles were completely demolished; the remains of the old keep were blown up. The foundation stone for the palace was laid on September 5, 1869; in 1872 its cellar was completed and in 1876, everything up to the first floor, the gatehouse being finished first. At the end of 1882 it was completed and fully furnished, allowing Ludwig to take provisional lodgings there and observe the ongoing construction work. The topping out ceremony for the Palas was in 1880, and in 1884, the king was able to move into the new building.

Despite its size, New Swanstone Castle did not have space for the royal court, but contained only the king's private lodging and servants' rooms. The court buildings served decorative, rather than residential purposes: The palace was intended to serve Ludwig II as a kind of inhabitable theatrical setting. As a temple of friendship it was also dedicated to the life and work of Richard Wagner, who died in 1883 before he had set foot in the building. In the end, Ludwig II only lived in the palace for a total of 172 days.

2. Description of New Swanstone Castle

The effect of the New Swanstone ensemble is highly stylistic, both externally and internally. The king's influence is apparent throughout, and he took a keen personal interest in the design and decoration. The suite of rooms within the Palas contains the Throne Room, Ludwig's suite, the Singers' Hall, and the Grotto. Throughout, the design pays homage to the German legends of the Swan Knight. Hohenschwangau, where Ludwig spent much of his youth, had decorations of these sagas. These themes were taken up in the operas of Richard Wagner. Many rooms bear a border depicting the various operas written by Wagner, including a theater permanently featuring the set of one such play. Many of the interior rooms remain undecorated, with only 14 rooms finished before Ludwig's death. With the palace under construction at the king's death, one of the major features of the palace remained unbuilt. (See Fig. 4. 8)

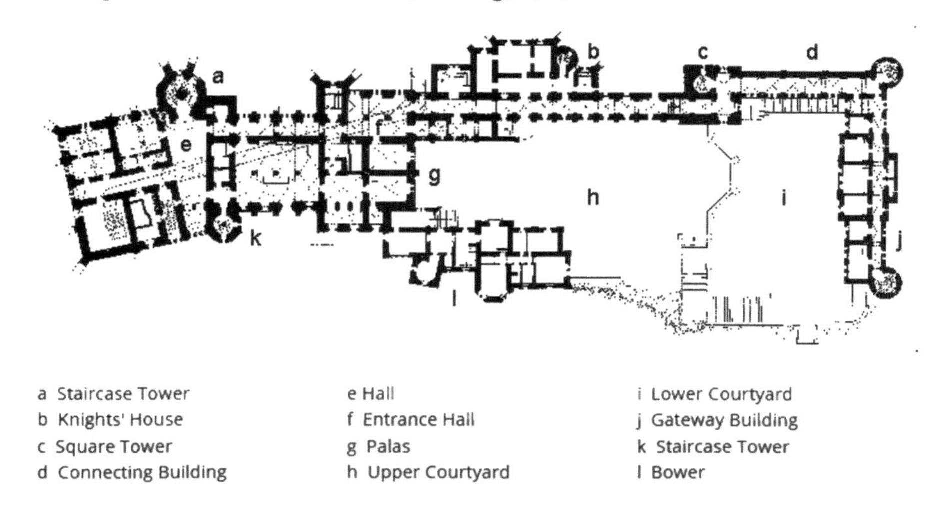

Fig. 4. 8 The layout of New Swanstone Castle

New Swanstone Castle consists of several individual structures which were erected over a length of 150 meters on the top of a cliff ridge. The elongate building is furnished with numerous towers, ornamental turrets, gables, balconies, pinnacles and sculptures. Following Romanesque style, most window openings are fashioned as bi-and triforia[6]. Before the backdrop of the Tegelberg and the Pöllat Gorge in the south and the Alpine foothills with their lakes in the north, the ensemble of individual buildings provides varying picturesque views of the palace from all directions. It was designed as the romantic ideal of a knight's castle. Unlike "real" castles, whose building stock is in most cases the result of centuries of building activity, New Swanstone was planned from the inception as an intentionally asymmetric building, and erected in consecutive stages. Typical attributes of a castle were included, but real fortifications—the most important feature of a medieval aristocratic estate—were dispensed with.

A massive keep, which would have formed the highest point and central focus of the ensemble, was planned for the middle of the upper courtyard but was never built, at the decision of the King's family. The foundation for the keep is visible in the upper courtyard.

The palace complex is entered through the symmetrical Gatehouse flanked by two stair towers. The eastward-pointing gate building is the only structure of the palace whose wall area is fashioned in high-contrast colors; the exterior walls are cased with red bricks, the court fronts with yellow limestone. The roof cornice is surrounded by pinnacles. The upper floor of the Gatehouse is surmounted by a crow-stepped gable and held Ludwig II's first lodging at New Swanstone, from which he occasionally observed the building work before the hall was completed. The ground floors of the Gatehouse were intended to accommodate the stables.

3. The Main Buildings of the Castle

The passage through the Gatehouse, crowned with the royal Bavarian coat of arms, leads directly into the courtyard. The courtyard has two levels, the lower one being defined to the east by the Gatehouse and to the north by the foundations of the so-called Rectangular Tower and by the gallery building. The southern end of the courtyard is open, imparting a view of the surrounding mountain scenery. At its western end, the courtyard is delimited by a bricked embankment, whose polygonally protracting bulge marks the choir of the originally projected chapel; this three-nave church, never built, was intended to form the base of a 90-metre keep, the planned centerpiece of the architectural ensemble. A flight of steps at the side gives access to the upper level.

3.1 Upper Court

The most striking structure of the upper court level is the so-called Rectangular Tower. Like most of the court buildings, it mostly serves a decorative purpose as part of the ensemble. Its viewing platform provides a vast view over the Alpine foothills to the north. The northern end of the upper courtyard is defined by the so-called Knights' House. The three-storey building is connected to the Rectangular Tower and the Gatehouse by means of a continuous gallery fashioned with a blind arcade. From the point of view of castle romanticism the Knights' House was the abode of a stronghold's men folk; at New Swanstone, estate and service rooms were envisioned here.

The western end of the courtyard is delimited by the hall. It constitutes the real main and residential building of the castle and contains the king's stateroom and the servants' rooms. The hall is a colossal five-story structure in the shape of two huge cuboids that are connected in a flat angle and covered by two adjacent high gable roofs. The building's shape follows the course of the ridge. In its angles there are two stair towers, the northern one surmounting the palace roof by several storeys with its height of 65 meters. With their polymorphic roofs, both towers are reminiscent of the Château de Pierrefonds. The western hall front supports a two-storey balcony with view on the Alpsee, while northwards a low chair tower and the conservatory protract from the main structure. The entire hall is spangled with numerous decorative chimneys and ornamental turrets, the court front with colorful frescos. The court-side gable is crowned with a copper lion, the western gable with the likeness of a knight.

3.2 The Hall of Singers

Had it been completed, the palace would have had more than 200 interior rooms, including premises for guests and servants, as well as for service and logistics. Ultimately, no more than about 15 rooms and halls were finished. The king's staterooms are situated in the upper stories: The anterior structure accommodates the lodgings in the third floor, above them the Hall of the Singers. The upper floors of the west-facing posterior structure are filled almost completely by the Throne Hall. The total floor space of all floors amounts to nearly 6,000 squaremeters.

The largest room of the palace by area is the Hall of the Singers, followed by the Throne Hall. The 27-by-10-metre Hall of the Singers is located in the eastern, court-side wing of the Palas, in the fourth floor above the king's lodgings. It is designed as an amalgamation of two rooms of the Wartburg: The Hall of the Singers and the Ballroom. It was one of the king's favorite projects for his palace. The rectangular room was decorated with themes from Lohengrin and Parzival[7]. Its longer side is terminated by a gallery that is crowned by a tribune, modeled after the Wartburg. The eastern narrow side is terminated by a stage that is structured by arcades and known. The Hall of the Singers was never designed for court festivities of the reclusive king. Rather, like the Throne Hall it served as a walkable monument in which the culture of knights and courtly love of the Middle Ages was represented.

3.3 Grotto

The eastward drawing room is adorned with themes from the Lohengrin legend. The furniture—sofa, table, armchairs and seats in a northward alcove—is comfortable and homelike. Next to the drawing room is a little artificial grotto that forms the passage to the study. The unusual room, originally equipped with an artificial waterfall and a so-called rainbow machine, is connected to a little conservatory. Depicting the Hörselberg grotto, it relates to Wagner's Tannhäuser, as does the décor of the adjacent study. In the park of Linderhof Palace the king had installed a similar grotto of greater dimensions. Opposite the study follows the dining room, adorned with themes of courtly love.

Notes:

1. New Swanstone Castle：新天鹅堡(Neuschwanstein)，是19世纪晚期的建筑，位于德国巴伐利亚西南方，建于1869年。这座城堡是巴伐利亚国王路德维希二世的行宫之一，共有360个房间，其中只有14个房间依照设计完工，其他的346个房间则因为国王在1886年逝世而未完成。新天鹅堡设计主要交付给剧场布景设计师克里斯蒂安·扬克(Christian Jank)设计。城堡中随处可见典型的哥特式建筑细节，而所有门窗、列柱回廊则呈现巴洛克风格。也许因为新天鹅堡的名字，整个城堡中所有的水龙头以及家具和房间配饰都是形态各异、栩栩如生的天鹅造型。
2. Romanesque Revival：罗马风复兴，是19世纪初期流行的一种建筑风格，受到11世纪到12世纪的欧洲罗马风建筑影响。与历史上的罗马风建筑有所不同，罗马风复兴风格建筑取向于使用更为简洁的拱形窗。
3. Ludwig II of Bavaria：路德维希二世(1845年8月25日—1886年6月13日)，维特尔斯巴赫王朝的巴伐利亚国王，以对艺术的狂热追求而著称。他兴建了包括新天鹅堡在内的数座城堡，同时也是瓦格纳的忠实崇拜者和资助人，修建了拜罗伊特节日剧院，专门上演瓦格纳的歌剧。路德维希二世沉浸在个人幻想中的行为引起了王室保守派的不满，1886年6月被以精神病为由废黜，数日后神秘地死于斯坦恩贝格湖。
4. Richard Wagner：威尔海姆·理查德·瓦格纳，1813年5月22日生于萨克森王国莱比锡，德国作曲家，著名的古典音乐大师。他是德国歌剧史上一位举足轻重的巨匠，前面承接莫扎特的歌剧传统，后面开启了后浪漫主义歌剧作曲潮流。1864年，他应巴伐利亚国王路德维希二世之召，在拜罗伊特自建歌剧院，上演其歌剧《尼伯龙根的指环》等。他所作歌剧还有《漂泊的荷兰人》、《汤豪舍》、《罗恩格林》、《歌唱大师》、《帕西法尔》等。
5. neo-Gothic：新哥特建筑，就是指哥特复兴建筑，运用了哥特建筑的表现手法，例如尖塔、尖拱券等。
6. bi-and triforia：两联或者三联拱。在欧洲教堂建筑中，筒拱上面，屋顶下面，天窗下面，侧廊的旁边空间，典型的在每一开间包含三联洞口形成的立面。
7. Lohengrin and Parzival：罗恩格林和帕西法尔，是德国民间传说故事，讲述了来自天国的圣杯武士罗恩格林帮助一位无辜受人诽谤和陷害的公主爱尔莎，不但把企图抢夺爱尔莎继承权的仇人杀死，又使爱尔莎的弟弟摆脱魔法恢复了人形。罗恩格林的父亲是圣杯国王帕西法尔。

Insights: Arch

An arch is a curved structure that spans a space and may or may not support weight above it. Arches appeared as early as the 2nd millennium B. C. in Mesopotamian brick architecture, and their systematic use started with the Ancient Romans who were the first to apply the technique to a wide range of structures. An arch can span a large area by resolving forces into compressive stresses and, in turn eliminating tensile stresses. (See Fig. 4. 9)

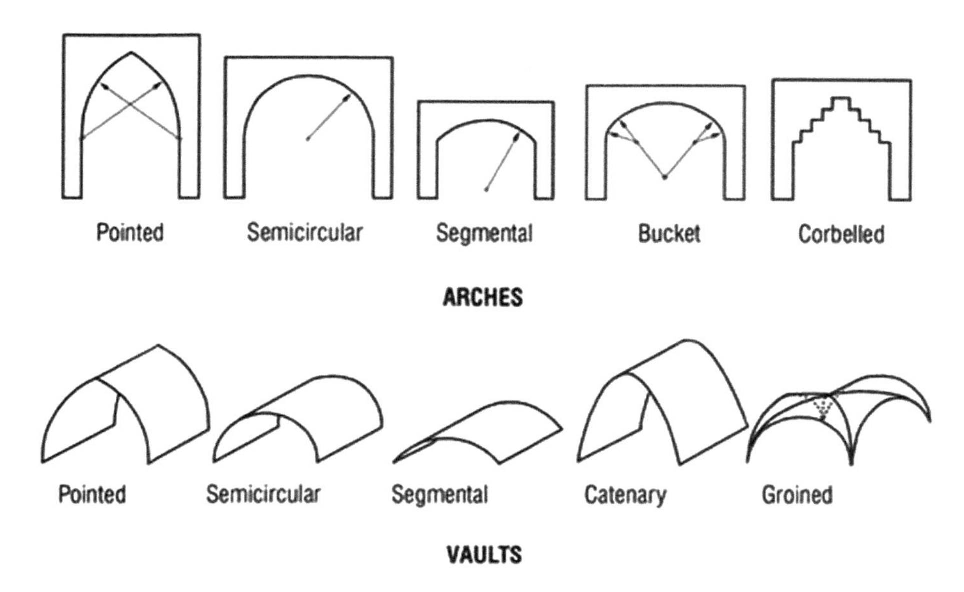

Fig. 4. 9 Arch-vault

Arches have many forms, but all fall into three basic categories: circular, pointed, and parabolic. Arches with a circular form were commonly employed by the builders of ancient heavy masonry arches. Ancient Roman builders relied heavily on the rounded arch to span large, open areas. Several rounded arches placed in-line, end-to-end, form an arcade; Pointed arches were most often used by builders of Gothic-style architecture. The advantage to using a pointed arch, rather than a circular one, is that the arch action produces less thrust at the base. This innovation allowed for taller and more closely spaced openings, typical of Gothic architecture; The parabolic arch produces the most thrust at the base, but can span the largest areas. It is commonly used in bridge design, where long spans are needed.

As the forces in the arch are carried to the ground, the arch will push outward at the base, called thrust. As the rise, or height of the arch decreases, the outward thrust increases. In order to maintain arch action and prevent the arch from collapsing, the thrust needs to be restrained, either with internal ties or external bracing, such as abutments.

Chapter Five

Market

A market, or market place, is an open space in town where people regularly gather for the purchase and sale of provisions, livestock, and other goods. Some markets operate on most days; others may be held once a week, or on less frequent specified days.

Markets have existed since ancient times. In ancient Greece it was called the agora, and in ancient Rome the forum. In Europe, market place is a public space with stalls selling goods in a public square or street market, with stalls along one or more public streets. In southeastern countries, chiefly found in Thailand, Indonesia and Vietnam, there are floating markets, where goods are sold from boats. In Greater China, there are wet markets, where traditionally live animals were sold.

The market square is a feature of many European and colonial towns. It is an open area where market stalls are traditionally set out for trading, commonly on one particular day of the week known as market day.

A typical market square consists of a square or rectangular area, or sometimes just a widening of the main street. It is usually situated in the centre of the town, surrounded by major buildings such as the parish church, town hall, important shops and hotels, and the post office, together with smaller shops and business premises. There is sometimes a permanent covered market building, and the entire area is a traditional meeting place for local people as well as a centre for trade.

Trajan's Market[1]

The Trajan's Market is the name given in the early 20th century to a complex of buildings in the imperial fora of Rome constructed in 107—110 A. D. during the reign of Trajan. Trajan's Market is located at the opposite end to the Colosseum[2]. The complex included a covered market, small shop fronts and a residential apartment block. The surviving buildings and structures, built as an integral part of Trajan's Forum and nestled against the excavated flank of the Quirinal Hill, present a living model of life in the Roman capital and a glimpse at the continuing restoration in the city, which reveals new treasures and insights about Ancient Roman architecture. (See Fig. 5. 1)

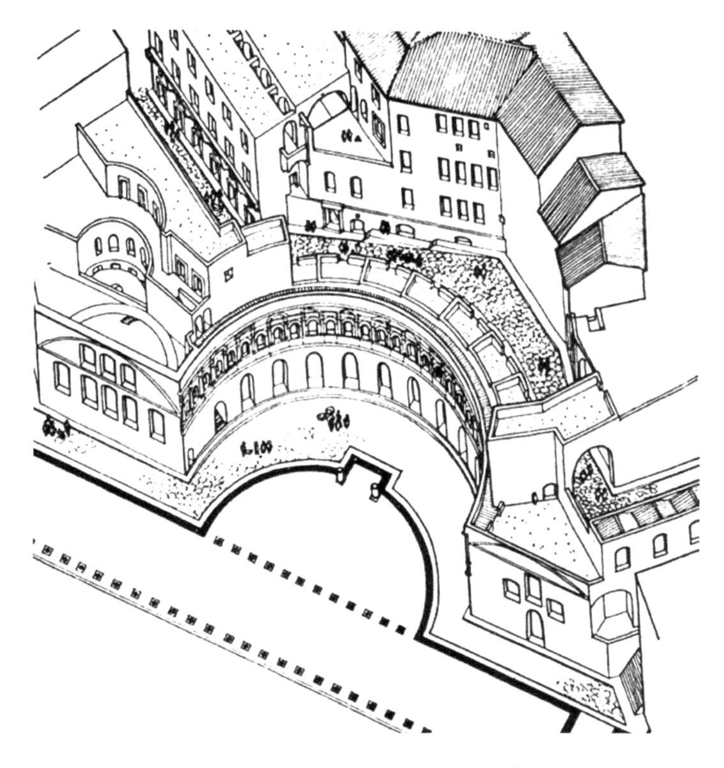

Fig. 5.1　Trajan's Market ①

1. History of Trajan's Market

In 106 A. D. in a remote town in the Carpatian Mountains, Decebalus, King of the Dacians killed himself; they were surrounded by Roman troops who had been chasing them for months after having conquered their capital Sarmizegetusa. It was the final event of the Second Dacian War. Emperor Trajan[3], who had personally led the campaign, was given an extraordinary triumphal procession in Rome. Dacia, today's Romania was rich in gold and silver and in 107 A. D. the emperor used these newly acquired resources to finance a major expansion of the Roman Forum[4].

Thought to be the world's oldest shopping mall, the arcades in Trajan's Market are now believed to be administrative offices for Emperor Trajan. The shops and apartments were built in a multi-level structure, and it is still possible to visit several of the levels. Highlights include delicate marble floors and the remains of a library.

Trajan's Market was probably built in 100—110 A. D. by Apollodorus of Damascus, an architect who always followed Trajan in his adventures and to whom Trajan entrusted the planning of his Forum, and inaugurated in 113 A. D. During the Middle Ages the

---

① studyblue. com

complex was transformed by adding floor levels which is still visible today, and defensive elements such as the Torre delle Milizie, the "militia tower" built in 1200. A convent, which was later built in this area, was demolished at the beginning of the twentieth century to restore Trajan's Markets to the city of Rome.

2. The Description of Trajan's Market

The market complex was built at one end of Trajan's Forum and includes buildings that had a number of different functions, predominantly commercial. Constructed on three different levels into the terraced hillside behind, access to the various parts was provided via connecting staircases. The ground level shallow alcoves opened onto a street and were used for small shops whilst there were more shops in the arcades above. The ground level alcoves are of uneven depth due to the fact that they were constructed following the bedrock of the hill. Originally, they would have all been framed with travertine, extending their capacity for displaying goods. One alcove has been restored to illustrate the original look and also has the typical window above the lintel.

On the upper level, access staircases were built at each end of via Biberatica running above the semicircular facade. The large uppermost central building was used as an apartment block, and the large structure further left functioned as a covered shopping arcade. In front of the whole complex, separated by a tufawall but accessible through a large central gateway, was the Exhedra and Porticus of Trajan's Forum. (See Fig. 5. 2)

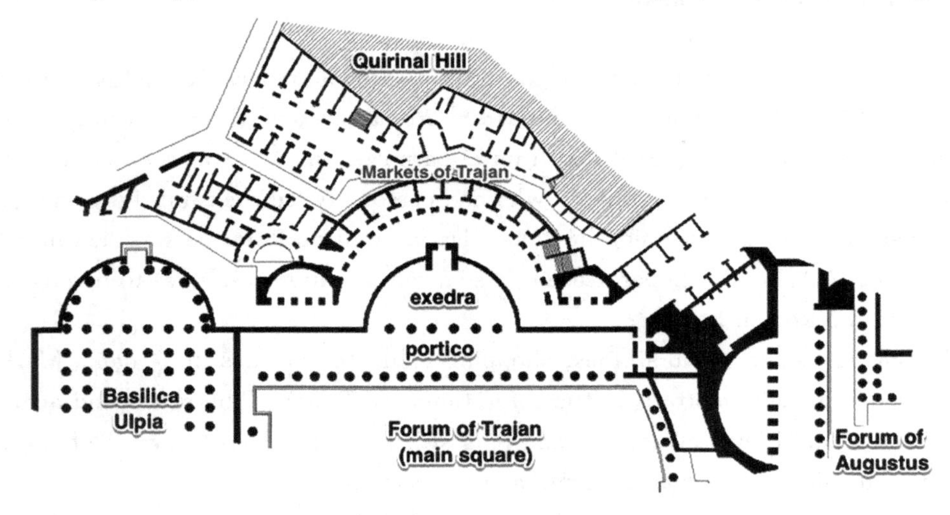

Fig. 5. 2 The plan of Trajan's Market

The buildings are largely constructed using concrete and faced with brick, and it is one of the finest examples of Imperial Roman architecture using these materials. The decorative semi-circular facade includes brick pilasters with travertine bases and capitals framing each archway on the second level. Decorative brick-work gives an added elegance, including entablatures of carved brick and alternate triangular and semi-circular pediments. White stucco would have once covered much of this brickwork and the pilasters, entablature, and pediments display evidence of having once been painted red.

The market is roofed by a concrete vault raised on piers, both covering and allowing air and light into the central space. The market itself is constructed primarily out of brick and concrete.

The name of the street on the upper level of the Market was via Biberatica, perhaps from the Latin biber—meaning drink—and hinting at the various drinking establishments which served the market shoppers. The principal products sold at the market would have come from across the empire and included fruit, vegetables, fish, wine, oil, and spices such as pepper.

The large square in front of the Great Hemicycle was paved with colored marbles, while the shops were paved with mosaics. Although the mosaics were designed according to very simple schemes yet these were different in each shop. A few years later a similar approach was followed in the decoration of the similar to many other ancient monuments, the markets were stripped of their marbles: the few remaining lintels show an elaborate design with a predominance of curved lines. The upper terraces offer interesting views over Trajan's Forum and the Capitoline Hill. Occasionally Modern Art exhibitions are arranged in Trajan's Markets.

3. Roman Forum

3.1 Trajan Forum

Forum is a Latin word meaning "public square surrounded by porticoes and buildings used for judicial and other businesses". In the lack of newspapers it was also the place where people went so they would know what was going on. The Roman Forum was located between the Capitoline and the Palatine hills. In 54 B. C. Julius Caesar[5] started enlarging its western part; in 2 B. C. Augustus expanded the Forum to the north. At this point the Forum was surrounded by hills: the Capitoline to the west, the Palatine to the south, the Quirinale to the north: an expansion to the east was blocked by the Colosseo and other buildings. Trajan asked Apollodorus of Damascus[6] to develop a project to make room for a further enlargement of the Forum.

Apollodorus decided to level to the ground the southern part of Quirinale: he then built on the flattened land a new forum dedicated to Trajan where he placed a gigantic column. Apollodorus realized that the hilly ground behind the new forum was both a poor background for the monuments he was erecting and a potential threat as it could slide down on them. He considered building a thick wall, similar to that which hid the view of the poor dwellings behind Foro di Augusto, but feared it could collapse. Eventually Apollodorus built a low wall to mark the boundaries of Trajan's Forum, but devised a more radical solution to prevent landslides. The slope of the hill was terraced and became the site of a commercial district—Trajan's Markets.

The lower terraces were given a round shape because in this way they better withstood the pressure of the hill. So the so called Great Hemicycle is both a finely designed monument and the evidence of an advanced knowledge of construction techniques. After 1900 years it still serves the purpose for which it was built.

The Great Hemicycle was strengthened by two halls at its end: they too had the shape of a hemicycle and were covered by a low dome, which had the same structure as the one which was built for the Pantheon[7] a few years later. Many experts credit that dome to Apollodorus.

The upper terraces have an irregular shape: the shops aligned along the market's main street were most likely taverns; Via Biberatica, the medieval name of the street is most likely a reference to biber, a Latin slang term for beverage, which is also the origin of beer. Most of the travertine blocks which frame the shops are the result of a 1930s reconstruction.

During the Middle Ages the solid buildings designed by Apollodorus were modified to serve other purposes such as monasteries and a series of fortifications which included a high tower.

The sixth and highest terrace is mainly occupied by a large rectangular building, the use of which is not fully understood yet. Its two storeys house a small museum providing information on how the emperors enlarged the Roman Forum.

3.2 Trajan's Column

Trajan's Columnis located in Trajan's Forum, built near the Quirinal Hill, north of the Roman Forum to commemorates Roman emperor Trajan's victory in the Dacian Wars. It was probably constructed under the supervision of the architect Apollodorus of Damascus at the order of the Roman Senate. It Completed in 113 A. D. , its design has inspired numerous victory columns, both ancient and modern.

The structure is about 30 meters in height, 35 meters including its large pedestal. The shaft is made from a series of 20 colossal Carrara marble drums, each weighing about 32 tons, with a diameter of 3. 7 meters. The 190-metre frieze winds around the shaft 23 times. It was in its time an architectural innovation. The narrative band expands from about 1 meter at the base of the column to 1. 2 meters at the top. The scenes unfold continuously. Often a variety of different perspectives are used in the same scene, so that more can be revealed. The relief portrays Trajan's two victorious military campaigns against the Dacians; the lower half illustrating the first (101—102), and the top half illustrating the second (105—106). These campaigns were contemporary to the time of the Column's building. Throughout, the frieze repeats standardized scenes of imperial address, sacrifice, and the army setting out on campaign . Scenes of battle are very much a minority on the column, instead it emphasizes images of orderly soldiers carrying out ceremony and construction.

The interior of Trajan's Column is hollow: entered by a small doorway at one side of the base, a spiral stair of 185 steps gives access to the platform above, having offered the visitor in antiquity a view over the surrounding Trajan's forum; 43 window slits illuminate the ascent.

Located immediately next to the large Basilica Ulpia, the column had to be constructed sufficiently tall in order to function as a vantage point and to maintain its own visual impact on the forum. Trajan's Column, especially its helical stairway design, exerted a considerable influence on subsequent Roman architecture. While spiral stairs were before still a rare sight in Roman buildings, this space-saving form henceforth spread gradually throughout the empire. Apart from the practical advantages it offered, the design also became closely associated with imperial power, being later adopted by Trajan's successors. In Napoleon's time, a similar column decorated with a spiral of relief sculpture was erected in the Place Vendôme[8] in Paris to commemorate his victory at Austerlitz.

Notes:

1. Trajan's Market：图拉真市场，是位于意大利罗马市内的一处古罗马时代的遗迹，也是古罗马建筑的代表作。市场修建于公元 100 年至 110 年期间，由大马士革的阿波罗多洛斯设计。图拉真市场是世界上最早的购物中心，建筑主要由砖块和混凝土建造。这座建筑得到持续修复，现在作为博物馆对外开放。
2. Colosseum：斗兽场，始建于公元 72 年弗拉维王朝，公元 82 年提图斯时代完成。这是罗马帝国于公元 70 年征服了耶路撒冷后，为纪念胜利驱使 8 万犹太俘虏修建的。斗兽场呈椭圆形，长轴为 188 米，短轴 156 米。周长 527 米，高 57 米，占地 2.6 公顷，可容 5～7 万观众。建筑外形单纯、明确；外观宏伟雄壮，立面分为 4 层，自下而上分别采用多利克柱式、爱奥尼亚柱式和科林斯柱式。斗兽场内部的看台，由低到高分为四组，观众的席位按等级尊卑地位之差别分区。
3. Emperor Trajan：图拉真皇帝(公元 53 年 9 月 18 日—117 年 8 月 9 日)，古代罗马帝国安敦尼王朝第二任皇帝，公元 98 年至 117 年在位。图拉真在位期间，对内巩固了经济和社会制度，对外发动战争，将罗马帝国的疆域扩张到历史上最大范围。由于其功绩卓著，获得了罗马元老院赠予的"最佳元首"称号。
4. Roman Forum：古罗马广场，是古罗马时代的城市中心，其中还残留了些许的古罗马时期的重要建筑的废墟，有提图斯凯旋门、奥古斯都凯旋门、塞维鲁凯旋门、凯撒神庙、灶神庙、维纳斯和罗马神庙，是古罗马政治、宗教、商业、娱乐等建筑的聚集地。古罗马城市一般都有广场，开始是作为市场和公众集会场所，后来也用于发布公告，进行审判，欢度节庆，甚至举行角斗。广场多为长方形。在自发形成的城市中，广场的位置因城而异；在按规划建造的营寨城市中，大多位于城中心交叉路口。

5. Julius Caesar：盖乌斯·尤利乌斯·恺撒，罗马共和国末期杰出的军事统帅、政治家，并且以其卓越的才能成为了罗马帝国的奠基者。恺撒出身贵族，历任财务官、祭司长、大法官、执政官、监察官、独裁官等职。公元前49年，他率军占领罗马，打败庞培，集大权于一身，实行独裁统治。公元前44年，恺撒遭元老院成员暗杀身亡。恺撒死后，其甥孙及养子屋大维开创罗马帝国并成为第一位帝国皇帝。
6. Apollodorus of Damascus：大马士革的阿波罗多洛斯，是一位在图拉真时代参与并完成了所有标志性公共建筑建设的建筑师。希腊人阿波罗多洛斯尽管出生在大马士革，却是所有建筑师中最具备罗马风格的建筑师。阿波罗多洛斯留下了大量的建筑遗迹，为现代考古学家们提供了极富价值的研究实物。
7. Pantheon：万神庙，位于意大利首都罗马圆形广场的北部，是罗马最古老的建筑之一，也是古罗马建筑的代表作。万神庙采用了穹顶覆盖的集中式形制，重建后的万神庙是单一空间、集中式构图的建筑物的代表，也是罗马穹顶技术的最高代表。它是古代建筑中最为宏大，保存近乎完美的，同时也是历史上最具影响力的建筑之一。
8. Place Vendôme：旺多姆广场，是巴黎的著名广场之一，位于巴黎老歌剧院与卢浮宫之间，呈切角长方形，长224米，宽213米。由于旺多姆公爵(1594—1665年)的府邸坐落于此，广场因而冠此名。旺多姆纪念铜柱坐落在广场的中央，由拿破仑皇帝下令建于1810年，是模仿罗马的图拉真柱修建的，柱高44米，用法国军队在奥斯特利兹战役中缴获的1250门大炮铸成，上面的螺旋形图案描绘着拿破仑征战的诸多场面，柱顶上立着拿破仑·波拿巴的铜像。

Grand Bazaar in Istanbul[1]

The Grand Bazaar is located inside the walled city of Istanbul, in the district of Fatih and in the neighborhood bearing the same name. It stretches roughly from west to east between the mosques of Beyazit and of Nuruosmaniye. The Grand Bazaar is one of the largest and oldest covered markets in the world, with 61 covered streets and over 3,000 shops which attract between 250,000 and 400,000 visitors daily. (See Fig. 5.3)

1. The History of Grand Bazaar

The construction of the Grand Bazaar's core started during the winter of 1455 or 1456, shortly after the Ottoman conquest of Constantinople. Sultan Mehmet II had an edifice erected devoted to the trading of textiles. The building was near the first sultan's palace, the Old Palace which was also in construction in those same years, and not far from the Artopoléia quarter, a location already occupied in Byzantine times by the bakers.

Fig. 5. 3　The Grand Bazaar

Analysis of the brickwork shows that most of the structure originates from the second half of the 15th century, although a Byzantine relief representing a Comnenian eagle[2] still enclosed on the top of the East Gate of the Bedesten has been used as proof that the edifice was a Byzantine structure.

In a market near the Bedesten is a covered market usually for haberdashery and craftsmanship. Roofed markets were built in Ottoman Empire and their design is based on the design of the mosques. Usually a Bedestan is the central building of the commercial part of the town. It has its origins in the Greco-Roman Basilica or Kaiserion, which served a similar purpose. During Ottoman times, the Bedestan was such an important building that cities were often classified under two categories, cities with a Bedestan and cities without a Bedestan. The slave trade was active in this zone also during the Byzantine Empire. Other important markets in the vicinity were the second-hand market, the "Long Market", corresponding to the Greek Long Arcade', a long porticoed mall stretching downhill from the Forum of Constantine to the Golden Horn, which was one of the main market areas of the city, while the old book market was moved from the Bazaar to the present picturesque location near the Beyazid Mosque only after the 1894 Istanbul earthquake.

Some years later—according to other sources, Mehmet II had another covered market built, the "Sandal Bedesten", which had the color of sandalwood. After the erection of the Sandal Bedesten the trade in textiles moved there while the Cevahir Bedesten was reserved for the trade in luxury goods. At the beginning the two buildings were isolated, between them and the Mosque of Beyazid stood the ruins of churches and a large cistern; However, soon many sellers opened their shops between and around them, so that a whole quarter was born, devoted exclusively to commerce.

At the beginning of the 17th century the Grand Bazaar had already achieved its final shape. The enormous extent of the Ottoman Empire in three continents, and the total control of road communication between Asia and Europe, rendered the Bazaar and the surrounding hans or caravanserais, the hub of the Mediterranean trade. The 19th-century growth of the textile industry in western Europe, introduction of mass production methods, capitulations signed between the Empire and many European countries, and the forestalling—always by European merchants—of the raw materials needed to produce goods in the Empire's closed economy were factors which all provoked the decadence of the Market. By 1850, rents in Bedesten were ten times lower than two to three decades before. Moreover, the birth of a West-oriented bourgeoisie and the commercial success of the Western products pushed the merchants belonging to the minorities moving out of the Bazaar, perceived as antiquated, and for opening new shops in quarters frequented by Europeans, as Pera and Galata.

2. Description of Grand Bazaar

According to several European travelers' survey in the late of 19th Century, in the Bazaar were 4,399 active shops, 2 bedesten, 2195 rooms, 1 hamam, one mosque, 10 medrese, 19 fountains, one mausoleum and 24 hans. In the 30.7 hectares of the complex, protected by 18 gates, there are 3,000 shops along 61 streets, the 2 bedesten. Until the first half of the 19th century, the market was unrivaled in Europe with regards to the abundance, variety and quality of the goods on sale. At that time we know from European travelers that the Grand Bazaar had a square plan, with two perpendicular main roads crossing in the middle and a third road running along the outer perimeter. In the Bazaar there were 67 roads, several squares used for the daily prayers, 5 mosques, 7 fountains, 18 gates which were opened each day in the morning and closed in the evening from these comes the modern name of the Market, "Closed Market". The number of shops amounted to 3,000, plus 300 located in the surrounding hans, large Caravanserais with two or three storeys round a porticoed inner courtyard, where goods could be stored and merchants could be lodged. In that period one tenth of the shops of the city were concentrated in the market and around it. For all that, at that time the market was not yet covered. (See Fig. 5.4)

The cevanir Bedesten has a rectangular plan (43.30 m×29.50 m). Two rows of stone piers, four in each row, sustain three rows of bays, five in each row. Each bay is surmounted by a brick dome with blind drum. In the inner and in the outer walls have been built 44 cellars, vaulted rooms without external openings. The sunlight in Bedesten comes from rectangular windows placed right under the roof: they can be accessed through a

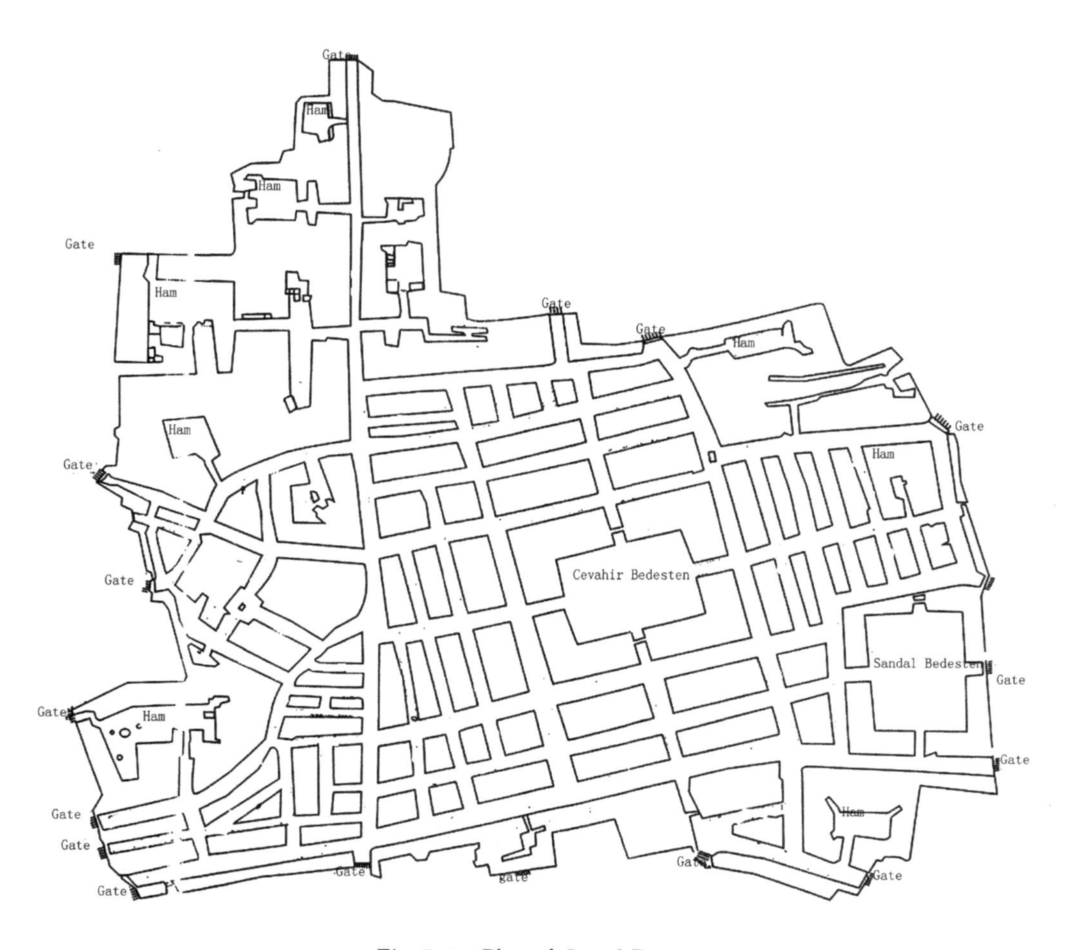

Fig. 5. 4 Plan of Grand Bazaar

wooden ambulatory. Due to the scarce illumination, the edifice was kept open only some hours each day, and was devoted to the trade of luxury goods, above all textiles. Moreover, the Bedesten's Mahzen(cellars) were also used as safes. The building can be accessed through four gates.

The Sandal Bedesten has also a rectangular plan (40. 20 m × 42. 20 m), with 12 stone piers bearing 20 bays surmounted by brick domes with blind drum. In this case shops are carved only in the outer walls. In both edifices, each bay is tied to the others through brick arches tied by juniper beams, and masonry is made with rubble. Both buildings were closed by Iron gates.

Aside the Bedesten, originally the Grand Bazaar structures were built with wood, and only after the 1700 fire, they were rebuilt in stone and brickwork, and covered. All the bazaar edifices, except the fur dealers market, a later addition which is two-story, are one story. The roofs are mainly covered with tiles, while the part burnt in 1954 uses nowtarmac. In the bazaar no artificial light was foreseen, also to prevent fires, and smoking was strictly prohibited. The roads outside the inner Bedesten are roughly parallel to it. Anyway, the damages caused by the many fires and quakes along the centuries, together with the repairs done without a general plan, gave to the market—especially in its western part—a picturesque appearance, with its maze of roads and lanes crossing each other at various angles.

3. Social Features of the Grand Bazaar

Until the restoration following the quake of 1894, the grand Bazaar had no shops: along both sides of the roads merchants sat on wooden divans in front of their shelves. Each of them got a space 1.8 to 2.4 m in width, and 0.91 to 1.22 m in depth. The name of this space was in Turkish dolap, meaning "stall". The most precious merchandise was not on display, but kept in cabinets. Only clothes were hung in long rows, with a picturesque effect. A prospective client could sit in front of the dealer, talk with him and drink a tea or a Turkish coffee, in a totally relaxed way. At the end of the day, each stall was closed with drapes.

Another peculiarity was the totally lack of advertising. Moreover, as everywhere in the East, traders of the same type of good were forcibly concentrated along one road, which got its name from their profession. The inner Bedesten hosted the most precious wares: jewelers, armourers, crystal dealers had their shops there. The Sandal Bedesten was mainly the center of the silk trade, but also other goods were on sale there. The most picturesque parts of the market were—aside the two Bedesten—the shoe market, where thousands of shoes of different colors were on display on high shelves, the spice and herbs market, which stood near the jewelers, the armour and weapon market, the old book market and the flea market. This kind of organization disappeared gradually, although nowadays a concentration of the same business along certain roads can be observed again.

Another peculiarity of the market during the Ottoman age was the total lack of restaurants. The absence of women in the social life and the nomadic conventions in the Turkish society made the concept of restaurant alien. Merchants brought their lunch in a food box called sefertas, and the only food on sale was simple dishes such as doner kebab and Turkish coffee.

Notes:

1. Grand Bazaar in Istanbul：伊斯坦布尔大巴扎，是世界上最大、最古老的巴扎之一，近4000家商店、60条街道纵横交错，身处其中，就仿佛置身于迷宫一般，极易迷失方向。这个大巴扎最初由苏丹穆罕默德二世下令修建，1455年开工到1461年建成，以首饰、陶瓷、香料、地毯店而闻名于世。许多摊位分类集中经营，例如皮衣、黄金首饰等。大巴扎在16

世纪苏丹苏莱曼一世在位期间大为扩展，形成现在的规模。1894 年经历了大地震后重建，是世界上历史悠久、规模巨大的室内集市之一，土耳其人称之为封闭式商场，由于黄金首饰店占据了大市场的大部分街道，所以它又称黄金大市场。

2. Comnenian eagle：科穆宁王朝徽记。科穆宁家族属于小亚细亚军事贵族家族，始祖伊萨克·科穆宁在 1057 年君士坦丁堡军队首领们发动的政变中被拥立为皇帝。后科穆宁王朝在曼努埃尔一世的统治下达到巅峰，恢复了东罗马东方霸主的地位。在其死后王朝却迅速衰落。

Covent Garden[1]

Covent Garden is a district in London on the eastern fringes of the West End, between St. Martin's Lane and Drury Lane. It is associated with the former fruit-and-vegetable market in the central square, now a popular shopping and tourist site, and with the Royal Opera House, which is also known as "Covent Garden". The district is divided by the main thorough fare of Long Acre, north of which is given over to independent shops centered on Neal's Yard and Seven Dials, while the south contains the central square with its street performers and most of the elegant buildings, theatres and entertainment facilities, including the Theatre Royal, Drury Lane and the London Transport Museum. (See Fig. 5. 5)

Fig. 5. 5 Covent Garden Market ①

1. History of Covent Garden

The area of Covent Garden was fields, settled in the 7th century when it became the heart of the Anglo-Saxon[2] trading town of Lundenwic, then returned to fields after Lundenwic was abandoned at the end of the 9th century. By 1201 part of it had been walled off by Westminster Abbey[3] for use as arable land and orchards.

By the 13th century this area had become a 40-acre quadrangle of mixed orchard, meadow, pasture and arable land, lying between modern-day St. Martin's Lane and Drury Lane, and Floral Street and Maiden Lane. The use of the name "Covent"—an Anglo-French term for a religious community, equivalent to "monastery" or "convent"—appears

① Etsy. com

in a document in 1515, when the Abbey, which had been letting out parcels of land along the north side of the Strand for inns and market gardens, granted a lease of the walled garden, referring to it as "a garden called Covent Garden". This is how it was recorded from then on.

In 1552, Edward VI, granted the land to John Russell, 1st Earl of Bedford. The Russell family had Bedford House and garden built on part of the land, with an entrance on the Strand, the large garden stretching back along the south side of the old walled-off convent garden. The church of St Paul's was the first building, begun in July 1631 on the western side of the square. The last house was completed in 1637.

By 1654 a small open-air fruit-and-vegetable market had developed on the south side of the fashionable square. Gradually, both the market and the surrounding area fell into disrepute, as taverns, theatres, coffee-houses and brothels opened up. By the 18th century it had become a well-known red-light district. An Act of Parliament was drawn up to control the area, and Charles Fowler's neo-classical building was erected in 1830 to cover and help organize the market. The market grew and further buildings were added: the Floral Hall, Charter Market, and in 1904 the Jubilee Market. By the end of the 1960s traffic congestion was causing problems, and in 1974 the market relocated to the New Covent Garden Market about three 5 km south-west at Nine Elms. The central building re-opened as a shopping centre in 1980, and is now a tourist location containing cafes, pubs, small shops, and a craft market called the Apple Market, along with another market held in the Jubilee Hall.

2. Description of Covent Garden

2.1 The Central Square

The central square in Covent Garden is simply called "Covent Garden", often marketed as "Covent Garden Piazza" to distinguish it from the eponymous surrounding area. Designed and laid out in 1630, with building work starting in 1631, it was the first modern square in London, and was originally a flat, open space or piazza with low railings. A casual market started on the south side, and by 1830 the present market hall was built. The space is popular with street performers, who audition with the site's owners for an allocated slot.

The square was originally laid out when the 4th Earl of Bedford, Francis Russell, commissioned Inigo Jones to design and build a church and three terraces of fine houses around the site of a former walled garden belonging to Westminster Abbey. Jones's design was informed by his knowledge of modern town planning in Europe, particularly Piazza d'Arme, in Leghorn, Tuscany, Piazza San Marco in Venice, Piazza Santissima Annunziata in Florence, and the Place des Vosges in Paris. The centerpiece of the project was the large square, the concept of which was new to London, and this had a significant influence on modern town planning as the metropolis grew, acting as the prototype for the laying-out of new estates, such as the Ladbroke Estate and the Grosvenor Estate. Isaac de Caus, the French Huguenot architect, designed the individual houses under Jones's overall design. (See Fig. 5.6)

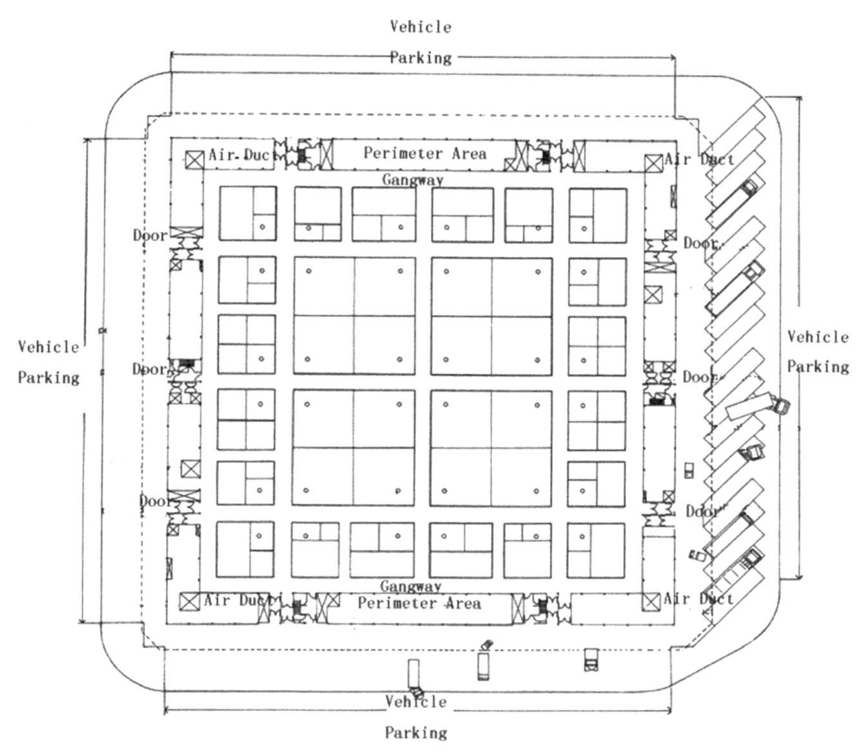

Fig. 5.6 The layout of Covent Garden

The church of St. Paul's was the first building, and was begun in July 1631 on the western side of the square. The last house was completed in 1637. Seventeen of the houses had arcaded portico walks organized in groups of four and six either side of James Street on the north side, and three and four either side of Russell Street. These arcades, rather than the square itself, took the name Piazza; the group from James Street to Russell Street became known as the "Great Piazza" and that to the south of Russell Street as the "Little Piazza". None of Inigo Jones's houses remain, though part of the north group was reconstructed in 1877—1879 as Bedford Chambers by William Cubitt to a design by Henry Clutton.

2.2 Market Building

The first record of a "new market in Covent Garden" is in 1654 when market traders set up stalls against the garden wall of Bedford House. The Earl of Bedford acquired a private charter from Charles II in 1670 for a fruit and vegetable market, permitting him and his heirs to hold a market every day except Sundays and Christmas Day. The original market, consisting of wooden stalls and sheds, became disorganized and disorderly, and the 6th Earl requested an Act of Parliament in 1813 to regulate it, then commissioned Charles Fowler in 1830 to design the neo-classical market building that is the heart of Covent Garden today. The contractor was William Cubitt and Company. Further buildings were added—the Floral hall, Charter Market, and in 1904 the Jubilee Market for foreign flowers was built by Cubitt and Howard.

By the end of the 1960s, traffic congestion was causing problems for the market, which required increasingly large lorries for deliveries and distribution. Redevelopment was considered, but protests from the Covent Garden Community Association in 1973 prompted the Home Secretary, Robert Carr, to give dozens of buildings around the square listed-building status, preventing redevelopment. The following year the market relocated to its new site, New Covent Garden Market, about 5 km south-west at Nine Elms. The central building re-opened as a shopping centre in 1980, with cafes, pubs, small shops and a craft market called the Apple Market. Another market, the Jubilee Market, is held in the Jubilee Hall on the south side of the square. The market halls and several other buildings in Covent Garden have been owned by the property company Capital & Counties Properties (CapCo) since 2006.

3. Cultural Connection of Covent Garden

The Covent Garden area has long been associated with both entertainment and shopping, and this continues. Covent Garden has 13 theatres, and over 60 pubs and bars, with most south of Long Acre, around the main shopping area of the old market.

The Royal Opera House, often referred to as simply "Covent Garden", was constructed as the "Theatre Royal" in 1732 to a design by Edward Shepherd. During the first hundred years or so of its history, the theatre was primarily a playhouse, with the Letters Patent granted by Charles II giving Covent Garden and Theatre Royal, Drury Lane exclusive rights to present spoken drama in London. In 1734, the first ballet was presented; a year later Handel's first season of operas began. Many of his operas and oratorios were specifically written for Covent Garden and had their premières here. It has been the home of The Royal Opera since 1945, and the Royal Ballet since 1946.

The current building is the third theatre on the site following destructive fires in 1808 and 1857. The facade, foyer and auditorium were designed by Edward Barry, and date from 1858, but almost every other element of the present complex dates from an extensive £178 million reconstruction in the 1990s. The Royal Opera House seats 2,268 people and consists of four tiers of boxes and balconies and the amphitheatre gallery. The stage performance area is roughly 15 meters square. The main auditorium is a Grade 1 listed building. The inclusion of the adjacent old Floral Hall, previously a part of the old Covent Garden Market, created a new and extensive public gathering place. In 1779 the pavement outside the playhouse was the scene of the murder of Martha Ray, mistress of the Earl of Sandwich, by her admirer the Rev. James Hackman.

Street entertainment at Covent Garden was noted in Samuel Pepys's diary in May 1662, when he recorded the first mention of a *Punch and Judy*[4] show in Britain. Covent Garden is licensed for street entertainment, and performers audition for timetabled slots in a number of venues around the market, including the North Hall, West Piazza, and South Hall Courtyard. The courtyard space is dedicated to classical music only. There are street performances at Covent Garden Market every day of the year, except Christmas Day.

Covent Garden, and especially the market, have appeared in a number of works. Eliza Doolittle, the central character in George Bernard Shaw's play, Pygmalion, and the musical adaptation by Alan Jay Lerner, My Fair Lady, is a Covent Garden flower seller. Alfred Hitchcock's 1972 film Frenzy about a Covent Garden fruit vendor who becomes a serial sex killer, was set in the market where his father had been a wholesale greengrocer. The daily activity of the market was the topic of a 1957 Free Cinema documentary by Lindsay Anderson, Every Day Except Christmas, which won the Grand Prix at the Venice Festival of Shorts and Documentaries.

Notes:

1. Covent Garden：科芬园，位于伦敦的苏豪区(SOHO)，在中世纪时期原为修道院花园，15世纪时重建为适合绅士居住的高级住宅区，同时造就了伦敦第一个广场，后来成为蔬果市场，目前以街头艺人和购物街区著称。这是一个由玻璃和钢铁管搭建顶棚的花园，具有典型的古罗马风情，建于18世纪。亨利八世之后，科芬园则转为市场，曾经是英国最大的蔬果花卉批发市场。
2. Anglo-Saxon：盎格鲁撒克逊人，是古代日耳曼人的部落分支，原居北欧日德兰半岛、丹麦诸岛和德国西北沿海一带。公元450年，盎格鲁、撒克逊两个部落大举移民大不列颠岛，此后两部落逐渐融合为盎格鲁撒克逊人。通过征服、同化，盎格鲁撒克逊人与大不列颠岛的"土著人"克尔特人，再加上后来移民的"丹人"、"诺曼人"经长时期融合，才形成近代意义上的英吉利人。
3. Westminster Abbey：威斯敏斯特修道院，也称西敏寺，坐落在伦敦泰晤士河北岸，原是一座天主教本笃会隐修院，始建于公元960年，1045年进行了扩建，1065年建成，1220年至1517年进行了重建。威斯敏斯特教堂在1540年英王创建圣公会之前，一直是天主教本笃会教堂。1540年之后，它成为圣公会教堂。威斯敏斯特教堂全系石造，包括教堂及修道院两大部分，由圣殿、翼廊、钟楼等组成。

4. *Punch and Judy*：英国著名木偶戏《潘奇与朱迪》，故事描述主人公潘奇生性残忍，因为看自己的孩子不顺眼，将自己还是婴儿的孩子扔出了窗外摔死。他的妻子朱迪十分生气，于是用木棍追打潘奇，不料被潘奇夺回木棍，被毒打之后也死了。接着潘奇又杀死了追捕他的警察，即使入狱被判绞刑后，刽子手也被他哄骗，他让刽子手为他演示如何绞刑，却趁机绞杀了刽子手。最后魔鬼前来缉拿，也遭他百般戏弄，棍打致死。

Central Market in Phnom Penh[1]

The Central Market in Phnom Penh which the local people call Phsar Themy is a remarkable monument, not only because of its architecture but also by its influence on the shape of urban space. Inaugurated in 1937, nearly seventy years ago, it still remains an architectural model for the covered markets. When the Central Market first opened in 1937, it was said to be the biggest market in Asia; today it still operates as a market. From 2009 to 2011, it underwent a US $4. 2 million renovation funded by the French Development Agency. (See Fig. 5. 7)

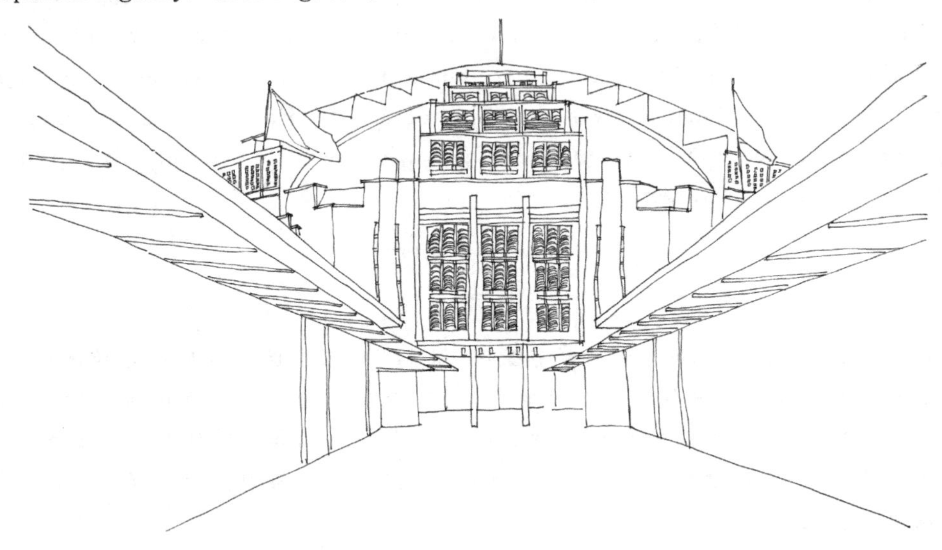

Fig. 5. 7 Central Market in Phnom Penh

1. The Description of the Central Market

The Central Market is a large market in the shape of a dome with four arms branching out into vast hallways with countless stalls of goods. Initial design and layouts are from French architect Louis Chauchon. Construction works were supervised by French architects Jean Desbois and Wladimir Kandaouroff.

The Central Market is a Phnom Penh landmark for its unique Art Deco[2] building. Before 1935, the area was a lake that received runoff during the rainy season. In 1434, when the King Pohea Yat first moved the capital of Cambodia to Phnom Penh, he dug up the lake using the earth to erect the hill at Wat Phnom. This lake was used to receive runoff water in rainy season. During the French colonial period, the lake was drained to build the market. But, wet season flooding around the market has remained a problem and is vestigial evidence of the old lake.

The construction began in 1935 and was completed in 1937. The market features an Art Deco style, with a big dome in the middle and four wings extending. The dome of the market is 30 meters wide and 26 meters high. The length of each wing is 44 meters and their height is 12 meters.

Central Market is truly an engineering marvel that largely reflects traditional Southeast Asian architecture featuring an enormous yellow-painted central dome with four wings extending to huge hallways, each of which teems with an array of shopping stalls. In fact, the major plus point of this lively market is its well shielded, properly ventilated structure that enables both sellers and buyers to engage in the trade while not being affected by monsoon rain or blazing heat.

The architect and official of Cambodian said "Phsar Themy is not just a nicely designed building, but its plan is also economical. The vendors do not need to spend much on the electric bill for lighting, fans or air conditioners" . He explained that "even though the market has no air-conditioning as contemporary supermarkets, people still can get fresh air from the many small windows in the dome of the building."

2. The Renovation of Central Market

In 2011 after an intensive rehabilitation of its emblematic architecture the Central Market in Phnom Penh was reopened on May 25th, 2011. The market was repainted and added some more vendor booths that were built in concrete around the four wings. In the project, four main ideas guided the design of the renovation: Integrate the market in its urban environment, Enhance the historic building, Provide spaces with a good level of comfort, hygiene and safety and Maintain all merchants on the site. (See Fig. 5. 8)

The reconciliation of the market with its environment required a reorganization of its surroundings in order to provide better continuity of pedestrian traffic. This was completed in cohesion with the local transport plan for the center of Phnom Penh developed by the municipality. This work was made with the support of experts of the City of Paris.

Parking spaces and drop-off areas are now clearly defined and allow easy access to the market without being a nuisance to pedestrian traffic.

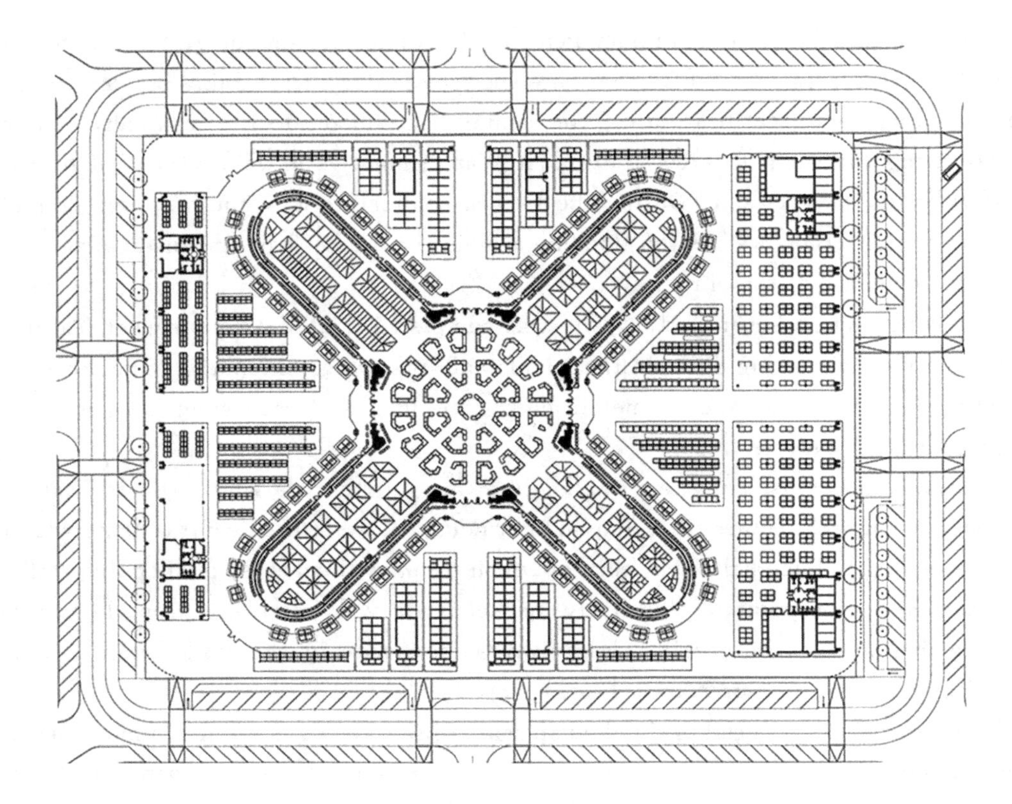

Fig. 5. 8　The plan of Central Market

The streamlining effort has also focused on the organization of the market itself. Four wide open lanes lead to the central dome of the market which is thus emphasized. Inside the market, the project offers fluid circulation from the street to the historic building.

To meet the merchants needs, extensions are implanted, they are independent from the historic building on the east and west squares of the Central Market. These new structures are designed to provide a modern and functional protection to merchants and passers-by while respecting the architecture of the traditional market. The same architectural design is reproduced on the north and south squares with extensions incorporating smaller gauge of existing houses to be retrofitted. These buildings also house the premises of services dedicated to the proper functionning of the market, such as offices, toilets, waste management, etc.

The architectural integration of this new market is in compliance with the general composition of Central Market. The halls around the market restore a new urban uniform facade, which consists f a succession of arcades overlooking wider sidewalks and allowing commercial animation on the outside. This circular walkway thus becomes a trading area opened out into the city.

The entrance to the market is lined with souvenir merchants hawking everything from T-shirts and postcards to silver curios and kramas. Inside is a dazzling display of jewels and gold. Electronic goods, stationery, secondhand clothes and flowers are also sold.

3. The French Colonial Buildings Around

Beginning in 1870, the French Colonialists turned Phnom Penh from a riverside village into a city where they built hotels, schools, prisons, barracks, banks, public works offices, telegraph offices, law courts, and health services buildings. In 1872, the first glimpse of a modern city took shape when the colonial administration employed the services of a French contractor Le Faucheur, to construct the first 300 concrete houses for sale and rental to the Chinese traders.

Hotel Le Royal is a luxury five star boutique hotel in Phnom Penh. It was first established in 1929, and today operated by the Raffles Hotels & Resorts. It was not until 1997 that the hotel re-opened, after a careful restoration and refurbishment; the historic hotel was upgraded to fulfill the requirements of a modern world-class hotel with its facilities and amenities, but has an old-world charm through its style and décor. It was designed in French style and luxuriously decorated with Cambodian and French furnishings. The building is typical of French colonial style, but the roof is in traditional Khmer style. The restaurant of the hotel offers traditional Khmer cuisine under a ceiling of lotus and honeysuckle paintings by painter and commissioned by His Majesty King Sihanouk. The artwork is based on the ceiling of the Dance pavilion at the Royal Palace in Phnom Penh. Elephant Bar has paintings on the ceiling of this watering hole that lend it its name.

The National Archive of Cambodia was built in 1926 by French colonial government. Unlike the National Library of Cambodia which integrated traditional Cambodian elements in its design, this building is more French in character. In 1995, the building was innovated with the help of international contributors.

The Post Office is also a building from the French colonial period. It is at the heart of French quarter. There are a number of other buildings from the French Colonial period around the square in front of the Post Office. They including the former police headquarters, the Bank of Indochina[3], and former Hotel Manolis. The post Office still serves its original purpose and was re-opened after renovation in 2004.

The building used to have a central squat tower surmounted by acupola roof, but this was removed in the late 1930's. The building mixed French influences with Asian style, especially in regard to how the air gets in and out easy. Besides the ventilation, they also want to make the building look more comfortable and attractive. During the renovation, the original windows have been replaced.

Cine Luxis the only old cinema survived in Phnom Penh was built in a splendid modern Art Deco style, Cine Lux was also designed by French architect in 1935. The cinema was used until the early 1990s for a variety of activities including film screenings and theater performances. It has total 650 seat, and reopened after the renovation in 2001.

Notes:

1. Central Market in Phnom Penh：柬埔寨金边中心市场，为金边的标志性建筑始建于1937年，于2011年3月修复后重新开业。该建筑由2005年在竞赛中获奖的柬埔寨当地建筑公司建造，由法国国际建筑设计机构负责设计。
2. Art Deco：艺术装饰风格，演变自19世纪末的新艺术运动。Art Nouveau是当时的欧洲中产阶级追求的一种艺术风格，它的主要特点是感性的自然界的优美线条，称为有机线条，比如花草动物的形体，尤其喜欢用藤蔓植物的颈条以及东方文化图案，如日本浮世绘。Art Deco不排斥机器时代的技术美感，机械式的、几何的、纯粹装饰的线条也被用来表现时代美感，比较典型的装饰图案，如扇形辐射状的太阳光、齿轮或流线型线条、对称简洁的几何构图等等；色彩运用方面以明亮且对比强烈的的颜色来彩绘，具有强烈的装饰意图。随着考古发现，远东、中东、希腊、罗马、埃及与玛雅等古老文化的物品或图腾，也都成了Art Deco装饰的素材来源。
3. Indochina：印度支那，亦称中南半岛或中印半岛，指东南亚半岛，东临南海，西濒印度洋，因位于印度、中国之间，而被近代欧洲人方便记忆式命名。中南半岛通常特指曾经是法国殖民地的法属印度支那，包括今日的越南、柬埔寨（旧称高棉）、老挝三国；广义的中南半岛则指东南亚大陆，包括越南、老挝、柬埔寨三国及缅甸、泰国、马来西亚的马来亚地区及新加坡、槟城、马六甲等地。

Insights: Arcade

An arcade is a succession of arches, each counter-thrusting the next, supported by columns, piers, or a covered walkway enclosed by a line of such arches on one or both sides. In ancient Greek, it is called stoa commonly for public use. Early stoas were open at the entrance with columns, usually of the Doric order, lining the side of the building; they created a safe, enveloping, protective atmosphere. Later examples were built as two stories, with a roof supporting the inner colonnades where shops or sometimes offices were located. They followed Ionic architecture. These buildings were open to the public; merchants could sell their goods, artists could display their artwork, and religious gatherings could take place. Stoas usually surrounded the market places or agora of large cities and were used as a framing device. In warmer or wet climate regions, exterior arcades provide shelter for pedestrians. The walkway may be lined with stores. Blind arcades are a feature of Romanesque architecture that was taken into Gothic architecture. In Gothic architecture, the arcade can be located in the interior, in the lowest part of the wall of the nave, supporting the triforium and the clerestory in a cathedral, or on the exterior, in which they are usually part of the walkways that surround the courtyard and cloisters. One of the earliest examples of a European shopping arcade, the Covered Market, Oxford, England was officially opened on 1st November 1774 and is still active today. (See Fig. 5. 9)

Fig. 5. 9 Arcade

Chapter Six

Square

A town square is an open public space commonly found in the heart of a traditional town used for community gatherings. Most town squares are hardscapes suitable for open markets, music concerts, political rallies, and other events that require firm ground. Being centrally located, town squares are usually surrounded by shops. At their center is often a fountain, well, monument, or statue. Many of those with fountains are actually called fountain square. In urban planning, a city square is a planned open area in a city, usually or originally rectangular in shape.

The German word for square"Platz" is a common term for central squares in German-speaking countries. These have been focal points of public life in towns and cities from the Middle Ages to today. A piazza is a term for city square in Italy and in surrounding regions. A piazza is commonly found at the meeting of two or more streets. Most Italian cities have several piazzas with streets radiating from the center. Shops and other small businesses are found on piazzas as it is an ideal place to set up a business.

Squares are often called "market" because of the usage of the square as a market place. Almost every town in west European countries has a market where the town hall is situated and therefore the centre of the town. In Mainland China, People's Square is a common designation for the central town square of modern Chinese cities, established as part of urban modernization. These squares are the site of government buildings, museums and other public buildings.

Red Square[1]

Red Square is a city square in Moscow, Russia. It separates the Kremlin[2], the former royal citadel and currently the official residence of the President of Russia. The name Red Square neither originates from the pigment of the surrounding bricks nor from the link between the colour red and communism. Rather, the name came about because the Russian word "krasnaya", which means both "red" and "beautiful", was applied to a small area between St. Basil's Cathedral, the Spassky Tower of the Kremlin. Red Square is often considered the central square of Moscow since it serve as Moscow's main market place. It was also the site of various public ceremonies and proclamations, and occasionally a coronation for Russia's Tsars would take place. (See Fig. 6.1)

Fig. 6.1 The Red Square of Moscow ①

1. History of Red Square

Before the 16th century, the East side of the Kremlin triangle, lying adjacent to Red Square and situated between the rivers Moskva and the now underground Neglinnaya River was deemed the most vulnerable side of the Kremlin to attack, since it was neither protected by the rivers, nor any other natural barriers, as the other sides were. Therefore, the Kremlin wall was built to its greatest height on this side, and the Italian architects involved in the building of these fortifications convinced Ivan the Great[3] to clear the area outside of the walls to create a field for shooting.

From 1508 to 1516, the Italian architect Aloisio arranged for the construction of a moat in front of the Eastern wall, which would connect the Moskva and Neglinnaya and be filled in with water from Neglinnaya. This moat was lined with limestone and, in 1533, fenced on both sides with low, 4 metre thick cogged brick walls.

Red Square was the landing stage and trade centre for Moscow. After a fire in 1547, Ivan the Terrible reorganized the lines of wooden shops on the Eastern side into market

① cityillustration. com

lines. The streets were divided into the Upper lines, now the GUM department store, Middle lines and Bottom lines.

After a few years, the Cathedral of Intercession of the Virgin, commonly known as Saint Basil's Cathedral, was built on the moat under the rule of Ivan IV. This was the first building which gave the square its present-day characteristic silhouette. In 1595, the wooden market lines were replaced with stone.

In the late 17th century the square was cleared of all wooden structures. Then all Kremlin towers received tent roofs. One tent was erected on the wall above Red Square, so that the tsar could watch from this spot the ceremonies in the square. In 1804, at the request of merchants, the square was paved in stone. In 1806 Nikolskaya Tower was reconstructed in the Gothic style and received a tent roof.

In the early 19th century, the Spassky Gate was the main front gate of the Kremlin and used for royal entrances. From this gate, wooden and stone bridges stretched across the moat. Books were sold on this bridge and stone platforms were built nearby for guns. The Tsar Cannon was located on the platform. The new phase of improvement of the square began after the Napoleonic invasion[4] and fire in 1812. The moat was filled in 1813 and in its place, rows of trees were planted.

During the Soviet era, Red Square maintained its significance, becoming a focal point for the new state. Besides being the official address of the Soviet government, it was renowned as a showcase for military parades from 1919 onward. Lenin's Mausoleum would from 1924 onward be a part of the square complex, and also as the grandstand for important dignitaries in all national celebrations. Two of the most significant military parades on Red Square were November 7, 1941, when the city was besieged by Germans and troops were leaving Red Square straight to the front lines, and the Victory Parade in 1945, when the banners of defeated Nazi armies were thrown at the foot of Lenin's Mausoleum.

2. The Description of Red Square Buildings

The square itself is around 330 meters long and 70 meters wide. The buildings surrounding the Square are all significant in some respect. Lenin's Mausoleum, for example, contains the embalmed body of Vladimir Ilyich Lenin, the founder of the Soviet Union. Nearby to the South is the elaborate brightly domed Saint Basil's Cathedral and the palaces and cathedrals of the Kremlin. On the Eastern side of the square is the GUM department store, and next to it the restored Kazan Cathedral. The Northern side is occupied by the State Historical Museum, whose outlines echo those of Kremlin towers. The Iberian Gate and Chapel have been rebuilt to the Northwest. (See Fig. 6. 2)

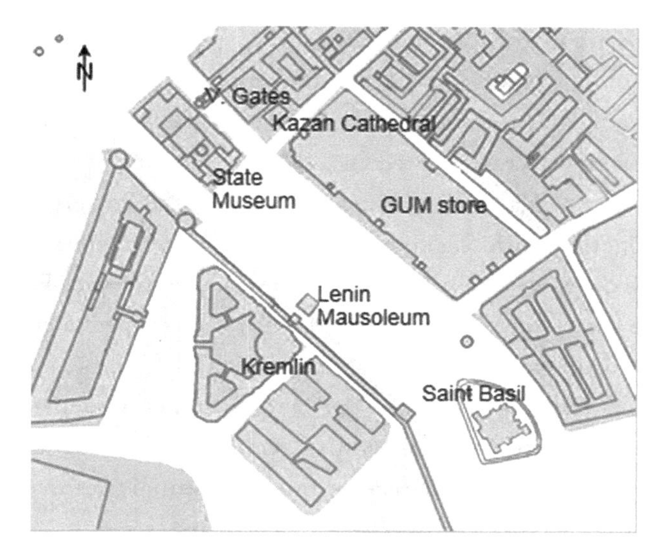

Fig. 6. 2 The layout of Red Square

2. 1 The Saint Basil's Cathedral

Cathedral of Vasily the Blessed, commonly known as Saint Basil's Cathedral, is now a museum. It was built from 1555-61 on orders from Ivan the Terrible and commemorates the capture of Kazan and Astrakhan. The original building, known as Trinity Church, contained eight side churches arranged around the ninth, central church of Intercession; the tenth church was erected in 1588 over the grave of venerated local saint Vasily (Basil).

The building is shaped as a flame of a bonfire rising into the sky, a design that has no analogues in Russian architecture. Dmitry Shvidkovsky states that "it is like no other Russian building. Nothing similar can be found in the entire millennium of Byzantine tradition from the fifth to fifteenth century ... a strangeness that astonishes by its unexpectedness, complexity and dazzling interleaving of the manifold details of its design." The cathedral foreshadowed the climax of Russian national architecture in the 17th century.

Instead of following the original layout, as seven churches around the central core, Ivan's architects opted for a more symmetrical floor plan with eight side churches around the core, producing "a thoroughly coherent, logical plan" . The central core and the four larger churches placed on the four major compass points are octagonal; the four diagonally placed smaller churches are cuboid, although their shape is barely visible through later additions. The larger churches stand on massive foundations, while the smaller ones were each placed on a raised platform, as if hovering above ground.

2. 2 Kremlin Towers

The Kremlin Wall is a defensive wall that surrounds the Moscow Kremlin, recognizable by the characteristic notches and its towers. The original walls were likely a simple wooden fence with guard towers built in 1156. The Kremlin is flanked by 19 towers with a 20th, the Koutafia Tower, not part of its walls.

The Borovitskaya Tower is a corner tower with a through-passage on the west side of the Kremlin. The tower was constructed in 1490 on the spot of an old Kremlin gate by Italian architect by order of Vasili III of Russia. In 1935, the Soviets installed a red star on top of the tower. Together with the star, its height is 54.05 meters.

The Spasskaya Tower was built in 1491 by the Italian architect Pietro Antonio Solari. The Tower was the first one to be crowned with the hipped roof, the clock on the Spasskaya Tower is usually referred to as the Kremlin chimes and designates official Moscow Time. The tower gate was once the main entrance into the Kremlin. In tsarist times, anyone passing through the gates had to remove their headgear and dismount their horses. Nowadays, the gate opens to receive the presidential motorcades on inauguration day, for the victory parades, and to receive the new years tree.

3. GUM

Gum,known as State Department Store facing Red Square, is currently a shopping mall. With the facade extending for 242 meters along the eastern side of Red Square. The trapezoidal building features a combination of elements of Russian medieval architecture and a steel framework and glass roof. William Craft Brumfield described the GUM building as "a tribute both to Shukhov's design and to the technical proficiency of Russian architecture toward the end of the 19th century".

The glass-roofed design made the building unique at the time of construction. The diameter of the roof is 14 meters, looks light, but it is a firm construction made of more than 50,000 metal pods, capable of supporting snowfall accumulation. Illumination is provided by huge arched skylights of iron and glass, each weighing some 820 short tons and containing in excess of 20,000 panes of glass. The facade is divided into several horizontal tiers, lined with red Finnish granite, Tarusa marble, and limestone. Each arcade is on three levels, linked by walkways of reinforced concrete.

Notes:

1. Red Square：红场，位于俄罗斯首都莫斯科市中心，临莫斯科河，是莫斯科最古老的广场，是重大历史事件的见证场所，也是俄罗斯重要节日举行群众集会、大型庆典和阅兵活动的地方。红场是著名旅游景点，是世界上著名的广场之一。红场南北长695米，东西宽130米，总面积9.035万平方米，呈不规则的长方形。红场地面全部由古老的条石铺成。红场的北面为俄罗斯国家历史博物馆，东面是莫斯科国立百货商场，南部为圣瓦西里大教堂，西侧是列宁墓和克里姆林宫的红墙及三座高塔。列宁墓上层，修建有主席台。每当俄罗斯举行重要仪式时，领导人就站在列宁墓上观礼指挥。

2. Kremlin：克里姆林宫，为红场最主要建筑，初建于12世纪中期，15世纪莫斯科大公伊凡三世时初具规模，以后逐渐扩大。16世纪中叶起它成为沙皇的宫堡，17世纪逐渐失去城堡的性质而成为莫斯科的市中心建筑群。克里姆林宫南临莫斯科河，西北接亚历山大罗夫斯基花园，东南与红场相连，呈三角形，周长2000多米。20多座塔楼、参差错落地分布在三角形宫墙边，宫墙上有5座城门塔楼和箭楼。宫殿的核心部分是宫墙之内的一系列宫殿，另有政府大厦和各种博物馆。它由俄著名建筑师巴尔马和波斯尼设计，不同于欧洲的哥特式与罗马式。克里姆林宫的几幢主要建筑都是由意大利设计师设计。
3. Ivan the Great：伊凡·瓦西里耶维奇(1530年8月25日—1584年3月18日)，又被称为伊凡雷帝、恐怖的伊凡、伊凡四世，是俄罗斯历史上的第一位沙皇。伊凡四世在领土扩张方面很有作为，伊凡四世内政方面的策略是加强中央集权。在伊凡四世在位期间，"沙皇"成为俄国君主的正式称谓。
4. Napoleonic invasion：俄法战争(1812年)。法皇拿破仑一世借口沙皇亚历山大一世破坏《蒂尔西特和约》，率军60万侵入俄境，企图歼灭俄军20余万。后俄军主动撤离莫斯科，并组织军民"坚壁清野"，袭扰法军。法军饥寒交迫，只好撤出莫斯科，沿南方撤回。俄军转入反攻，追歼大量法军。法军丧失了全部骑兵和几乎所有炮兵，只剩三万人退出国境。

Piazza San Marco[1]

Piazza San Marco is the principal public square of Venice, Italy, where it is generally known just as la Piazza, meaning "the Square". The "little Piazza" is an extension of the Piazza towards the lagoon in its south east corner. The two spaces together form the social, religious and political centre of Venice and are commonly considered together. A remark usually attributed to Napoleon calls the Piazza San Marco "the drawing room of Europe". (See Fig. 6. 3)

①

Fig. 6. 3 Piazza San Marco

① shuttershock. com

1. The History of Piazza San Marco

The history of the Piazza San Marco can be conveniently covered in four periods, but the only pre-renaissance buildings and monuments still standing there are Church of St. Marco[2], the Doge's Palace[3] and the two great columns in the Piazzetta.

In 828/9 relics of St. Mark[4] were brought to the city from Alexandria[5] and the Venetians and the Doge[6] adopted the apostle as their new patron. The first church of St. Mark was begun on the south side of the existing chapel. The design of the church was based on the Church of the Twelve Apostles[7] in Constantinople and it seems to have covered the same area as the central part of the present church. The Doge's palace, in the same area as its modern successor, was at that time surrounded by water.

In 1063 a new church was finished and its main structure of the present church, though the west front facing the Piazza in the Romanesque style[8] was in undecorated brickwork. It had five domes, but their exterior profile was low, unlike the present high, onion-shaped structures.

In 1204 Constantinople was captured in the course of the 4th Crusade[9] and, at that time and later during the 13th century, much valuable material was taken from the city and shipped back for the adornment of Venice.

The original 9th century Doge's palace was rebuilt in 1340. Because of the great expense involved, nothing more was done for many years, but in 1422 the Doge Tomaso Mocenigo insisted that for the honor of the city the remaining part of the old palace should be demolished and the new part extended.

In 1493 an astronomical clock was commissioned by Venice and it was decided to install it in a new clock tower in the Piazza with a high archway beneath it leading into the street known as the Merceria. The building was completed with the clock installed by February 1499.

2. The Description of the Piazza San Marco

The Square is dominated at its eastern end by the great church of St. Mark. The Piazzetta dei Leoncini is an open space on the north side of the church named after the two marble lions, the neo-classic building on the east side adjoining the Basilica is the Palazzo Patriarcale, the seat of the Patriarch of Venice.

Beyond that is the Clock Tower, completed in 1499, above a high archway where the street known as the Merceria leads through shopping streets to the Rialto, the commercial and financial center. To the right of the clock-tower is the closed church of San Basso, sometimes open for exhibitions.

To the left is the long arcade along the north side of the Piazza, the buildings on this side are formerly the homes and offices of the Procurators of St. Mark, high officers of state in the days of the republic of Venice. They were built in the early 16th century. The arcade is lined with shops and restaurants at ground level, with offices above.

Turning left at the end, the arcade continues along the west end of the Piazza, which was rebuilt by Napoleon about 1810 and is known as the Napoleonic Wing. It holds, behind the shops, a ceremonial staircase which was to have led to a royal palace but now forms the entrance to the Correr Museum.

Turning left again, the arcade continues down the south side of the Piazza. The buildings on this side are known as the new procuracies. At the far end the Procuratie meet the north end of mid-16th century Sansovino's Libreria, whose main front faces the Piazzetta. The arcade continues round the corner into the Piazzetta.

Opposite to this, standing free in the Piazza, is the Campanile of St. Mark's church, rebuilt in 1912 after the collapse of the former campanile on 14 July 1902. Adjacent to the Campanile, facing towards the church, is the elegant small building known as the Loggetta del Sansovino, and used as a lobby by patricians waiting to go into a meeting of the Great Council in the Doge's Palace and by guards when the Great Council was sitting. (See Fig. 6. 4)

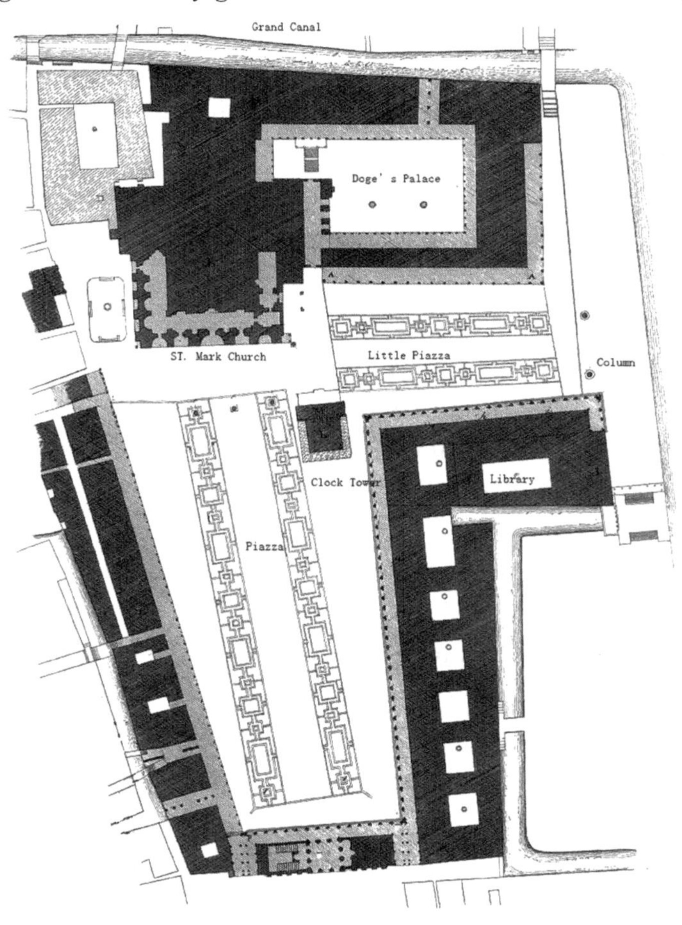

Fig. 6. 4 The plan of Piazza San Marco

3. The Key Buildings around Piazza San Marco

3.1 Church of Saint Mark

The Cathedral Basilica of Saint Mark is the most famous of the city's churches and one of the best known examples of Italo-Byzantine architecture. It lies at the eastern end of the Piazza San Marco, adjacent and connected to the Doge's Palace. For its opulent design, gold ground mosaics, and its status as a symbol of Venetian wealth and power, from the 11th century on the building has been known by the nickname Church of gold.

The exterior of the west facade of the basilica is divided in three registers: lower, upper, and domes. In the lower register of the facade, five round-arched portals, enveloped by polychrome marble columns, open into the narthex through bronze-fashioned doors. The upper level of mosaics with scenes showing the history of the relics of Saint Mark from right to left fill the lunettes of the lateral portals; the first on the left is the only one on the facade still surviving from the 13th century.

The stone sculpture is relatively limited at the lower level, where a forest of columns and patterned marble slabs are the main emphases. It includes relatively narrow bands of Romanesque work on the portals, richly carved borders of foliage mixed with figures to the ogee arches and other elements, and large shallow relief saints between the arches. Along the roofline, by contrast, there is a line of statues, many in their own small pavilions, culminating in Saint Mark flanked by six angels in the centre, above a large gilded winged lion.

In the upper register, from the top of ogee arches, statues of The ological and Cardinal Virtues, four Warrior Saints, Constantine, Demetrius, George, The odosius and St. Mark watch over the city. Above the large central window of the facade, under St. Mark, the Winged Lion holds the book quoting "Peace to you Mark my evangelist" . In the centre of the balcony the famous bronze horses face the square.

The interior is based on a Greek cross, with each arm divided into three naves with a dome of its own as well as the main dome above the crossing. The dome above the crossing and the western dome are bigger than the other three. This is based on Constantine's Church of the Holy Apostles in Constantinople. The marble floor is entirely tessellated in geometric patterns and animal designs. One particular panel in the pavement shows two cocks carrying a trussed-up fox, has been interpreted politically by some, as a reference to the French conquest of Milan in the Italian Wars. Others see it as a sacred symbol of the faithful wish for immortality, with the victory of the cross, and "analogous to the hope of resurrection, the victory of the soul over death". The lower register of walls and pillars is completely covered with polychrome marble slabs. The transition between the lower and the upper register is delimited all around the basilica by passageways which largely substituted the former galleries.

3.2 The Doge's Palace

The Doge's Palace is built in Venetian Gothic style, and one of the main landmarks of the city of Venice. The palace was the residence of the Doge of Venice, the supreme authority of the Republic of Venice. By the end of the 19th century, the structure was showing clear signs of decay, and the Italian government set aside significant funds for its restoration and all public offices were moved elsewhere. In 1923, the Italian State, owner of the building, entrusted the management to the Venetian municipality to be run as a museum.

The oldest part of the palace is the wing overlooking the lagoon, the corners of which are decorated with 14th-century sculptures. The ground floor arcade and the loggia above are decorated with 14th-and 15th-century capitals, some of which were replaced with copies during the 19th century.

Flanked by Gothic pinnacles, with two figures of the Cardinal Virtues per side, the gateway is crowned by a bust of St. Mark over which rises a statue of Justice with her traditional symbols of sword and scales. In the space above the cornice, there is a sculptural portrait of the Doge Francesco Foscari kneeling before the St. Mark's Lion.

The north side of the courtyard is closed by the junction between the palace and St. Mark's Basilica, which used to be the Doge's chapel. At the center of the courtyard stand two well-heads dating from the mid-16th century. Since 1567, the Giants' Staircase is guarded by two colossal statues of Mars and Neptune, which represents Venice's power by land and by sea, and therefore the reason for its name. Members of the Senate gathered before government meetings in the Senator's Courtyard, to the right of the Giants' Staircase.

The rooms in which the Doge lived were always located in this area of the palace, between the water entrance to the building-the present-day Golden Staircase and the apse of St. Mark's Basilica. The core of these apartments forms a prestigious, though not particularly large, residence, given that the rooms nearest the Golden Staircase had a mixed private and public function. In the private apartments, the Doge could set aside the trappings of office to retire at the end of the day and dine with members of his family amidst furnishings that he had brought from his own house.

3.3 The Clock Tower

The Clock Tower is an early renaissance building and it comprises a tower, which contains the clock, and lower buildings on each side. Both the tower and the clock date from the last decade of the 15th century, though the mechanism of the clock has subsequently been much altered. It was placed where the clock would be visible from the waters of the lagoon and give notice to everyone of the wealth and glory of Venice. The lower two floors of the tower make a monumental archway into the main street of the city, the Merceria, which linked the political and religious centre with the commercial and financial centre.

On a terrace at the top of the tower are two great bronze figures, hinged at the waist, which strike the hours on a bell. One is old and the other young, to show the passing of time and, although said to represent shepherds or giants they are always known as "the Moors" because of the dark patina acquired by the bronze.

Below this level is the winged lion of Venice with the open book, before a blue background with gold stars. Below again, is a semi-circular gallery with statues of the Virgin and Child seated, in gilt beaten copper. On either side are two large blue panels showing the time: the hour on the left in Roman numerals and the minutes on the right in Arabic numerals. Twice a year, at Epiphany (6 January) and on Ascension Day the three Magi, led by an angel with a trumpet, emerge from one of the doorways normally taken up by these numbers and pass in procession round the gallery, bowing to the Virgin and child, before disappearing through the other door.

Notes:

1. Piazza San Marco：圣马可广场，又称威尼斯中心广场，是威尼斯的政治、宗教和传统节日的公共活动中心。圣马可广场初建于9世纪，当时只是圣马可大教堂前的一座小广场。马可是圣经中《马可福音》的作者，威尼斯人将他奉为守护神。相传828年，两个威尼斯商人从埃及亚历山大将耶稣圣徒马可的遗骨偷运到威尼斯，并在同一年为圣马可兴建教堂。教堂内有圣马可的陵墓，大教堂以圣马可的名字命名，大教堂前的广场也因此得名"圣马可广场"。1797年拿破仑进占威尼斯后，赞叹圣马可广场是"欧洲最美的客厅"和"世界上最美的广场"。圣马可广场是由公爵府，圣马可大教堂，圣马可钟楼，新、旧行政官邸大楼，连接两大楼的拿破仑翼大楼，圣马可大教堂的四角形钟楼和圣马可图书馆等建筑及威尼斯大运河所围成的长方形广场，长约170米，东边宽约80米，西侧宽约55米。广场四周的建筑从中世纪到文艺复兴时代都有。
2. Church of Saint Mark：圣马可大教堂，始建于公元829年，重建于1043—1071年，曾是中世纪欧洲最大的教堂，也是一座收藏丰富的艺术品的宝库。大教堂是东方拜占庭艺术、古罗马艺术、中世纪哥特式艺术和文艺复兴艺术等多种艺术式样的结合体。大教堂有五个圆形大屋顶，是典型的东方拜占庭风格。大教堂内外有400根大理石柱子，内外有4000平方米面积的马赛克镶嵌画。
3. Doge's Palace：威尼斯公爵府，始建于9世纪。由于当时威尼斯与地中海东部的伊斯兰国家密切的文化贸易往来，大量阿拉伯人定居威尼斯，所以总督府立面的席纹图案明显受到了伊斯兰建筑的影响。

4. St. Mark：圣马可，基督教《圣经》故事人物，《马可福音》的作者，其耶路撒冷的住所后成为基督教徒聚会之所。据传，圣徒保罗被拘送罗马后，他去照顾保罗；待彼得至罗马，又为彼得助手。他曾至亚历山大里亚传教，后于该地被杀。传说他根据彼得的叙述撰写了《马可福音》。据说在威尼斯还是一片荒芜海滩的时候，马可到意大利各地传教，当时风暴骤起，把船刮到荒凉的沼泽地带搁浅了。马可似乎听到天使在召唤："愿你平安，马可！你和威尼斯共存。"这样，马可成为威尼斯的护城神，其标志为狮子。
5. Alexandria：亚历山大，位于尼罗河口以西一条狭长地带上，是埃及第二大城市，最大海港。亚历山大建城至今已有2000多年的历史。公元前332年，希腊马其顿国王亚历山大一世征服埃及后，在北临地中海的尼罗河三角洲上建立一座都城，并以自己的名字给城市命名。亚历山大曾经为欧洲与东方贸易和文化交流的枢纽，创造了"亚历山大文化"，即"希腊后期文化"，对人类的科学与文化发展做出了杰出的贡献。被视为古代世界七大建筑奇迹之一的亚历山大灯塔便是这一时期科学和文化高度发达的典型代表。
6. Doge：威尼斯共和国总督。威尼斯共和国是位于意大利北部的城市共和国，始建于公元687年，1797年为拿破仑·波拿巴所灭，1866年并入意大利王国。公元687年威尼斯产生第一任总督，建立共和国。建国初期隶属东罗马帝国。公元10世纪末获独立，威尼斯成为富庶的商业国。
7. Twelve Apostles：耶稣十二门徒，其名录在圣经里记载了四次，次序如下：西门彼得、安得烈、雅各布、约翰、腓力、巴多罗买、多马、马太、亚勒腓的儿子雅各布、达太、西门、犹大。
8. Romanesque style：罗马式建筑，是公元10—12世纪欧洲基督教流行地区的一种建筑风格，多见于修道院和教堂，因采用古罗马式的券、拱而得名。罗马式建筑给人以雄浑庄重的印象，对后来的哥特式建筑影响很大。
9. Crusade：十字军东征(1096—1291年)，是一系列在罗马天主教教宗的准许下进行的、持续近200年的、有名的宗教性军事行动，由西欧的封建领主和骑士，对地中海东岸的国家，以收复阿拉伯入侵占领的土地名义发动的战争，前后共计有八次，第四次十字军东征就是针对信奉东正教的拜占庭帝国。

Tian'anmen Square[1]

Tian'anmen Square is a large city square in the center of Beijing, China, named after the Gate of Heavenly Peace(天安门), located to its North, separating it from the Forbidden City. Tian'anmen Square is the fourth largest city square in the world. It has great cultural significance as it was the site of several important events in Chinese history. With the towering Monument to the People's Heroes(人民英雄纪念碑) at the center, Tian'anmen Tower in the north, the Mao Zedong Memorial Hall in the south, the National Museum of China in the east and the Great Hall of the People in the west. (See Fig. 6. 5)

Fig. 6. 5 Tian'anmen Square①

1. History of Tian'anmen square

The Tian'anmen square was designed and built in 1651, and has since enlarged four times its original size in the 1950s. Near the center of today's square, stood the "Great Ming Gate"(大明门), the southern gate to the Imperial City, renamed "Great Qing Gate"(大清门) during the Qing Dynasty, and "Gate of China"(中华门) later. It was a purely ceremonial gateway, with three arches but no ramparts, this gate had a special status as the "Gate of the Nation", as can be seen from its successive names. It normally remained closed, except when the Emperor passed through, the traffic was diverted to two side gates at the western and eastern ends of today's square, respectively. Because of this diversion in traffic, a busy market place, called Chessgrid Streets(棋盘街) developed in the big, fenced square to the south of this gate.

In 1954, the Gate of China was demolished, allowing for the enlargement of the square. In November 1958 a major expansion of Tian'anmen Square started, which was completed after only 11 months. This followed the vision of Mao Zedong to make the square the largest and most spectacular in the world, and intended to hold over 500,000 people. On its southern edge, the Monument to the People's Heroes has been erected. As part of the Ten Great Buildings[2] constructed between 1958—1959 to commemorate the ten-year anniversary of the People's Republic of China, the Great Hall of the People(人民大会堂) and the Revolutionary History Museum(革命历史博物馆), now National Museum of China(国家博物馆) were erected on the western and eastern sides of the square.

The year after Mao's death in 1976, a Mausoleum was built near the site of the former Gate of China, on the main north-south axis of the square. In connection with this project, the square was further increased in size to become fully rectangular and being able to accommodate 600,000 persons.

The urban context of the square was altered in the 1990s with the construction of National Grand Theatre(国家大剧院) in its vicinity and the expansion of the National Museum.

---

① 申伟.《乾坤方圆——天安门广场》

2. The Landmarks around Tian'anmen Square

2.1 Tian'anmen Tower(天安门)

The Tian'anmen is widely used as a national symbol, first built During the Ming Dynasty in 1420, Tian'anmen is often referred to as the front entrance to the Forbidden City. However, the Meridian Gate (午门) is the first entrance to the Forbidden City proper, while Tian'anmen was the entrance to the Imperial City, within which the Forbidden City was located. Tian'anmen is located to the north of Tian'anmen Square, separated from the plaza by Chang'an Avenue(长安街).

The Tian'anmen gate was originally named Chengtianmen (承天门), and the original building was first constructed in 1420 and was based on a gate of an imperial building in Nanjing with the same name and hence inherited the name. The gate was damaged by lightning in July, 1457, and was completely burnt down. In 1465, the Chenghua(成化) Emperor of the Ming Dynasty ordered the Minister of Works to rebuild the gate, and the design was changed from the original paifang(牌坊) form to the gatehouse that is seen today. Following the establishment of the Qing Dynasty and the Manchu conquest of China proper, the gate was once again rebuilt, beginning in 1645, and was given its present name upon completion in 1651.

The gate building is 66 meters long, 37 meters wide and 32 meters high. Like other official buildings of the empire, the gate has unique imperial roof decorations. In front of the gate are two lions standing in front of the gate and two more guarding the bridges.

Two stone columns, called huabiao(华表), each with an animal (hou 吼) on top of it, also stand in front of the gate.

The gate has five arched gates and nine principle hall columns. With the delicately carved white marbles on its base and yellow tiles on the roof, the tower is quite resplendent. Under the tower flows the limpid Jinshui River(金水河), across which seven exquisite bridges are perched, named the Golden Water Bridges(金水桥). (See Fig. 6.6)

2.2 Monument to the People's Heroes (人民英雄纪念碑)

In the center of the Square stands the Monument to the People's Heroes, which commemorates the martyrs who devoted their lives to the Chinese people. The Monument is a square building, covering an area of 3,000 square meters. It is composed of three parts: the body, the Buddhist-style base, and the pedestal, reaching as tall as 37.94m. The body of the monument is made up of 413 pieces of granite 32 layers deep. In the center of the north side of the monument, a single complete piece of stone, 14.7m long, 2.9m wide and 1m thick, is inscribed with large, glazed words by Mao Zedong which read: "Eternal Glory to the People's Heroes." The south side of the monument is composed of 7 pieces of stone

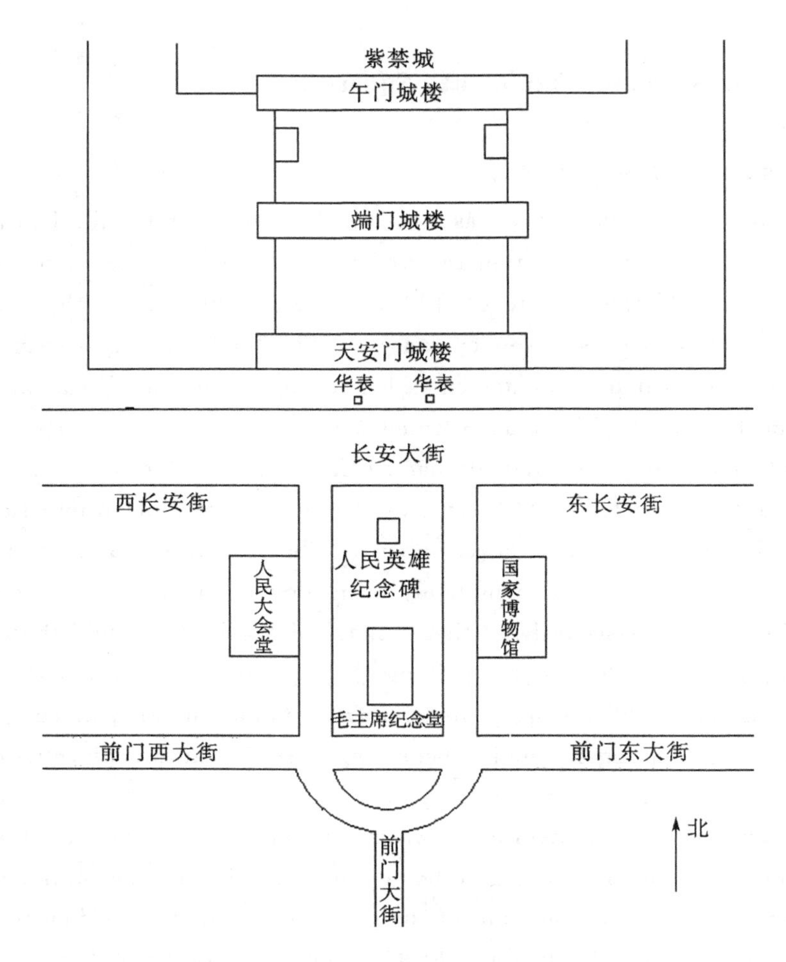

Fig. 6. 6 The layout of Tian'anmen Square

with a draught of an epigraph by Mao Zedong and inscribed by Zhou Enlai, the first premier of the People's Republic of China who served from October 1949 until his death in January 1976. The east and west sides of the monument are carved with patterns of five-pointed stars, pine trees and flags.

It is surrounded by white balusters, the eight reliefs depict the crucial events which took place in modern times, showing a part of the centuries-old history of China.

2. 3 Mao Zedong Memorial Hall(毛主席纪念堂)

In the south of the Square is the Mao Zedong Memorial Hall. Mao Zedong, the first chairman of the People's Republic of China, is greatly admired by the Chinese People. The hall was built in 1977. The remains of Mao Zedong are laid in a crystal coffin in the main hall. From the exhibit in the north hall, you can learn something about Chairman Mao, Zhou Enlai and other founders of the state.

The memorial hall is about 260 meters long, 220 meters wide and 33. 5 meters high.

The remains of the Chairman Mao Ze Dong in a crystal coffin are on display. The Chairman Mao's Mausoleum has five parts—the Memorial Hall, the North Hall, the South Hall, the Viewing Hall and the Courtyard. People throughout China were involved in the design and construction of the mausoleum, with 700,000 people from different provinces, autonomous regions, and nationalities doing symbolic voluntary labor. Materials from all over China were used throughout the building: granite from Sichuan province, porcelain plates from Guangdong province, pine trees from Yan'an in Shaanxi province, saw-wort seeds from the Tian Shan mountains in the Xinjiang Autonomous Region, earth from quake-stricken Tangshan, colored pebbles from Nanjing, milky quartz from the Kunlun Mountains, pine logs from Jiangxi province, and rock samples from Mount Everest. Water and sand from the Taiwan Straits were also used to symbolically emphasize the People's Republic of China's claims over Taiwan.

2.4 National Museum of China(国家博物馆)

The museum was established in 2003 by the merging of the two separate museums that had occupied the same building since 1959: the Museum of the Chinese Revolution in the northern wing and the National Museum of Chinese History in the southern wing.

The building was completed in 1959 as one of the Ten Great Buildings celebrating the ten-year anniversary of the founding of the People's Republic of China. It complements the opposing Great Hall of the People that was built at the same time. The structure sits on 6.5 hectares and has a frontal length of 313 meters, a height of four stories totaling 40 meters, and a width of 149 meters. The front displays eleven square pillars at its center.

A large whitish interior space with a very high ceiling lit by many windows on its left stretches off into the far background. There are people walking around within. At left in the foreground is a large dark wooden model of a round three-tiered pagoda

After four years of renovation, the museum reopened on March 17, 2011, with 28 new exhibition halls, more than triple the previous exhibition space, and state of the art exhibition and storage facilities. It has a total floor space of nearly 200,000 m^2 to display.

2.5 The Great Hall of the People(人民大会堂)

The Great Hall of the People is a state building located at the western edge of Tian'anmen Square. This building, erected in 1959, is used for legislative and ceremonial activities by the People's Republic of China government and Communist Party of China. It is also the meeting place of the National Congress of the Communist Party of China, which, since 1982, has occurred once every five years. The Great Hall is also used for many special events, including national level meetings of various social and political organizations, large anniversary celebrations, as well as the memorial services for former leaders.

The central section principally includes the Great Auditorium, the Main Auditorium, the Congress Hall, the Central Hall, the Golden Hall and other main halls. The northern section consists of the State Banquet Hall, the Salute State Guest Hall, the North Hall, the East Hall, the West Hall and other large halls.

Each province, special administrative region, autonomous region of China has its own hall in the Great Hall, such as Beijing Hall, Hong Kong Hall and Taiwan Hall. Each hall has the unique characteristics of the province and is furnished according to the local style.

The Great Auditorium, with volume of 90,000 cubic meters, seats 3,693 in the lower auditorium, 3,515 in the balcony, 2,518 in the gallery and 300 to 500 on the dais. Government leaders make their speeches; and the representatives do much of their business. It can simultaneously seat 10,000 representatives. The ceiling is decorated with a galaxy of lights, with a large red star is at the centre of the ceiling, and a pattern of a water waves nearby represents the people. Its facilities equipped with audio-visual and other systems adaptable to a variety of meeting types and sizes. A simultaneous interpretation system is also provided with a language booth.

The State Banquet Hall with an area of 7,000 square meters can entertain 7,000 guests, and up to 5,000 people can dine at one time. Smaller gatherings can be held in the Main Auditorium, with larger groups having the use of one or more of the conference halls, such as Golden Hall and North Hall, and the smallest assemblies accommodated in one or more of the over 30 conference halls that are named after provinces and regions in China.

Notes:

1. Tian'anmen Square：天安门广场，位于中国北京东长安街，北起天安门，南至正阳门，东起中国国家博物馆，西至人民大会堂，南北长880米，东西宽500米，面积达44万平方米，可容纳100万人举行盛大集会，是世界上最大的城市广场。天安门原名承天门，建于明永乐十五年(1417年)。清顺治八年(1651年)改建后称天安门。此时的天安门前只是一块封闭的T形宫廷广场，为明清两朝举办重大庆典和向全国发布政令的重要场所。1914年5月，北洋政府的朱启钤启动改造旧都城计划。拆除天安门前千步廊、修筑沥青路、瓮城等，原本封闭的宫廷广场变成可自由穿行和逗留的开放空间，神秘的皇权被消解，天安门开始成为现代意义上的广场。1958年，为迎接十周年国庆，天安门广场开始了史上最大规模的一次扩建，拆除了中华门、棋盘街及广场上的红墙。人民英雄纪念碑和两侧建起的人民大会堂、国家博物馆和军事博物馆，奠定了广场作为政治中心的基调。
2. Ten Great Buildings：十大建筑。为迎接中华人民共和国建国10周年，中央人民政府决定在首都北京大兴土木，建设包括人民大会堂在内的国庆工程。由于这项计划大体上包括10个大型项目，故又称新中国"十大建筑"，包括人民大会堂、国家博物馆、军事博物馆、北京火车站、北京工人体育场、全国农业展览馆、北京民族文化宫、中国美术馆、北京饭店等10座建筑，总面积达67.3万平方米。这个浩大的工程采用的是边设计、边勘察、边施工的做法。十大工程采取的是折中主义的古典风格，将古今中外的建筑风格为我所用。

Place de la Concorde[1]

The Place de la Concorde at the eastern end of the Champs-Élysées[2], is one of the major public squares in Paris, France. Measuring 8.64 hectares in area, it is the largest square in the French capital. The Place was designed by Ange-Jacques Gabriel in 1755 as a moat-skirted octagon between the Champs-Élysées to the west and the Tuileries Garden to the east. Decorated with statues and fountains, the area was named Place Louis XV to honor the king at that time. The square showcased an equestrian statue of the king, which had been commissioned in 1748 by the city of Paris. (See Fig. 6.7)

Fig. 6.7 The Place de la Concorde

In 1792, during the French revolution, the statue of Louis XV was replaced by another large statue called Liberte or freedom, and the square was renamed the Place de la Revolution. A guillotine was installed at the centre of the square and during the following couple of years, many people were beheaded here, including King Louis XVI, Marie Antionette, and eventually the revolutionary Robespierre. After the revolution the square was renamed several times, until 1830, when it was once again named the Place de la Concorde.

1. The Orientation of the Place de la Concorde

To the west of the Place is the famous Champs-Élysées; to the east of the Place are the Tuileries Gardens. The Galerie nationale du Jeu de Paume and the Musée de l'Orangerie, both in the Tuileries Gardens, border the Place. At the north end, two magnificent identical stone buildings were constructed. Separated by the rue Royale, these remain among the best examples of architecture from that period. Initially they served as government offices, with the eastern one being the home of the French Naval Ministry. Shortly after its construction, the western building was made into the luxurious Hotel de Crillon, which is still operating today, this is here that Marie Antoinette spent afternoons relaxing and taking piano lessons. The hotel also served as the headquarters of the occupying German army during World War II.

The Rue Royale leads to the Église de la Madeleine. The Embassy of the United States is located in the corner of the Place at the intersection of Avenue Gabriel and Rue Boissy d'Anglas. The northeastern corner of the Place is the western end of the Rue de Rivoli. South of the Place: the River Seine, crossed by the Pont de la Concorde, built by Jean-Rodolphe Perronnet between 1787—1790 and widened in 1930—1932. The Palais Bourbon, home of the French National Assembly, is across the bridge, on the opposite bank of the river. At each of the eight angles of the octagonal Place is a statue, initiated by architect Jacques-Ignace Hittorff, representing a French city. That same year a bronze fountain, called La fontaine des Mers was added to the square. Later in 1839 a second fountain, the Elevation of the Maritime fountain, was installed. This fountain, like the first, was designed by Hittorf. (See Fig. 6. 8)

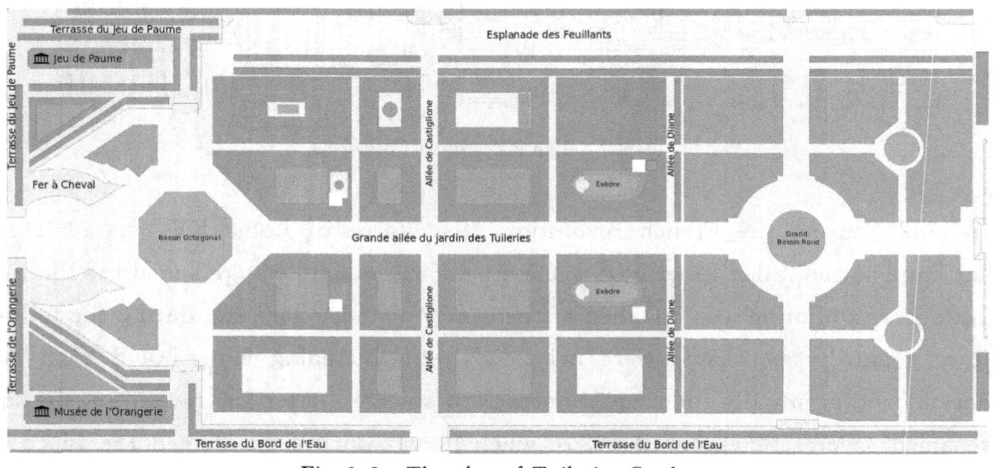

Fig. 6. 8 The plan of Tuileries Garden

In the 19th century the 3200 years old obelisk from the temple of Ramses II at Thebes, now modern day Luxor, was installed at the centre of the Place dela Concorde.

2. Important monuments on the Place de la Concorde

2.1 Obelisk

The center of the Place is occupied by a giant Egyptian obelisk decorated with hieroglyphics exalting the reign of the pharaoh Ramesses II. It is one of two obelisks the Egyptian government gave to the French in the 19th century. The other one stayed in Egypt, too difficult and heavy to move to France with the technology at that time. In the 1990s, President Francois Mitterrand gave the second obelisk back to the Egyptians.

The obelisk once marked the entrance to the Luxor Temple. The Ottoman viceroy of Egypt, Mehmet Ali, offered the 3,300-year-old Luxor Obelisk to France in 1829. It arrived in Paris on 21 December 1833. Three years later, on 25 October 1836, King Louis Philippe had it placed in the center of Place de la Concorde, where a guillotine used to stand during the Revolution.

The obelisk, a yellow granite column, rises 23 meters high, including the base, and weighs over 250 metric tons. Given the technical limitations of the day, transporting it was no easy feat—on the pedestal are drawn diagrams explaining the machinery that was used for the transportation. The obelisk is flanked on both sides by fountains constructed at the time of its erection on the Place.

The obelisk is covered with hieroglyphs picturing the reign of pharaohs Ramses II and Ramses III. Pictures on the pedestal describe the transportation to Paris and its installation at the square in 1836. One story that is often told about the installation of the obelisk, is that during the final stage of its erection, it was found that due to the placement of the winches, they had met there mechanical limits before the obelisk was fully upright, then a voice out of an estimated 200,000 onlookers, shouted, "moisten the ropes". It was, it is claimed, a sailor who knew that hemp ropes would shrink while drying.

Missing its original cap, believed stolen in the 6th century B. C., the government of France added a gold-leafed pyramid cap to the top of the obelisk in 1998.

2.2 Two Fountains

The two fountains in the Place de la Concorde have been the most famous of the fountains built during the time of Louis-Philippe, and came to symbolize the fountains in Paris. They were designed by Jacques Ignace Hittorff, a student of the Neoclassical designer Charles Percier at the École des Beaux-Arts. The German-born Hittorff had served as the official Architect of Festivals and Ceremonies for the deposed King, and had spent two years studying the architecture and fountains of Italy.

Hittorff's two fountains were on the theme of rivers and seas, in part because of their proximity to the Ministry of Navy, and to theSeine. Their arrangement, on a north-south axis aligned with the Obelisk of Luxor and the Rue Royale, and the form of the fountains themselves, were influenced by the fountains of Rome, particularly Piazza Navona and the Piazza San Pietro, both of which had obelisks aligned with fountains.

Both fountains had the same form: a stone basin; six figures of tritons or naiads holding fish spouting water; six seated allegorical figures, their feet on the prows of ships, supporting the pedestal, of the circular vasque; four statues of different forms of genius in arts or crafts supporting the upper inverted upper vasque; whose water shot up and then cascaded down to the lower vasque and then the basin.

The north fountain was devoted to the Rivers, with allegorical figures representing the Rhone and the Rhine, the arts of the harvesting of flowers and fruits, harvesting and grape growing; and the geniuses of river navigation, industry, and agriculture.

The south fountain, closer to the Seine, represented the seas, with figures representing the Atlantic and the Mediterranean; harvesting coral; harvesting fish; collecting shellfish; collecting pearls; and the geniuses of astronomy, navigation and commerce.

2.3 The Tuileries Gardens

The Tuileries Garden is a public garden located between the Louvre Museum and the Place de la Concorde. Created by Catherine de Medici as the garden of the Tuileries Palace in 1564, it was eventually opened to the public in 1667, and became a public park after the French Revolution. In the 19th and 20th century, it was the place where Parisians celebrated, met, promenaded, and relaxed.

In July 1559, Queen Catherine de Medicis decided that she would build a new palace for herself, separate from the Louvre, with a garden modeled after the gardens of her native Florence. Catherine commissioned a landscape architect from Florence to build an Italian Renaissance garden, with fountains, a labyrinth, and a grotto, decorated with faience images of plants and animals.

The garden of Catherine de Medicis was an enclosed space five hundred meters long and three hundred meters wide, separated from the new chateau by a lane. It was divided into rectangular compartments by six alleys, and the sections were planted with lawns, flower beds, and small clusters of five trees and, more practically, with kitchen gardens and vineyards.

Notes:

1. Place de la Concorde：巴黎协和广场，位于巴黎市中心、塞纳河北岸，是法国最著名广场和世界上最美丽的广场之一，18 世纪由国王路易十五下令营建。广场呈八角形，中央矗立着埃及方尖碑，是由埃及总督赠送给查理五世的。方尖碑是由整块的粉红色花岗岩雕出来的，上面刻满了埃及象形文字，赞颂埃及法老的丰功伟绩。广场的四周有 8 座雕像，象征着法国的八大城市。
2. Champs-Élysées：香榭丽舍大街，位于巴黎市中心商业繁华区，是闻名世界的大街。香榭丽舍大街横贯首都巴黎的东西主干道，全长 1800 米，最宽处约 120 米，为双向八车道，东起协和广场，西至戴高乐广场。东段以自然风光为主；西段是高级商业区，世界一流品牌、服装店、香水店都集中在这里。每年 7 月 14 日的法国国庆大阅兵在这条大道上举行。

Insights: Triumphal Arch

A triumphal arch is a monumental structure in the shape of an archway with one or more arched passage ways, often designed to span a road. In its simplest form a triumphal arch consists of two massive piers connected by an arch, crowned with a flat entablature or attic on which a statue might be mounted or which bears commemorative inscriptions. The main structure is often decorated with carvings, sculpted reliefs, and dedications. (See Fig. 6. 9)

Fig. 6. 9 Constantine triumphal arch ①

① Deviantart. com

Triumphal arches are one of the most influential and distinctive types of architecture associated with ancient Rome. The triumphal arch was used to commemorate victorious generals or significant public events such as the founding of new colonies, the construction of a road or bridge, the death of a member of the imperial family or the accession of a new emperor.

The ornamentation of an arch was intended to serve as a constant visual reminder of the triumph and triumphator. The facade was ornamented with marble columns, and the piers and attics with decorative cornices. Sculpted panels depicted victories and achievements, the deeds of the triumphator, the captured weapons of the enemy or the triumphal procession itself. The spandrels usually depicted flying Victories, while the attic was often inscribed with a dedicatory inscription naming and praising the triumphator. The piers and internal passageways were also decorated with reliefs and free-standing sculptures.

Garden

A garden is a planned space, usually outdoors, set aside for the display, cultivation, and enjoyment of plants and other forms of nature. The garden can incorporate both natural and man-made materials. Gardens may exhibit structural enhancements, sometimes called follies, including water features such as fountains, ponds, waterfalls or creeks, dry creek beds, statuary, arbors, trellises and more. Some gardens are for ornamental purposes only, while some gardens also produce food crops, sometimes in separate areas, or sometimes intermixed with the ornamental plants.

Gardens may be designed by garden owners themselves, or by professionals. Professional garden designers tend to be trained in principles of design and horticulture, and have a knowledge and experience of using plants. Garden design can be roughly divided into two groups, formal and naturalistic gardens. Elements of garden design include the layout of hard landscape, such as paths, rockeries, walls, water features, sitting areas and decking, as well as the plants themselves, with consideration for their horticultural requirements, their season-to-season appearance, lifespan, growth habit, size, speed of growth, and combinations with other plants and landscape features.

Persian Gardens[1]

The tradition and style in the design of Persian gardens, known as Iranian gardens in Iran, has influenced the design of gardens from Andalusia[2] to India and beyond. The gardens of the Alhambra show the influence of Persian Garden philosophy and style in a Moorish Palace scale, from the era of Al-Andalus in Spain. From the time of the Achaemenid Dynasty the idea of an earthly paradise spread through Persian literature and example to other cultures, both the Hellenistic gardens of the Seleucids and the Ptolemies in Alexandria. Such gardens would have been enclosed. The garden's purpose was, and is, to provide a place for protected relaxation in a variety of manners: spiritual, and leisurely such as meetings with friends, essentially a paradise on earth. The Common Iranian word for enclosed space, a term that was adopted by Christianity to describe the garden of Eden[3] or Paradise on earth. (See Fig. 7. 1)

Fig. 7.1 Persian gardens

1. History of Persian Gardens

Persian gardens may originate as early as 4000 BCE. Decorated pottery of that time displays the typical cross plan of the Persian garden. The outline of the Pasargad Garden, built around 500 B. C., is viewable today.

During the reign of the Sassanids[4], and under the influence of Zoroastrianism[5], water in art grew increasingly important. This trend manifested itself in garden design, with greater emphasis on fountains and ponds in gardens.

During the Arab occupation, the aesthetic aspect of the garden increased in importance, overtaking utility. During this time, aesthetic rules that govern the garden grew in importance. An example of this is the chahār bāgh[6], a form of garden that attempts to emulate Eden, with four rivers and four quadrants that represent the world. The design sometimes extends one axis longer than the cross-axis, and may feature water channels that run through each of the four gardens and connect to a central pool. (See Fig. 7.2)

The invasion of Persia by the Mongols in the thirteenth century led to a new emphasis on highly ornate structure in the garden. Examples of this include tree peonies and chrysanthemums. The Mongol empire then carried a Persian garden tradition to other parts of their empire notably India.

Babur introduced the Persian garden to India. The now unkempt Aram Bāgh garden in Agra was the first of many Persian gardens he created. The Taj Mahal embodies the Persian concept of an ideal, paradise-like garden.

The Safavid Dynasty[7] built and developed grand and epic layouts that went beyond a simple extension to a palace and became an integral aesthetic and functional part of it. Nowadays, traditional forms and style are still applied in modern Iranian gardens. They also appear in historic sites, museums and affixed to the houses of the rich.

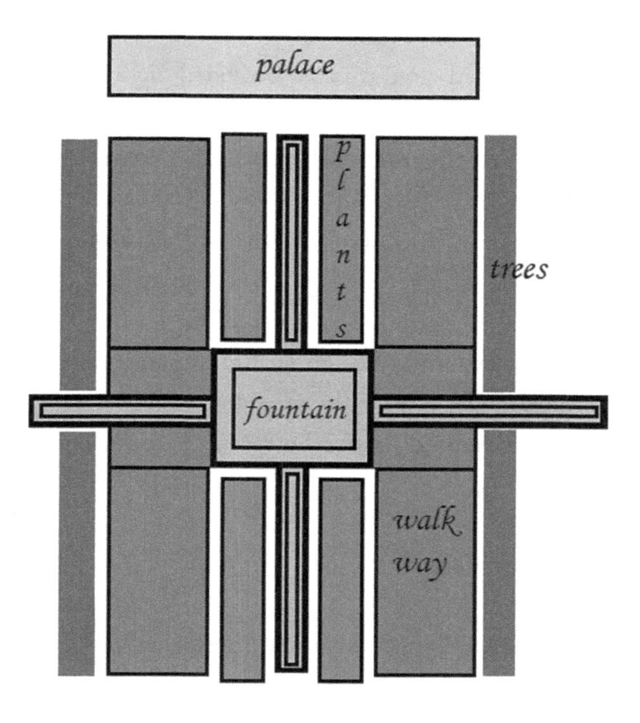

Fig. 7. 2 The layout of Persian gardens

2. Elements of Persian Gardens

The oldest representational descriptions and illustrations of Persian gardens come from travelers who reached Iran from the west. An early description of a Persian garden from the first half of the fourth century B. C. is found in Xenophon's Oeconomicus[8] in which he has Socrates relate the story of the Spartan general Lysander's visit to the Persian prince Cyrus[9] the Younger, who shows the Greek his "paradise at Sardis". In this story Lysander is "astonished at the beauty of the trees within, all planted at equal intervals, the long straight rows of waving branches, the perfect regularity, the rectangular symmetry of the whole, and the many sweet scents which hung about them as they paced the park".

These descriptions show chahar bāgh type gardens that featured an enclosing wall, rectangular pools, an internal network of canals, garden pavilions and lush planting. There are surviving examples of this garden type at Yazd and at Kashan. Sunlight and its effects were an important factor of structural design in Persian gardens. Textures and shapes were specifically chosen by architects to harness the light.

Iran's dry heat makes shade important in gardens, which would be nearly unusable without it. Trees and trellises largely feature as biotic shade; pavilions and walls are also structurally prominent in blocking the sun.

The heat also makes water important, both in the design and maintenance of the garden. Irrigation may be required, and may be provided via a form of underground tunnel called a qanat[10], that transports water from a local aquifer. Well-like structures then connect to the qanat, enabling the drawing of water. Alternatively, an animal-driven Persian well would draw water to the surface. Such wheel systems also moved water around surface water systems, such as those in the chahar bāgh style. Trees were often planted in a ditch called a juy, which prevented water evaporation and allowed the water quick access to the tree roots.

The Persian style often attempts to integrate indoors with outdoors through the connection of a surrounding garden with an innercourtyard. Designers often place architectural elements such as vaulted arches between the outer and interior areas to open up the divide between them.

3. The Styles of Persian Gardens

The Persian garden's construction may be formal which is with an emphasis on structure or casual which is with an emphasis on nature, following several simple design rules. This allows a maximization, in terms of function and emotion, of what may be done in the garden.

There are six primary styles of the Persian gardens according to their functions, but they are not limited to a particular style, but often integrate different styles, or have areas with different functions and styles.

Hayāt is a public, classical Persian layout with heavy emphasis on aesthetics over function. Man-made structures in the garden are particularly important, with arches and pools which may be used to bathe. The ground is often covered in gravel flagged with stone. Plantings are typically very simple—such as a line of trees, which also provide shade.

Privately, these gardens are often pool-centered and, again, structural. The pool serves as a focus and source of humidity for the surrounding atmosphere. There are few plants, often due to the limited water available in urban areas.

Meidān is a public, formal garden that puts more emphasis on the biotic element than the hayāt and that minimizes structure. Plants range from trees, to shrubs, to bedding plants, to grasses. Again, there are elements such as a pool and gravel pathways which divide the lawn. When structures are used, they are often built, as in the case of pavilions, to provide shade.

Chahar Bāgh are private and formal. The basic structure consists of four quadrants divided by waterways or pathways. Traditionally, the rich used such gardens in work-related functions such as entertaining ambassadors. These gardens balance structure with greenery, with the plants often around the periphery of a pool and path based structure.

Park is much like many other parks, the Persian park serves a casual public function with emphasis on plant life. They provide pathways and seating, but are otherwise usually limited in terms of structural elements. The purpose of such places is relaxation and socialization.

Bāgh is a casual garden, the bāgh emphasizes the natural and green aspect of the garden. Unlike the park, it is a private area often affixed to houses and often consisting of lawns, trees, and ground plants. The waterways and pathways stand out less than in the more formal counterparts and are largely functional. The primary function of such areas is familial relaxation.

Bāgh-e Eram is a historic Persian garden in Shiraz[11], Iran. Eram is the Personalized version of the Arabic word "Iram" meaning heaven in the Qur'an. Eram Garden therefore is so called for its aesthetic attractions resembling "heaven." The garden, and the Qavam House within it, are located on the northern shore of the Khoshk river in the Fars province.

Both the Qavam House pavilion and the garden were built during the middle of nineteenth century by the Ilkhanate or a paramount chief of the Qashqai tribes of Pars. The original layout of the garden however, with its quadripartite Persian Paradise garden structure was most likely laid in 18th century by the Seljuqs, and was then referred to as the "Bāgh-e Shāh" which means "the king's garden" in Persian and was much less complicated or ornamental.

Over its 150 years the structure has been modified, restored or stylistically changed by various participants. The Qavam House faces south along the long axis. It was designed by a local architect, Haji Mohammad Hasan. The structure housed 32 rooms on two stories, decorated by tiles with poems from the poetHafez written on them. The structure underwent renovation during the Zand and Qajar dynasties.

Today, Eram Garden and Qavam House are within Shiraz Botanical Garden of Shiraz University. They are open to the public as a historic landscape garden and house museum. They are World Heritage Site, and protected by Iran's Cultural Heritage Organization.

Notes:

1. Persian Gardens：波斯园林，主要设计理念突出了对伊甸园及琐罗亚斯德教四大元素——天空、水、大地、植物的象征意象，所有园林都分为四个部分，并且水在园林的灌溉与装饰中发挥了重要的作用。波斯园林最早的可以追溯到公元前 6 世纪。楼台、亭榭、墙垣以及精密的水流灌溉系统是园林的重要特征。波斯园林对印度及西班牙园林艺术都产生了影响。波斯庭园的布局多以位于十字形道路交叉点上的水池为中心，象征伊甸园四条河，这一手法为阿拉伯人继承，成为伊斯兰园林的传统，流布于北非、西班牙、印度，传入意大利后，演变成各种水法，成为欧洲园林的重要内容。波斯园林多规则地布局树木，大量配置鲜花，视花园为天上人间，并高筑围墙以防风。
2. Andalusia：安达卢西亚，是西班牙最南的历史地理区，也是西班牙南部一富饶的自治区，意思是“汪达尔人的土地”。该地区南临大西洋、直布罗陀海峡和地中海，出产甘蔗、香蕉、葡萄和棉花，有许多8—15世纪摩尔人统治时代的遗址。

3. Eden：伊甸园。伊甸园在《圣经》的原文中含有乐园的意思。根据《旧约·创世纪》记载，《圣经》记载伊甸园在东方，有四条河从伊甸流出滋润园子。这四条河分别是幼发拉底河、底格里斯河、基训河和比逊河。上帝耶和华照自己的形像造了人类的祖先男人亚当，再用亚当的一个肋骨创造了女人夏娃，并安置第一对男女住在伊甸园中。

4. Sassanids：萨珊王朝，也称波斯第二帝国，始自公元224年，651年亡。萨珊王朝统治时期的领土包括当今伊朗、阿富汗、伊拉克、叙利亚、高加索地区、中亚西南部、土耳其部分地区、阿拉伯半岛海岸部分地区、波斯湾地区、巴基斯坦西南部，控制范围甚至延伸到印度。古典时代晚期的萨珊王朝被认为是伊朗或波斯最具重要性和影响力的历史时期之一，是在伊斯兰对波斯的征服及伊斯兰教流行之前最后一个伊朗大帝国。萨珊王朝统治时期见证了古波斯文化发展至巅峰状态，它在很大程度上影响了罗马文化。萨珊王朝的文化影响力远远超出了它的边界，影响力遍及西欧、非洲、中国及印度，对欧洲及亚洲中世纪艺术的成形起着显著的作用。

5. Zoroastrianism：琐罗亚斯德教，是流行于古代波斯(今伊朗)及中亚等地的宗教，中国史称袄教、火祆教、拜火教。琐罗亚斯德教是基督教诞生之前中东和西亚最有影响的宗教，古代波斯帝国的国教，曾被伊斯兰教徒贬称为“拜火教”。琐罗亚斯德教的教义一般认为是神学上的一神论和哲学上的二元论。

6. chahar bāgh：四分花园，波斯园林的平面布局方式。花园用道路或者水渠分隔为四个小部分。

7. Safavid Dynasty：萨法维帝国，是由波斯人建立的从1501年至1736年统治伊朗的帝国。它实现了波斯的复兴，并以伊斯兰教什叶派为国教，团结国民，凝聚人心，奠定波斯特色的伊斯兰教和民族精神的统一。帝国强盛时，疆域东起霍拉桑，西至幼发拉底河，北抵卡拉库姆沙漠与咸海，南达波斯湾与阿拉伯海，囊括伊朗全境，伊拉克大部，高加索部分地区，土克曼斯坦，阿富汗斯坦西部，乌兹别克斯坦南部。在阿巴斯一世时期，版图甚至远达库尔德斯坦与土耳其东部的迪亚巴克尔。

8. Xenophon's Oeconomicus：色诺芬(公元前430年左右—公元前355年)，古希腊历史学家。他的著作16世纪就被译成多种欧洲文字。直到近代，他仍然具有崇高的声望。他一生的重大经历是在波斯王子小居鲁士的希腊雇佣兵团中服役。公元前355年写作《财源论》(*Ways and Means*)，向各城邦鼓吹和平政策。在他的著作中，最富有他个人色彩的是《远征记》。

9. Cyrus：居鲁士二世(约公元前600年或公元前576年—公元前530年)，即居鲁士大帝(Cyrus the Great)，古代波斯帝国的缔造者，公元前550—公元前529年在位。他以伊朗西南部的一个小首领起家，经过一系列的胜利，打败了3个帝国，即米底、吕底亚和巴比伦，统一了大部分的古中东，建立了从印度到地中海的大帝国。今天，伊朗人将居鲁士尊称为“国父”。

10. qanat：坎儿井，是“井穴”的意思，其结构是由竖井、暗渠、明渠、涝坝(积水潭)四部分组成。在高山雪水潜流处，寻其水源，在一定间隔打一深浅不等的竖井，然后再依地势高下在井底修通暗渠，沟通各井，引水下流。地下渠道的出水口与地面渠道相连接，把地下水引至地面灌溉桑田。坎儿井的分布区域辽阔，坎儿井可能起源于亚洲中部地区，非常可能起源于亚美尼亚或波斯高原。

11. Shiraz：设拉子，伊朗南部最大城市，伊朗最古老的城市之一，以玫瑰和夜莺之城及诗人的故乡闻名于世。2500 年前，波斯人居鲁士以此为中心创建了波斯帝国。大流士将首都迁至距该城 60 千米处的波斯波利斯，后被马其顿的亚历山大所毁，现留有居鲁士大帝的陵墓和宫殿。波斯波利斯是波斯帝国阿契美尼德王朝的首都，公元前 6 世纪时是波斯帝国的中心地区，公元 10 世纪时仍为波斯首都，18 世纪时曾为赞德王朝首都，有 13 世纪初期，蒙古人建的清真寺和塔赫特城堡。

Chinese Classical Garden[1]

The Chinese Classical garden is a landscape garden style which has evolved over three thousand years. It includes both the vast gardens of the Chinese emperors and members of the Imperial Family, built for pleasure and to impress, and the more intimate gardens created by scholars, poets, former government officials, soldiers and merchants, made for reflection and escape from the outside world. They create an idealized miniature landscape, which is meant to express the harmony that should exist between man and nature. (See Fig. 7.3)

①

Fig. 7.3 Humble administritive garden

① 彭一刚. 中国古典园林分析[M]. 北京：中国建筑工业出版社，1984 年.

1. The Elements of Chinese Classical Garden

A typical Chinese garden is enclosed by walls and includes one or more ponds, rock works, trees and flowers, and an assortment of halls and pavilions within the garden, connected by winding paths and zig-zag galleries. By moving from structure to structure, visitors can view a series of carefully composed scenes, unrolling like a scroll of landscape paintings.

A Chinese classical garden was not meant to be seen all at once; the plan of a classical Chinese garden presented the visitor with a series of perfectly composed and framed glimpses of scenery; a view of a pond, or of a rock, or a grove of bamboo, a blossoming tree, or a view of a distant mountain peak or a pagoda. Jesuit priest Jean Denis Attiret[2], who lived in China from 1739 and was a court painter for the Chinese Emperor, observed there was a "beautiful disorder, an anti-symmetry" in the Chinese garden.

The wall of the Chinese classical garden was usually painted white, which served as a pure backdrop for the flowers and trees. A pond of water was usually located in the center. Many structures, large and small, were arranged around the pond. In a scholar garden the central building was usually a library or study, connected by galleries with other pavilions which served as observation points of the garden features. These structures also helped divide the garden into individual scenes or landscapes. The other essential elements of a scholar garden were plants, trees, and rocks, all carefully composed into small perfect landscapes.

According to Ji Cheng's 16th century book "*The Craft of Gardens*"[3], "borrowed scenery" was the most important element of a garden. This could mean using scenes outside the garden, such as a view of distant mountains or the trees in the neighboring garden, to create the illusion that garden was much bigger than it was. He recommended locating a pavilion near a temple, so that the chanted prayers could be heard; planting fragrant flowers next to paths and pavilions, so visitors would appreciate their aromas; that bird perches be created to encourage birds to come to sing in the garden, that streams be designed to make pleasant sounds, and that banana trees be planted in courtyards so the rain would patter on their leaves.

The Chinese classical garden was laid out to present a series of scenes. Visitors moved from scene to scene either within enclosed galleries or by winding paths which concealed the scenes until the last moment. The scenes would suddenly appear at the turn of a path, through a window, or hidden behind a screen of bamboo. They might be revealed through round "moon doors" or through windows of unusual shapes, or windows with elaborate lattices that broke the view into pieces. So Ji Cheng instructed garden builders to "hide the vulgar and the common as far as the eye can see, and include the excellent and the splendid."

2. The Main Architecture in Chinese Classical Garden

Chinese classical gardens are filled with architecture; halls, pavilions, temples, galleries, bridges, kiosks, and towers, occupying a large part of the space. The garden structures are not designed to dominate the landscape, but to be in harmony with it. Classical gardens traditionally have these structures as follow:

The ceremony hall (堂) which is a building used for family celebrations or ceremonies, usually with an interior courtyard, not far from the entrance gate. The principal pavilion (厅) is for the reception of guests, for banquets and for celebrating holidays. It often has a veranda around the building to provide cool and shade. The pavilion of flowers (花厅) is located near the residence, where the building has a rear courtyard filled with flowers, plants, and a small rock garden.

In addition to the larger halls and pavilions, the garden is filled with smaller pavilions (亭), which are designed for providing shelter from the sun or rain, for contemplating a scene, reciting a poem, taking advantage of a breeze, or simply resting. These kinds of pavilions might be located where the dawn can best be watched, where the moonlight shines on the water, where autumn foliage is best seen, where the rain can best be heard on the banana leaves, or where the wind whistles through the bamboo stalks. They are sometimes attached to the wall of another building or sometimes stood by themselves at view points of the garden, by a pond or at the top of a hill. Some gardens have a picturesque stone pavilion in the form of a boat, located in the pond which generally had three parts; a kiosk with winged gables at the front, a more intimate hall in the center, and a two story structure with a panoramic view of the pond at the rear.

Galleries (廊) are narrow covered corridors which connect the buildings, protect the visitors from the rain and sun, and also help divide the garden into different sections. The galleries are rarely straight; they zigzag or are serpentine, following the wall of the garden, the edge of the pond, or climbing the hill of the rock garden. They have small windows, sometimes round or in odd geometric shapes, to give glimpses of the garden or scenery to those passing through.

Like the galleries, bridges are another common feature of the Chinese classical garden. they are rarely straight, but zigzag or arch over the ponds, to provide view points of the garden. Bridges are often built from rough timber or stone-slab raised pathways. Some gardens have brightly painted or lacquered bridges, which give a lighthearted feeling to the garden.

3. The Specific Symbolism of Garden Elements

3.1 The Artificial Mountain

The artificial mountain or rock is an integral element of Chinese classical gardens. The mountain peak was a symbol of virtue, stability and endurance in the philosophy. A mountain peak on an island was also a central part of the legend of the Isles of the Immortals, and thus became a central element in many Chinese classical gardens. In smaller classical gardens, a single scholar rock represents a mountain, or a row of rocks represents a mountain range. (See Fig. 7.4)

Fig. 7.4 Rock garden in humble administrative garden ①

3.2 Water

The lake or pond has an important symbolic role in the garden. In the *Book of Transformations* (易经) water represents lightness and communication, and carried the food of life on its journey through the valleys and plains. It is also the complement to the mountain, the central element of the garden, and represents dreams and the infinity of spaces. The shape of the garden pond often hides the edges of the pond from viewers on the

① 彭一刚. 中国古典园林分析[M]. 北京:中国建筑工业出版社,1984 年.

other side, giving the illusion that the pond goes on to infinity. The softness of the water contrasts with the solidity of the rocks. The water reflects the sky, and therefore is constantly changing, but even a gentle wind can soften or erase the reflections.

Small gardens have a single lake, with a rock garden, plants and structures around its edge. Middle-sized gardens will have a single lake with one or more streams coming into the lake, with bridges crossing the streams, or a single long lake divided into two bodies of water by a narrow channel crossed by a bridge. Streams come into the lake, forming additional scenes. Numerous structures give different views of the water, including a stone boat, a covered bridge, and several pavilions by the side of or over the water. The streams always follow a winding course, and are hidden from time to time by rocks or vegetation.

3.3 Flowers and trees

Flowers and trees, along with water, rocks and architecture, are essential elements of the Chinese garden. They represent nature in its most vivid form, and contrast with the straight lines of the architecture and the permanence, sharp edges and immobility of the rocks. They change continually with the seasons, and provide both sounds and aromas to please the visitor.

Each flower and tree in the garden had its own symbolic meaning. The pine, bamboo and Chinese plum were considered the "Three Friends of Winter" (岁寒三友). by the scholars who created classical gardens, prized for remaining green or blooming in winter. They were often painted together by artists. For scholars, the pine was the emblem of longevity and tenacity, as well as constance in friendship.

In Chinese folk culture, the peach tree in the Chinese garden symbolized longevity and immortality; Pear trees were the symbol of justice and wisdom, the apricot tree symbolized the way of the Mandarin, or the government official, the pomegranate tree was offered to young couples so they would have male children and numerous descendants. The willow tree represented the friendship and the pleasures of life, guests were offered willow branches as a symbol of friendship.

Of the flowers in the Chinese classical garden, the most appreciated were the orchid, peony, and lotus. During the Tang Dynasty, the peony, the symbol of opulence and a flower with a delicate fragrance, was the most celebrated flower in the garden. The orchid was the symbol of nobility, and of impossible love, as in the Chinese expression "a faraway orchid in a lonely valley." The lotus was admired for its purity, and its efforts to reach out of the water to flower in the air made it a symbol of the search for knowledge.

Notes:

1. Chinese Classical Garden：中国古典园林艺术，是指以江南私家园林和北方皇家园林为代表的中国山水园林形式。在中国传统建筑中，古典园林是独树一帜有重大成就的建筑形式。中国古典园林共由六大要素构成：筑山，理池，植物，动物，建筑，匾额，楹联与刻石。为表现自然，筑山是造园的最重要的因素之一。水是最富有生气的因素，无水不活。园林一定要凿池引水。古代园林的理水方法，一般有掩、隔、破三种。植物是另一个重要的因素。

2. Jean Denis Attiret：王致诚(1702—1768 年)，天主教耶稣会传教士，法国人，自幼学画于里昂，后留学罗马，工油画人物肖像。清乾隆三年(1738 年)，他来到中国，献《三王来朝耶稣图》，乾隆时受召供奉内廷。初绘西画，后学中国绘画技法，参酌中西画法，别立中西折中之新体，曲尽帝意，乃得重视，与郎世宁、艾启蒙、安德义合称四洋画家。
3. *The Craft of Gardens*：《园冶》，中国古代造园专著，也是中国第一本园林艺术理论的专著。明末造园家计成著，崇祯四年(公元 1631 年)成稿，崇祯七年刊行。全书共 3 卷，附图 235 幅。《园冶》是计成将园林创作实践总结提高到理论的专著，全书论述了宅园、别墅营建的原理和具体手法，反映了中国古代造园的成就，总结了造园经验，是一部研究古代园林的重要著作，为后世的园林建造提供了理论框架以及可供模仿的范本。

Japanese Karesansui

Karesansui or Japanese rock garden[1] (枯山水) or "dry landscape" garden, often called a Zen[2] garden, creates a miniature stylized landscape through carefully composed arrangements of rocks, water features, moss, pruned trees and bushes, and uses gravel or sand that is raked to represent ripples in water. A zen garden is usually relatively small, surrounded by a wall, and is usually meant to be seen while seated from a single viewpoint outside the garden, such as the porch of the hojo, the residence of the chief monk of the temple or monastery. Classical zen gardens were created at temples of Zen Buddhism in Kyoto during the Muromachi period[3]. They were intended to imitate the intimate essence of nature, not its actual appearance, and to serve as an aid to meditation about the true meaning of life. (See Fig. 7. 5)

Fig. 7. 5 Japanese garden ①

① Keyword suggest. org

1. The History of Karesansui

Karesansui or Rock gardens existed in Japan at least since the Heian period (784—1185). These early gardens were described in the first manual of Japanese gardens, Sakuteiki (作庭记), written at the end of the 11th century by Tachibana no Toshitsuna[4] (1028—1094). They were largely copied from the Chinese gardens of the Song Dynasty (960—1279), where groups of rocks symbolized Mount Penglai (蓬莱), the legendary mountain-island home of the Eight Immortals (八仙) in Chinese mythology, known in Japanese as Horai. The Sakuteiki described exactly how rocks should be placed. In one passage, he wrote: "In a place where there is neither a lake or a stream, one can put in place what is called a kare-sansui, or dry landscape". This kind of garden featured either rocks placed upright like mountains, or laid out in a miniature landscape of hills and ravines, with few plants. He described several other styles of rock garden, which usually included a stream or pond, including the great river style, the mountain river style, and the marsh style. The ocean style featured rocks that appeared to have been eroded by waves, surrounded by a bank of white sand, like a beach.

White sand and gravel had long been a feature of Japanese gardens. In the Shinto religion[5], it was used to symbolize purity, and was used around shrines, temples, and palaces. In zen gardens, it represents water, or, like the white space in Japanese paintings, emptiness and distance. They are places of meditation.

Zen Buddhism was introduced into Japan at the end of the 12th century, and quickly achieved a wide following, particularly among the Samurai class[6] and war lords, who admired its doctrine of self-discipline. The gardens of the early zen temples in Japan resembled Chinese gardens of the time, with lakes and islands. But in Kyoto in the 14th and 15th century, a new kind of garden appeared at the important zen temples. These zen gardens were designed to stimulate meditation. "Nature, if you made it expressive by reducing it to its abstract forms, could transmit the most profound thoughts by its simple presence", Michel Baridon wrote. "The compositions of stone, already common in China, became in Japan, veritable petrified landscapes, which seemed suspended in time, as in a certain moments of Noh theater[7], which dates to the same period."

2. The Elements of Japanese Rock Garden

2.1 The Selection and Placement of Rocks

The selection and placement of rocks is the most important part of making a Japanese rock garden. In Sakuteiki, it is expressed as "setting stones", the "act of setting stones upright." It laid out very specific rules for choice and the placement of stones, and warned that if the rules were not followed the owner of the garden would suffer misfortune. In Japanese gardening, rocks are classified as either tall vertical, low vertical, arching, reclining, or flat.

For creating "mountains", usually igneous volcanic rocks, rugged mountain rocks with sharp edges, are used. Smooth, rounded sedimentary rocks are used for the borders of gravel "rivers" or "seashores." In Japanese gardens, individual rocks rarely play the starring role; the emphasis is upon the harmony of the composition. For arranging rocks, there are many rules in the Sakuteiki. For example: "Make sure that all the stones, right down to the front of the arrangement, are placed with their best sides showing. If a stone has an ugly-looking top you should place it so as to give prominence to its side. Even if this means it has to lean at a considerable angle, no one will notice. There should always be more horizontal than vertical stones. If there are 'running away' stones there must be 'chasing' stones. If there are 'leaning' stones, there must be 'supporting' stones."

Rocks are rarely if ever placed in straight lines or in symmetrical patterns. The most common arrangement is one or more groups of three rocks. One common triad arrangement has a tall vertical rock flanked by two smaller rocks, representing Buddha and his two attendants. Other basic combinations are a tall vertical rock with a reclining rock; a short vertical rock and a flat rock; and a triad of a tall vertical rock, a reclining rock and a flat rock. Other important principles are to choose rocks which vary in color, shape and size, to avoid rocks with bright colors which might distract the viewer, and make certain that the grains of rocks run in the same direction.

At the end of the Edo period[8], a new principle was invented; the use of "discarded" or "nameless" rocks, placed in seemingly random places to add spontaneity to the garden. Other important principles of rock arrangement include balancing the number of vertical and horizontal rocks.

2.2 Sand and Gravel

Gravel is usually used in zen gardens, rather than sand, because it is less disturbed by rain and wind. The act of raking the gravel into a pattern recalling waves or rippling water, known as samon (砂纹), has an aesthetic function. Zen priests practice this raking also to help their concentration. Achieving perfection of lines is not easy. Rakes are according to the patterns of ridges as desired and limited to some of the stone objects situated within the gravel area. Nonetheless often the patterns are not static. Developing variations in patterns is a creative and inspiring challenge.

Stone arrangements and other miniature elements are used to represent mountains and natural water elements and scenes, islands, rivers and waterfalls. Stone and shaped shrubs are used interchangeably. In most gardens moss is used as a ground cover to create "land" covered by forest.

2.3 Symbolism of Rocks

In the Japanese rock garden, rocks sometimes symbolize mountains, particularly Horai, the legendary home of the Eight Immortals in Buddhist mythology; or they can be boats or a living creature, usually a turtle, or a carp. In a group, they might be a waterfall or a crane in flight.

In the earliest rock gardens of the Heian period, the rocks in a garden sometimes had a political message. As the Sakutei-ki wrote: "Sometimes, when mountains are weak, they are without fail destroyed by water. It is, in other words, as if subjects had attacked their emperor. A mountain is weak if it does not have stones for support. An emperor is weak if he does not have counselors. That is why it is said that it is because of stones that a mountain is sure, and thanks to his subjects that an emperor is secure. It is for this reason that, when you construct a landscape, you must at all cost place rocks around the mountain."

Some classical zen gardens have symbolism that can be easily read; it is a metaphorical journey on the river of life. Others, like Ryōan-ji, resist easy interpretation. Many different theories have been put forward about what the garden is supposed to represent, from islands in a stream to swimming baby tigers to the peaks of mountains rising above the clouds to theories about secrets of geometry or of the rules of equilibrium of odd numbers.

3. Ryōan-ji Garden

Ryōan-ji(龙安寺) is a Zen temple located in northwest Kyoto, Japan. The Ryōan-ji garden is considered one of the finest surviving examples of kare-sansui, a refined type of Japanese Zen temple garden design generally featuring distinctive larger rock formations arranged amidst a sweep of smooth pebbles.

In 1450, Hosokawa Katsumoto[9], a powerful warlord, acquired the land and founded a Zen temple, Ryōan-ji. There is controversy over who built the garden and when. Most sources date the garden to the second half of the 15th century. First descriptions of a garden, clearly describing one in front of the main hall, date from 1680—1682. It is described as a composition of nine big stones laid out to represent Tiger Cubs Crossing the Water. As the garden has fifteen stones at present, it was clearly different from the garden that we see today.

The garden is a rectangle of 248 square meters in twenty-five meters by ten meters. Placed within it are fifteen stones of different sizes, carefully composed in five groups; one group of five stones, two groups of three, and two groups of two stones. The stones are surrounded by white gravel, which is carefully raked each day by the monks. The only vegetation in the garden is some moss around the stones. (See Fig. 7. 6)

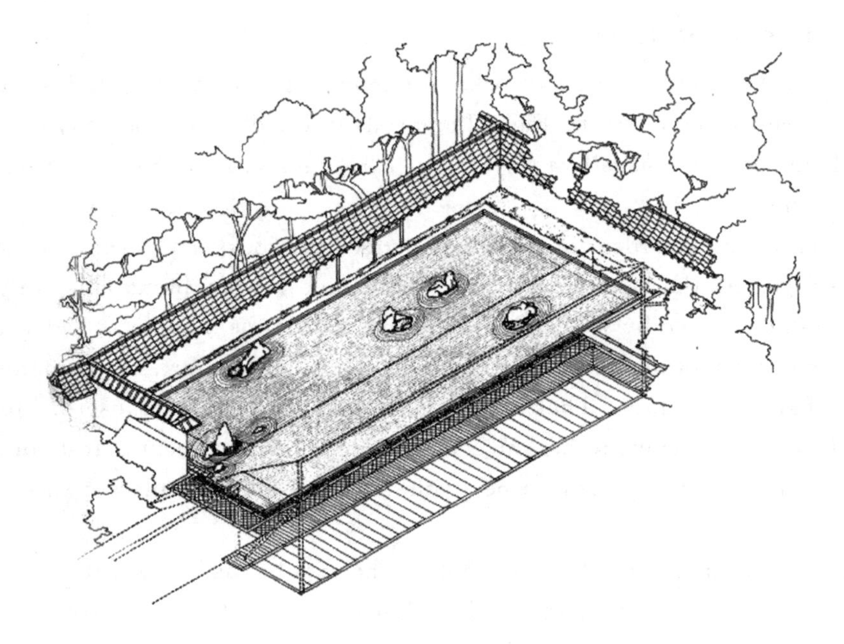

Fig. 7. 6　Plan of Ryōan-ji garden

The garden is meant to be viewed from a seated position on the veranda of the hōjō, the residence of the abbot of the monastery. The stones are placed so that the entire composition cannot be seen at once from the veranda. They are also arranged so that when looking at the garden from any angle, other than from above, only fourteen of the boulders are visible at one time. It is traditionally said that only through attaining enlightenment would one be able to view the fifteenth boulder.

The wall behind the garden is an important element of the garden. It is made of clay, which has been stained by age with subtle brown and orange tones. In 1977, the tile roof of the wall was restored with tree bark to its original appearance.

Like any work of art, the artistic garden of Ryōan-ji is also open to interpretation, or scientific research into possible meanings. Many different theories have been put forward inside and outside Japan about what the garden is supposed to represent, from islands in a stream, to swimming baby tigers to the peaks of mountains rising above to theories about secrets of geometry or of the rules of equilibrium of odd numbers. Garden historian Gunter Nitschke wrote: "The garden at Ryōan-ji does not symbolize anything, or more precisely, to avoid any misunderstanding, the garden of Ryōan-ji does not symbolize, nor does it have the value of reproducing a natural beauty that one can find in the real or mythical world. I consider it to be an abstract composition of 'natural' objects in space, a composition whose function is to incite meditation."

Notes:

1. Japanese rock garden：枯山水，是中国古典园林在传入日本后为适应日本地理条件限制而改造的缩微式园林景观，现多见于小巧、静谧、深邃的禅宗寺院。日本在园林艺术设计上探索精巧、细致，注重景观形式的象征和心理的感受，枯山水用石块象征山峦，用白沙象征湖海，用线条表示水纹，如一幅留白的山水画卷。在其特有的环境气氛中，因其无水而喻水，因无山无水而得名。日本园林使用一些如常绿树、苔藓、沙、砾石等静止不变的元素，营造枯山水庭园，园内几乎不使用任何开花植物，以期达到自我修行的目的。禅宗园林风格成熟后，庭园面积压缩，由早先的"园"转化为"庭"。最严格意义的枯山水是京都府龙安寺方丈楠庭和大仙院方丈北庭和东庭。
2. Zen：禅，是一种基于"静"的行为。比较有代表性的有：佛教的四禅八定；中国的禅宗；道家闭关；印度各路瑜伽。
3. Muromachi period：室町时代(1336—1573 年)，是日本史中世时代的一个划分，名称源自于幕府设在京都的室町，大约相当于中国的明朝。
4. Tachibana no Toshitsuna：橘俊岗(1028—1094 年)，相传为日本古代造园专著《作庭记》的作者。《作庭记》是世界上最早的一部关于造园的专著书籍，大约成于日本的藤原时代(894—1185 年)，相当于中国的唐代末期，此书在日本造园史上有不可忽视的地位，很受日本学界重视，被视为"国宝"。《作庭记》比中国明代计成的《园冶》早五百多年，但《园冶》在东方乃至世界的知名度要更大。《作庭记》主要重在论述理水，这与日本岛国多山的地理特征有关系，从日本枯山水景观中可以看出。
5. Shinto religion：神道，是日本的传统民族宗教，以自然崇拜为主，源于日本本土的崇神传统，属于泛灵多神信仰，视自然界各种动植物为神祇。神道教起初没有正式的名称，一直到公元 5 至 8 世纪，佛教经朝鲜传入日本，渐渐在当时的日本扩张开来，为了与"佛法"一词分庭抗礼，创造了"神道"一词来区分日本固有的神道与外国传入的佛法。日本人一般在出生 30 至 100 天内，都会被父母带领参拜神社，在 3、5、7 岁的 11 月 15 日所谓七五三节要参拜神社，升学、结婚要到神社祈求神佑。神社不设立灵牌，具有重要任务的神社，一般称为神宫。
6. Samurai class：武士，是 10—19 世纪日本的一个社会阶级，一般指通晓武艺、以战斗为职业的军人。武士遵守不畏艰难、忠于职守、精干勇猛的信念。
7. Noh theater：能剧。日本能剧的产生可以追溯到 8 世纪，随后的发展又融入了多种艺术表现形式，如杂技、歌曲、舞蹈和滑稽戏。今天，它已经成为了日本最主要的传统戏剧。这类剧主要以日本传统文学作品为脚本，在表演形式上辅以面具、服装、道具和舞蹈组成。
8. Edo period：江户时代，是德川幕府统治日本的年代，由 1603 年创立到 1867 年的大政奉还，江户时代是日本封建统治的最后一个时代。
9. Hosokawa Katsumoto：细川胜元(？—1473 年)，室町幕府末期的武将，室町幕府管领。细川胜元信仰禅宗，兴建了龙安寺、龙兴寺等寺庙，并且还是个通晓和歌的文人。

French Formal Garden[1]

The French formal garden is a style of garden based on symmetry and the principle of imposing order over nature. It reached its apogee in the 17th century with the creation of the Gardens of Versailles, designed for Louis XIV by the landscape architect André Le Nâtre. The style was widely copied by other courts of Europe. (See Fig. 7. 7)

Fig. 7. 7 The Garden of Versailles

1. History of French Formal Garden

The French formal garden evolved from the French Renaissance garden, a style which was inspired by the Italian Renaissance garden[2] at the beginning of the 16th century. The Italian Renaissance garden, typified by the Boboli Gardens[3] in Florence and the Villa Medici in Fiesole, was characterized by planting beds, or parterres, created in geometric shapes, and laid out symmetrical patterns; the use of fountains and cascades to animate the garden; stairways and ramps to unite different levels of the garden; grottos, labyrinths, and statuary on mythological themes. The gardens were designed to represent harmony and order, the ideals of the Renaissance, and to recall the virtues of Ancient Rome.

Beginning in 1528, King Francis I of France created new gardens at the Château de Fontainebleau[4], which featured fountains, parterres, a forest of pine trees brought from Provence and the first artificial grotto in France. While the gardens of the French Renaissance were much different in their spirit and appearance than those of the Middle Ages, they were still not integrated with the architecture of the châteaux, and were usually enclosed by walls. The different parts of the gardens were not harmoniously joined together, and they were often placed on difficult sites chosen for terrain easy to defend, rather than for beauty. All this was to change in the middle of the 17th century with the development of the Chateau of Vaux-le-Vicomte, created by Nicolas Fouquet, the superintendent of Finances to Louis XIV, beginning in 1656. Fouquet commissioned Louis Le Vau to design the chateau, Charles Le Brun to design statues for the garden, and André Le Nâtre to create the gardens. For the first time, that garden and the chateau were perfectly integrated. A grand perspective of 1500 meters extended from the foot of the chateau to the statue of the Hercules of Farnese; and the space was filled with parterres of evergreen shrubs in ornamental patterns, bordered by colored sand, and the alleys were decorated at regular intervals by statues, basins, fountains, and carefully sculpted topiaries. "The symmetry attained at Vaux achieved a degree of perfection and unity rarely equaled in the art of classic gardens. The chateau is at the center of this strict spatial organization which symbolizes power and success."

In the middle of the 18th century, the influence of the new English garden created by British aristocrats and landowners, and the popularity of the Chinese style, brought to France by Jesuit priests from the Court of the Emperor of China, a style which rejected symmetry in favor of nature and rustic scenes, brought an end to the reign of the symmetrical garden *à* la francaise. In many French parks and estates, the garden closest to the house was kept in the traditional *à* la francaise style, but the rest of the park was transformed into the new style, called variously the English garden. This marked the end of the age of the garden *à* la francaise and the arrival in France of landscape garden, which was inspired not by architecture but by painting, literature and philosophy.

2. The Elements of the French Formal Garden

The form of the French garden was largely fixed by the middle of the 17th century. It had the following elements, which became typical of the formal French garden: A geometric plan using the most recent discoveries of perspective and optics. A terrace overlooking the garden, allowing the visitor to see all at once the entire garden. As the French landscape architect Olivier de Serres wrote in 1600, It is desirable that the gardens should be seen from above, either from the walls, or from terraces raised above the parterres.

All vegetation is constrained and directed, to demonstrate the mastery of man over nature. Trees are planted in straight lines, and carefully trimmed, and their tops are trimmed at a set height.

A central axis, or perspective, perpendicular to the facade of the house, on the side opposite the front entrance. The axis extends either all the way to the horizon or to piece of statuary or architecture . The axis faces either South or east-west . The principal axis is composed of a lawn, or a basin of water, bordered by trees. The principal axis is crossed by one or more perpendicular perspectives and alleys.

The residence serves as the central point of the garden, and its central ornament. No trees are planted close to the house; rather, the house is set apart by low parterres and trimmed bushes. The most elaborate parterres, or planting beds, in the shape of squares, ovals, circles or scrolls, are placed in a regular and geometric order close to the house, to complement the architecture and to be seen from above from the reception rooms of the house. The parterres near the residence are filled with broderies, designs created with low boxwood to resemble the patterns of a carpet, and given a polychrome effect by plantings of flowers, or by colored brick, gravel or sand.

Farther from the house, the broderies are replaced with simpler parterres, filled with grass, and often containing fountains or basins of water. Beyond these, small carefully created groves of trees, serve as an intermediary between the formal garden and the masses of trees of the park. "The perfect place for a stroll, these spaces present alleys, stars, circles, theaters of greenery, galleries, spaces for balls and for festivities."

Bodies of water, such as canals and basins serve as mirrors, doubling the size of the house or the trees. The garden is animated with pieces of sculpture, usually on mythological themes, which either underline or punctuate the perspectives, and mark the intersections of the axes, and by moving water in the form of cascades and fountains.

Ornamental flowers were relatively rare in French gardens in the 17th century, and there was a limited range of colors; blue, pink, white and mauve. Brighter colors such as yellow, red and orange would not arrive until about 1730, because of botanical discoveries from around the world brought to Europe. Bulbs of tulips and other exotic flowers came from Turkey and the Netherlands; An important ornamental feature in Versailles and other gardens was the topiary, a tree or bush carved into geometric or fantastic shapes, which were placed in rows along the main axes of the garden, alternating with statues and vases.

Another trick used by French garden designers was the Ha-ha. This was a method used to conceal fences which crossed long alleys or perspectives. A deep and wide trench with vertical wall of stone on one side was dug wherever a fence crossed a view, or a fence was placed in bottom of the trench, so that it was invisible to the viewer.

The designers of the French garden saw their work as a branch of architecture, which simply extended the space of the building to the space outside the walls, and ordered nature according to the rules of geometry, optics and perspective. Gardens were designed like buildings, with a succession of rooms which a visitor could pass through following an established route, hallways, and vestibules with adjoining chambers. The "walls" were composed of hedges, and "stairways" of water. On the ground were carpets of grass, or embroidered with plants, and the trees were formed into curtains along the alleys. Just as architects installed systems of water into the chateaux, they laid out elaborate hydraulic systems to supply the fountains and basins of the garden. Long basins full of water replaced mirrors, and the water from fountains replaced chandeliers. The flowing water in the basins and fountains imitated water pouring into carafes and crystal glasses. The dominant role of architecture in the garden did not change until the 18th century, when the English garden arrived in Europe, and the inspiration for gardens began to come not from architecture but from romantic painting.

The French formal garden was often used as a setting for plays, spectacles, concerts, and displays of fireworks. In 1664, Louis XIV celebrated a six-day festival in the gardens, with cavalcades, comedies, ballets, and fireworks. Full-size ships were constructed for sailing on the Grand Canal, and the garden had an open-air ballroom, surrounded by trees; a water organ, a labyrinth, and a grotto.

3. Garden of Versailles

The Gardens of Versailles, created by André Le Nâtre between 1662 and 1700, were the greatest achievement of the Garden *à* la francaise. They were the largest gardens in Europe-with an area of 15000 hectares, and were laid out on an east-west axis followed the course of the sun: the sun rose over the Court of Honor, lit the Marble Court, crossed the Chateau and lit the bedroom of the King, and set at the end of the Grand Canal, reflected in the mirrors of the Hall of Mirrors. In contrast with the grand perspectives, reaching to the horizon, the garden was full of surprises-fountains, small gardens fill with statuary, which provided a more human scale and intimate spaces. (See Fig. 7. 8)

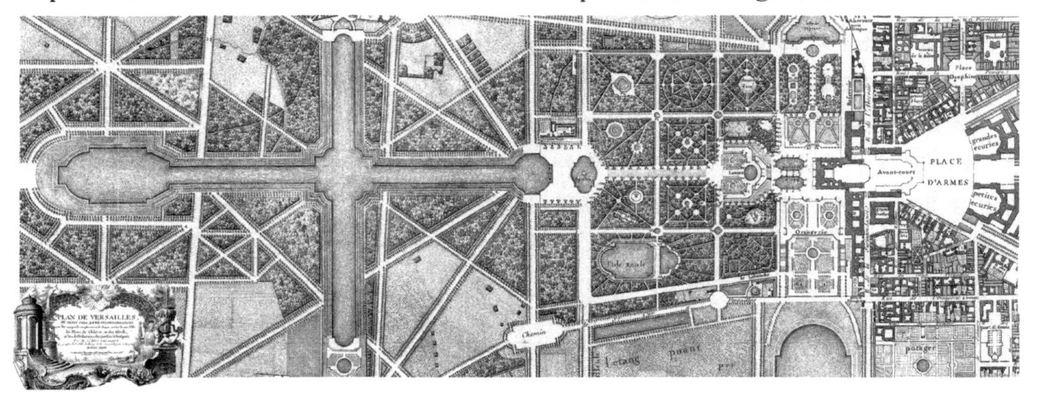

Fig. 7. 8 The layout of the Garden Versailles in 1746

Located on the east-west axis just west and below the Parterre d'Eau, is the Latona Fountain . The fountain depicted an episode from Ovid's Metamorphoses. Latona and her children, Apollo and Diana, being tormented with mud slung by Lycian peasants, who refused to let her and her children drink from their pond, appealed to Zeus who responded by turning the Lycians into frogs.

Further along the east-west axis is the Apollo Fountain which was constructed between 1668 and 1671, depicts the sun god driving his chariot to light the sky. The fountain forms a focal point in the garden and serves as a transitional element between the gardens of the Petit Parc and the Grand Canal.

With a length of 1,500 meters and a width of 62 meters, the Grand Canal, which was built between 1668 and 1671, physically and visually prolongs the east-west axis to the walls of the Grand Parc. During the Ancien Régime, the Grand Canal served as a venue for boating parties. Above and beyond the decorative and festive aspects of this garden feature, the Grand Canal also served a practical role. Situated at a low point in the gardens, it collected water it drained from the fountains in the garden above. Water from the Grand Canal was pumped back to the reservoir on the roof of the Grotte de Thétys via a network of windmill-powered and horse-powered pumps.

Notes:

1. French Formal Garden：法国几何园林。自 16 世纪以来，法国园林都采用了严格对称的形式，成为在欧洲有深远影响的一种形式。法国园林在学习意大利园林同时，结合本国特点，创作出一些独特的风格。其一，运用适应法国平原地区布局法，用一条道路将刺绣花坛分割为对称的两大块，有时图案采用阿拉伯式的装饰花纹与几何图形相结合。其二，用花草图形模仿衣服和刺绣花边，形成一种新的园林装饰艺术，称为“摩尔式”或“阿拉伯式”装饰。绿色植坛划分成小方格花坛，用黄杨做花纹，除保留花草外，使用彩色页岩细粒或砂子作为底衬，以提高装饰效果。其三，花坛是法国园林中最重要的构成因素之一。从把整个花园简单地划分成方格形花坛，到把花园当作一个整体，按图案来布置刺绣花坛，形成与宏伟建筑相匹配的整体构图效果，是法国园林艺术的重大飞跃。
2. Italian Renaissance garden：文艺复兴时期的意大利园林，当时的园林选址注意周围环境，可以远眺前景，多建在佛罗伦萨郊外风景秀丽的丘陵坡地上。多个台层相对独立，没有贯穿各台层的中轴线。建筑风格保留一些中世纪痕迹。建筑与庭园部分都比较简朴、大方，有很好的比例和尺度。喷泉、水池作为局部中心。绿丛植坛为常见的装饰，图案花纹简单。
3. Boboli Gardens：波波里花园，是享誉世界的古代罗马园艺花园。在 14 世纪初期是意大利佛罗伦萨最显赫贵族梅第奇家族的私家庭院，每逢节日梅第奇家族都会在庭院里举行盛大的音乐派对。波波里庭院自然地依山坡走向在树林中精心安排了小径和喷泉池水，颇具自然情趣，置身其中没有被环境雕饰的局促。山顶的庭院可以远眺四周的风景和佛罗伦萨城。自从由梅第奇家族兴建和特利波罗设计以来，波波里花园就成为意大利园林的经典杰作。
4. Château de Fontainebleau：枫丹白露宫，是法国最大的王宫之一，位于法国北部法兰西岛地区赛纳-马恩省，从 12 世纪起用作法国国王狩猎的行宫。“枫丹白露”的法文原义为“美丽的泉水”。宫殿花园狄安娜花园又称皇后花园或橙园，从 16 到 18 世纪花园内散布着花坛和雕塑。

Insights: Muqarnas

Muqarnas is a form of architectural ornamented vaulting, the "geometric subdivision of a corbel into a large number of miniature squinches, producing a sort of cellular structure", sometimes also called a "honeycomb" vault. It is used for domes, and especially half-domes in entrances, iwans and apses, mostly in traditional Islamic and Persian architecture. Where some elements project downwards, the style may be called mocárabe; these are reminiscent of stalactites, and are sometimes called "stalactite vaults".

Muqarnas developed around the middle of the 10th century in northeastern Iran and central North Africa. Examples can be found across Morocco and by extension, the Alhambra in Granada, Spain, the Abbasid Palace in Baghdad, Iraq, and Egypt. Large rectangular roofs in wood with muqarnas-style decoration adorn the 12th century Palatina in Palermo, Sicily, and other important buildings in Norman Sicily.

Muqarnas is a downward-facing shape; that is, a vertical line can be traced from the floor to any point on a muqarnas surface. It is also arranged in horizontal courses, as in a corbelled vault, with the horizontal joint surface having a different shape at each level. The edges of these surfaces can all be traced on a single plan view; architects can thus plan out muqarnas geometrically.

Muqarnas may be made of brick, stone, stucco, or wood, and clad with tiles or plaster. Since Muqarnas does not have a significant structural role, it need not be carved into the structural blocks of a corbelled vault and it can be hung from a structural roof as a purely decorative surface. (See Fig. 7. 9)

Fig. 7. 9 Decorative muqarnas

Chapter Eight

Theater

A theater, or playhouse, is a structure where theatrical works or plays are performed or other performances such as musical concerts may be produced. The facility is traditionally organized to provide support areas for performers, the technical crew and the audience members.

There are as many types of theaters as there are types of performance. Theaters may be built specifically for a certain types of productions, they may serve for more general performance needs or they may be adapted or converted for use as a theater. They may range from open-air amphitheaters to ornate, cathedral-like structures to simple, undecorated rooms or black box theaters.

The most important areas in the traditional theater is the acting space generally known as the stage. In some theaters, this area is permanent part of the structure. In addition to the acting space, there may be offstage spaces as well. These include wings on either side of a proscenium stage called "backstage" where props, sets and scenery may be stored as well as a place for actors awaiting an entrance.

All theaters provide a space for an audience. The audience is usually separated from the performers by the proscenium arch. In proscenium theaters and amphitheaters, the proscenium arch is a permanent feature of the structure. The seating areas can includestalls or arena, balconies or galleries, box and house seats.

Ancient Open Air Amphitheatre[1]

The dramatic performances were important to the Athenians—this is made clear by the creation of a tragedy competition and festival in the City Dionysia[2]. An indispensable element of every urban centre from the Classical period onwards, theatres were set in the centre of political, social and religious life: the acropolis, the agora, the stadium, the bouleuterion, and the sanctuaries. The theatre construction is a concept and an architectural achievement of Greek civilization: a plain structure in which coexist, in a balanced and complete manner, functionality and excellent aesthetics. (See Fig. 8.1)

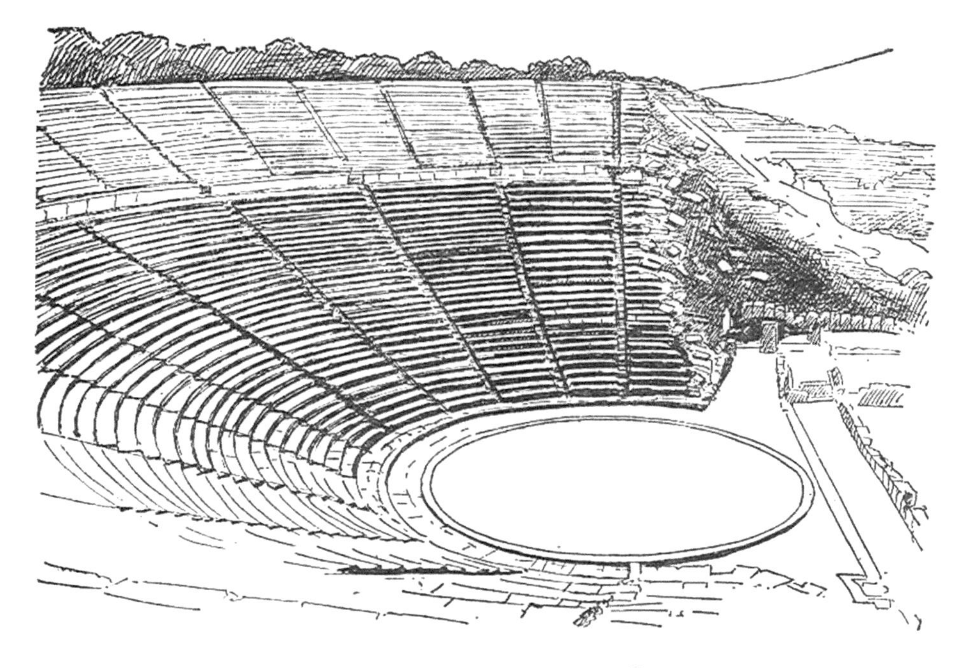

Fig 8. 1 Ancient theater [1]

1. The Description of Ancient Greek Amphitheatre

After the Great Destruction of Athens by the Persian Empire in 480 BCE, the town and acropolis were rebuilt, and theatre became formalized and an even greater part of Athenian culture and civic pride. The plays had a chorus from 12 to 15 people, who performed the plays in verse accompanied by music, beginning in the morning and lasting until the evening.

The performance space was a simple circular space, the orchestra, where the chorus danced and sang. The orchestra, which had an average diameter of 78 feet, was situated on a flattened terrace at the foot of a hill, the slope of which produced a natural theatron, literally "seeing place". Later, the term "theatre" came to be applied to the whole area of theatron, orchestra, and skené.

The orchestra was not only the site of the choral performances, but the religious rites, and, possibly, the acting. An altar was located in the middle of the orchestra; in Athens, the altar was dedicated to Dionysus. (See Fig. 8. 2)

Behind the orchestra was a large rectangular building called the skene meaning "tent" or "hut". At first, the skene was put up for the religious festival and taken down when it was finished. Later, the skene became a permanent stone structure. These structures were

① Levigilant. com

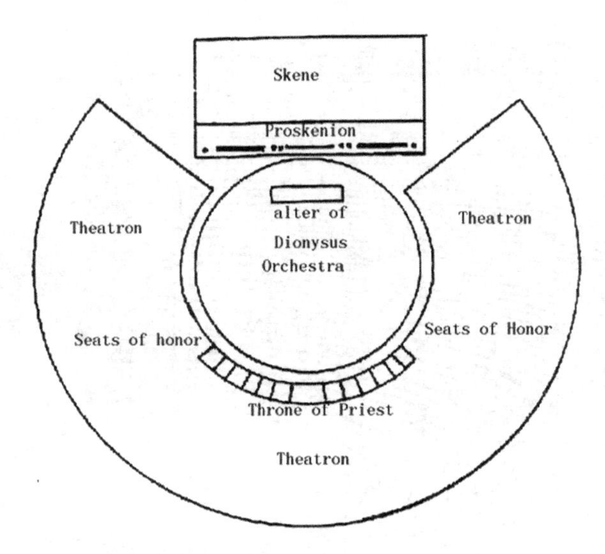

Fig 8. 2 Plan of Dionysus Theater

sometimes painted to serve as backdrops. Then, the skene was used as a "backstage" area where actors could change their costumes and masks, but also served to represent the location of the plays, which were usually set in front of a palace or house. Typically, there were two or three doors in the skene that led out onto orchestra, and from which actors could enter and exit. The death of a character was always heard behind the skênê, for it was considered inappropriate to show a killing in view of the audience.

In front of the skene there may have been a raised acting area called the proskenion, the ancestor of the modern proscenium stage. It is possible that the actors, as opposed to the chorus acted entirely on the proskenion.

Rising from the circle of the orchestra was the audience. The audience sat on tiers of benches built up on the side of a hill. Greek theaters, then, could only be built on hills that were correctly shaped. The theatres were originally built on a very large scale to accommodate the large number of people on stage, as well as the large number of people in the audience, around 15,000 viewers. The theaters were not enclosed; the audience could see each other and the surrounding countryside as well as the actors and chorus. Mathematics played a large role in the construction of these theatres, as their designers had to be able to create acoustics in them such that the actors' voices could be heard throughout the theatre, including the very top row of seats. The Greeks' understanding of acoustics compares very favorably with the current state of the art.

Greek theatres also had tall arched entrances, through which actors and chorus members entered and exited the orchestra. By the end of the 5th century B. C. , around the time of the Peloponnesian War[3], the skênê, the back wall, was two stories high. There were several scenic elements commonly used in Greek theatre, such as a crane that gave the impression of a flying actor; A wheeled platform was often used to bring dead characters into view for the audience trap doors, or similar openings in the ground to lift people onto the stage.

In Roman times, most Greek theatres were turned into arenas, adapted to the new types of spectacle which became popular during this period. Protective structures were added for the audience, while the orchestra area was enlarged to host gladiatorial combats and wild beast fights.

2. Theatre of Dionysus

The Theatre of Dionysus is a major theatre in Athens, built at the foot of the Athenian Acropolis. Dedicated to Dionysus, the god of plays and wine, the theatre could seat as many as 17,000 people with excellent acoustics, making it an ideal location for ancient Athens' biggest theatrical celebration, the Dionysia. It was the first stone theatre ever built, cut into the southern cliff face of the Acropolis, and supposedly birthplace of Greek tragedy. The site was used as a theatre since the sixth century B. C. The existing structure dates back to the fourth century B. C. but had many other later remodellings.

The Theatre of Dionysus on the south slope of the Athenian Acropolis have been dated to the 9th century B. C. , the early theatre must have been very simple, comprising a flat orchestra, with a few rows of wooden or stone benches set into the hill. The oldest orchestra in the theatre precinct is thought to have been circular with a diameter of around 27 meters, a wooden skene was apparently introduced at the back of the orchestra, serving for the display of artificial scenery and perhaps to enhance the acoustics. It was in this unpretentious setting that the plays of the great fifth century B. C. Attic tragedians were performed.

Alterations to the stage were made in the Hellenistic period, and 67 marble thrones were added around the periphery of the orchestra, inscribed with the names of the dignitaries that occupied them. The marble thrones that can be seen today in the theatre take the form of klismos chairs, and are thought to be Roman copies of earlier versions. At the center of this row of seats was a grand marble throne reserved for the priest of Dionysus.

The Theatre of Dionysus underwent a modernization in the Roman period, an entirely new stage was built in the first century, dedicated to Dionysus and the Roman emperor Nero. By this time, the floor of the orchestra had been paved with marble slabs, and new seats of honor were constructed around the edge of the orchestra.

The Theatre of Dionysus also sometimes hosted meetings. In the Roman period, "crude Roman amusements" that were ordinarily restricted to the amphitheatre replaced the sacred performances once held in the theatre, and by the Byzantine period, the entire complex had been destroyed.

3. Description of Ancient Roman amphitheatre

Ancient Roman amphitheatres were also the major public venues, circular or oval in shape, and used for events such as gladiator combats, chariot races, animal slayings and executions. About 230 Roman amphitheatres have been found across the area of the Roman Empire; the largest could accommodate 40, 000—60, 000 spectators, and the most elaborate theaters featured multi-storied, arcaded facades and were elaborately decorated with marble, stucco and statuary. After the end of gladiatorial games in the 5th century and of animal killings in the 6th century, most amphitheatres fell into disrepair, and their materials were mined or recycled. Some were razed, and others converted into fortifications. A few continued as convenient open meeting places; in some of these, churches were sited. (See Fig. 8. 3)

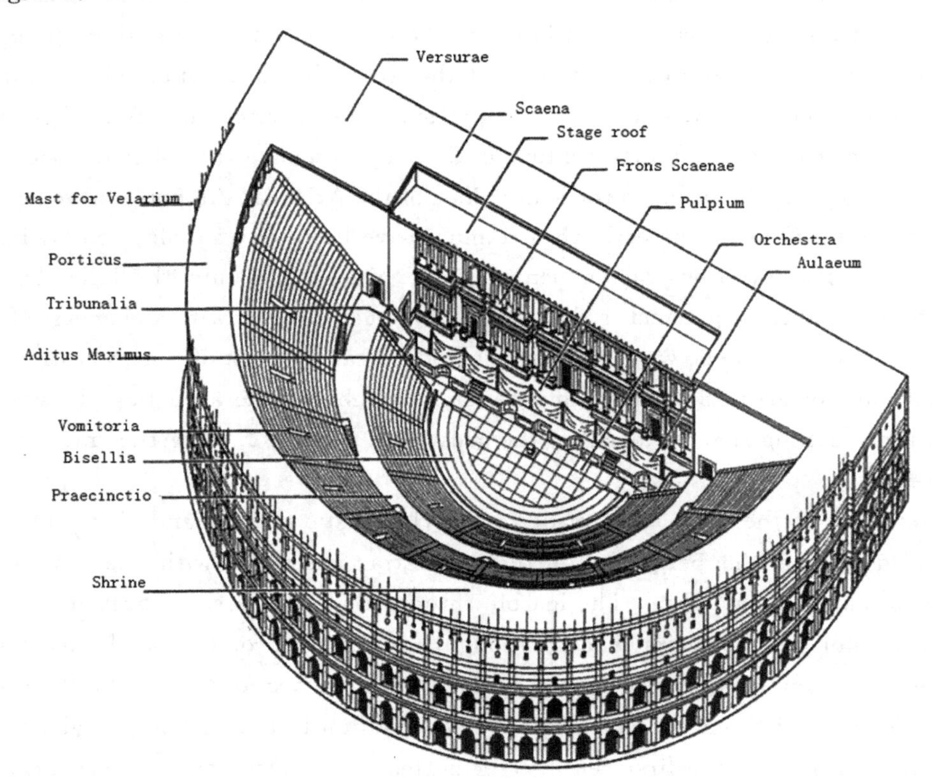

Fig 8. 3 Roman amphitheater standard plan

The characteristics of Roman theatres are due in large part to the influence of Ancient Greece . However, Roman theatres have specific differences, such as being built upon their own foundations instead of earthen works or a hillside and being completely enclosed on all sides. The amphitheatres of ancient Rome were constructed out of Roman concrete, and did not need superior acoustics.

These buildings were semi-circular and possessed certain inherent architectural structures, with minor differences depending on the region in which they were constructed. The Roman theatre also had a *podium*, which sometimes supported the columns of the *scaenae frons*. The *scaenae* was originally not part of the building itself, constructed only to provide sufficient background for the actors. Eventually, it became a part of the edifice itself, made out of concrete. The theatre itself was divided into the stage and the seating section. Entrances and exits were made available to the audience.

Romans tended to build their theatres regardless of the availability of hillsides. All theatres built within the city of Rome were completely man-made without the use of earthworks. The auditorium was not roofed; rather, awnings could be pulled overhead to provide shelter from rain or sunlight.

The stage buildings, in their fully developed form, almost always combine a stage, with a ground floor and first floor, with a proscenium. The proscenium usually takes the form of a small row of pillars, columns or semi-columns in the Doric or Ionic style. Paintings were placed in the spaces between the columns of the proscenium, while each of its three doorways, similarly painted, is conventionally thought to have led to the palace, the countryside or the port. The stage building always includes an upper storey, its floor level with the proscenium roof. Certain stages also included side rooms that served as outbuildings, while many stage buildings are connected to porticos. In some theatres, an underground passage from the stage to the orchestra allowed the gods of the netherworld to appear and intervene in the actions of the characters on stage.

Notes:

1. Amphitheatre：古希腊剧场，一般为露天的，包括三个部分：乐池、景屋以及观众席。乐池位于建筑的中央，通常也作为舞台，是一个周长为 150 米的圆形区域。乐池是演员表演和歌队吟唱的地方，也是举行宗教仪式的场所。在乐池中央往往有祭坛。在乐池后面是名为“景屋”(skene)的矩形建筑，意为“帐篷”或“棚屋”。景屋的功能相当于是古希腊剧场的后台，在这里演员可以换服装和面具。景屋外有三扇门通往乐池。有些剧场的景屋前方有一个凸出地面的表演区域，这便是现代剧场中舞台前部的雏形。环绕在乐池周围的坡形结构就是观众席。观众席通常依小山坡的坡度建造，观众们坐在整齐排列的长椅上观赏表演。古希腊剧场规模很大，能够容纳 1 万 5 千名观众。

2. City Dionysia：酒神节，也称狄俄尼索斯节，是希腊罗马宗教中为酒神狄俄尼索斯举行的祭仪。古希腊每年 3 月为表示对酒神狄俄尼索斯的敬意，都要在雅典举行这项活动，促进了古希腊戏剧、音乐艺术的发展。

3. Peloponnesian War：伯罗奔尼撒战争，公元前 431—前 404 年雅典及其同盟者与以斯巴达为首的伯罗奔尼撒同盟之间的战争。战争由于雅典势力的扩张引起斯巴达人的恐惧。斯巴达的同盟者科林斯与雅典的矛盾，在导致战争爆发的过程中起了重大作用。公元前 433 年，斯巴达盟邦忒拜进攻雅典盟邦普拉蒂亚，伯罗奔尼撒同盟对雅典宣战。雅典在西西里惨败，元气大伤。公元前 405 年，雅典舰队全军覆没，雅典海上霸权丧失殆尽。战争使参战双方的多数城邦蒙受人力和财力的巨大损失，国力下降。

Chinese Ancient Country Stage in Shanxi[1]

Chinese opera can be traced as far back as the Qin and Han Dynasties (221 B. C. —220 A. D.). However, it did not become a full-fledged art until the Song and Jin Dynasties (960—1234), and reached its first peak of prosperity in the Yuan Dynasty (1271—1368). As one of the cradles of Chinese civilization, local opera has an especially long history in Shanxi Province, and many varieties of operas from this region hold high status and have great importance in the development of traditional Chinese opera. In addition to countless opera pieces, passed down generation by generation, Shanxi is also home to a more tangible representation of this resplendent culture, its many ancient opera stages, of various shapes, sizes and types, scattered throughout its towns and villages. Shanxi Province alone contains more than 3000 of these ancient theater stages, 80% of all those in China, dating to the Jin, Yuan, Ming and Qing eras. (See Fig. 8. 4)

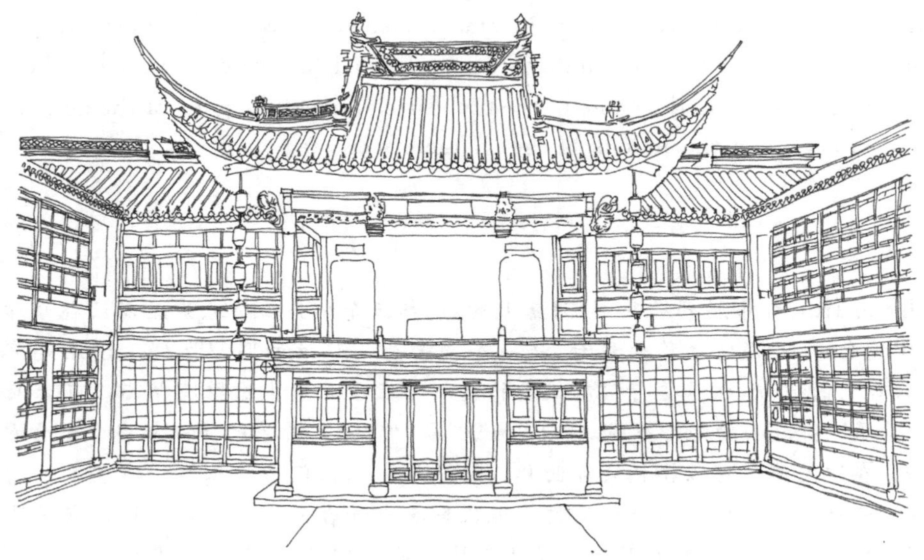

Fig. 8. 4 Shanxi theater stage

1. The History of Ancient Theater Stage

The earliest form of the opera stage, the open-air stage, appeared as early as the Han

Dynasty (202 B. C. —220 A. D.); it was made of either packed earth or stone, and constructed in courtyards or open spaces. Records show that in Western Han Dynasty, Emperor Wudi had platforms built for the purpose of worshiping the heavens, and on these platforms both sacrificial ceremonies and song and dance performances took place. It consists of a structure much like a Chinese pavilion, with a roof held up by four pillars, open on all four sides, and in the middle is a stone platform used for performances.

Later, the stages open on four sides evolved to only have three open sides, known as "three-sided opera viewing stages". Examples of this kind which can still be found include the Bull King Temple Stage and the Cui Manor Temple Stage . In ancient times, these performance venues, which closely resemble pavilions used for shade in the summer, are referred to as "dance pavilions (舞亭)", and although they look simple enough, they feature an ingenious "caisson" design. The deep ceiling would be decorated with patterns of aquatic plants, such as algae, water chestnuts and lotus flowers, but these plants were not merely as a decoration because in superstitious belief these plants could protect the architecture from fire. which imply the wish of constructor that the stage could fire. In fact, instead of showing off its fine mortise and tenon joint structure, a caisson was mainly used to improve the acoustics of the stage.

In ancient China opera was more than a form of entertainment, it was closely connected with religion, thus many of the stages were constructed as part of temple and ancestral hall complexes. The Bull King Temple Stage sits independently outside the temple, in an eye-catching location, and the front pillars are made of stone while the behind ones are wooden, all of which are carved with intricate designs.

Later stages added a wall on either side, so that only the front was left open to the audience. This appears to be a relatively basic innovation, but these additional walls reflect a very important step in the evolution of Chinese opera: whereas three-sided viewing stages are suitable for audiences of dance performances, for theatrical pieces with complex dramatic plots and longer running times, it would not be practical for viewers to be seated on all three sides of the stage, so Chinese opera scholars believe that the appearance of the one-sided viewing stage signifies the true maturation of Chinese theatrical arts.

In the Ming and Qing Dynasties, the propylaia stage is the most typical one among all kinds of stages, which is also called as the propylaia dance tower. Normal propylaia stage often has two floors, the first floor used for passages and the second floor is for performance. This kind of design can not only save the constructive cost but also the space. In 1904, Chen Duxiu (陈独秀) the forerunner of the Chinese New Culture Movement published a paper called On Chinese Opera, in which he said: "Theater venues are the classroom of the people, and the performers are their teachers; each and every one of us are students of these classrooms." This statement serves as an appropriate representation of the importance of theater in terms of its function of social education.

2. Types of the Ancient Theater Stages in Shanxi

The large number of Yuan Dynasty stages found in Shanxi serves as justification of this region as an important hub of Chinese opera culture. Due to the prosperity of Chinese opera, Shanxi has been scattered with all sorts of opera stages.

2.1 Double Floor Stage

Before the Ming Dynasty, most opera stages were single-floored. However, in the late Ming Dynasty, a new architecture appeared with its temple stage on both sides. The first floor was used as wing chamber and the upper floor was acted as an audience area. The innovation of this type of viewing tower was in order to adapt the appearance of the double-floor tower opera stage. The first floor of a double-floor stage could be used as office or public activity space, the second floor was the performance venue, and the wing chambers on the two sides of the stage were the dressing room for the performers.

2.2 品-shaped Propylaia Opera Stage

This complex propylaia opera stage often has a structure of"one big and two small", which means there are a jutting main stage and two rooms by its sides, forming a shape of the Chinese character "品". The appearance of this type of stage is an innovation through the developing history of Chinese ancient temple theaters, for the stage expands the audience area without occupying more area of the temple. The rooms beside the main stage are the dressing and storage room for the performers.

2.3 Double and Triple Stage

In the Ming and Qing Dynasties, there was a unique type of opera stage in Shanxi called "double stage", along with the closely related "triple stage" variety. Double stage is a type of opera stage with two same size stages stand side by side. Compared with normal opera stage, the double stage could make the performance more lively and vigorous, thus attracting more audience. Similarly, the triple stage is the stage that connected three small same-sized stages. Across from the triple stage are three temple halls, each facing a small stage, and shows would be played separately on different stages while worshiping different gods enshrined in the three halls. Sometimes, the shows will be played at the triple stages at the same time, which brings more exciting effect to the audience through the more fierce competition.

The two stages of double stage can be in the same size side by side, whereas the pair consists of a large one and a small one. Technically, stages of different size could be used for double opera performances, but this is not the case: in ancient times, due to the differences in the scale of temple festivals and various seasons of sacrificial ceremonies, the two stages would be used for different types of shows. Instead, the larger stage would usually be reserved for bangzi opera(梆子剧), whereas the smaller one could be used for puppet shows(木偶剧), which were performed by locally organized troupes.

The triple stage was an innovation made upon the foundation of the double stage. The design of the triple stage involves three side-by-side stages with the same length. The reason for this layout is so that performances may be underway at the same time on all three stages, as sort of a "battle" among the three shows. This is where in olden times a series of large-scale events held in worship of gods or to bring in the new year would often reach their climax.

2.4 Cross-street Propylaia Opera Stage

This type of opera stage is also called the "passage stage" by folks. Unlike the normal stages, there are passages under its propylaia, thus the stage would not get in the way of the passengers. This stage was innovated in Qing Dynasty, which often was built across a road, bringing the theater directly to the streets of ancient towns and commercial roads. The performance could start immediately after covering the top of the propylaia with wood plates. In addition to facilitating the transportation, the arches under the stage also played an important role in increasing resonance and expanding the sounds. (See Fig. 8.5)

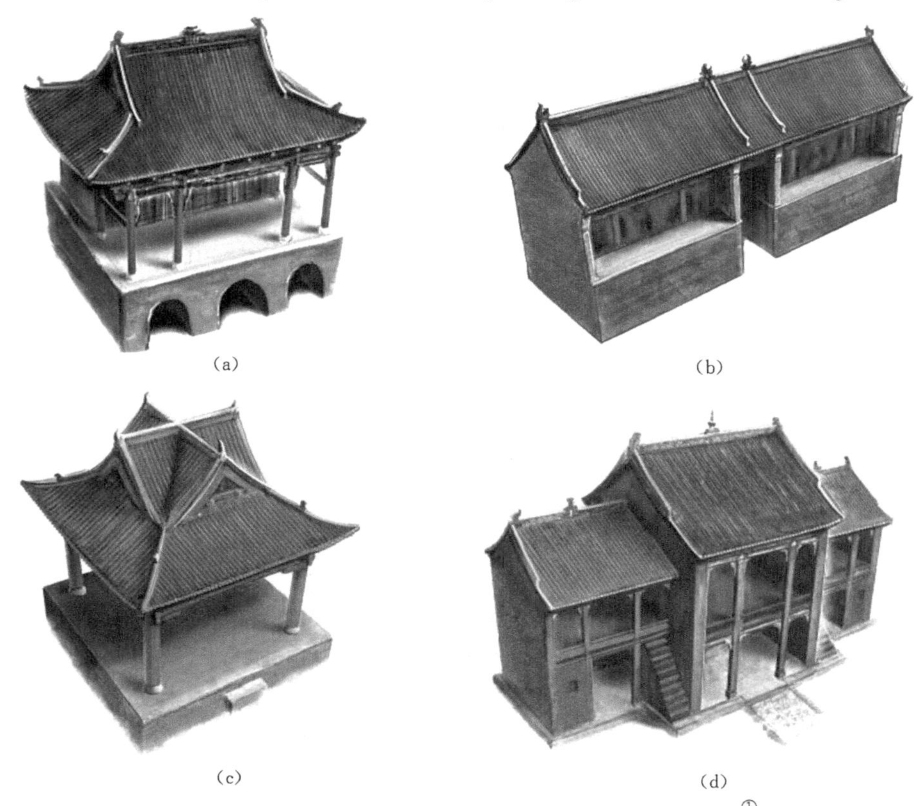

(a) (b) (c) (d)

Fig. 8.5 Samples of ancient theater stages in Shanxi[①]

① 乔延中. 山西古戏台[M]. 沈阳:辽宁人民出版社,2004 年.

3. Temple Theater

When visiting Shanxi you may notice that wherever there's a temple there's a opera stage, and vice-versa. In olden times, even for a village it had to have a temple, and each temple had to have its own opera stage, the temple grounds serving as the audience seating area during performances. In larger villages, and especially towns and cities, there were numerous temples, and a correspondingly larger number of opera stages.

The reason that these opera stages were all built within temples was that the performances were held for the viewing pleasure of the gods. Traditional Chinese opera originally developed from sacrificial ceremonies, eventually evolving into a form of entertainment for the gods. For every one of gods, performances would be given at different times, which would serve as a way for the locals to pay their respects to them. This is why the opera stage is always facing the main hall of the temple, so that the gods can have the best seat for the show.

In the downtown of Yuci(榆次) there is a beautiful ancient Chenghuang Temple(城隍庙); Chenghuang being a Taoist god is dedicated to protecting a specific town or city, and worshiped by the local residents. There is a Ming Dynasty opera stage inside the temple. When passing through the main gate of the temple you first see a tower called "xuanjian(玄鉴楼)" standing above you, on top of which is four-sided hip-and-gable roofing. Behind this tower is a "music tower (乐楼)", where shows were given during sacrificial ceremonies, and behind that is the "dance tower", as the locals call it. These three pagoda-style towers have different heights, for an overall very interesting aesthetic effect, and elaborate structures. It's been said that at the end of the Qing Dynasty, for the design plans of the opera stage inside the Summer Palace in the capital, the craftsmen made a trip to Shanxi to first view the layout of these structures as a reference.

Notes:

1. Chinese Ancient Country Stage in Shanxi：山西古戏台。山西被誉为"中国戏曲的摇篮"，早在汉代，山西就出现了戏曲的萌芽，到了元代，山西已是全国戏曲艺术的中心。北宋年间，当汴京的演出场所还被称作"勾栏"、"瓦舍"、"乐棚"的时候，山西早已有了固定的砖木建筑、被称作"舞亭"、"舞楼"、"乐楼"的正式戏台。目前，山西省现存元、明、清时期的旧戏台三千多座，在全国排名第一。全国仅存的6座元代戏台，都在山西省晋南一带，成为珍贵的"活历史"。山西古戏台的形式主要有三种，一种是单状舞台，一种是双状舞台，另有一种是三状并联式舞台。

Shakespeare's Globe[1]

Shakespeare's Globe is a reconstruction of the Globe Theatre, an Elizabethan playhouse in the London Borough of Southwark, on the south bank of the River Thames that was originally built in 1599, destroyed by fire in 1613, rebuilt in 1614, and then demolished in 1644. The modern reconstruction is an academic approximation based on available evidence of the 1599 and 1614 buildings. It was founded by American actor and director Sam Wanamaker and built about 230 meters from the site of the original theatre and opened to the public in 1997, with a production of *Henry V*. (See Fig. 8. 6)

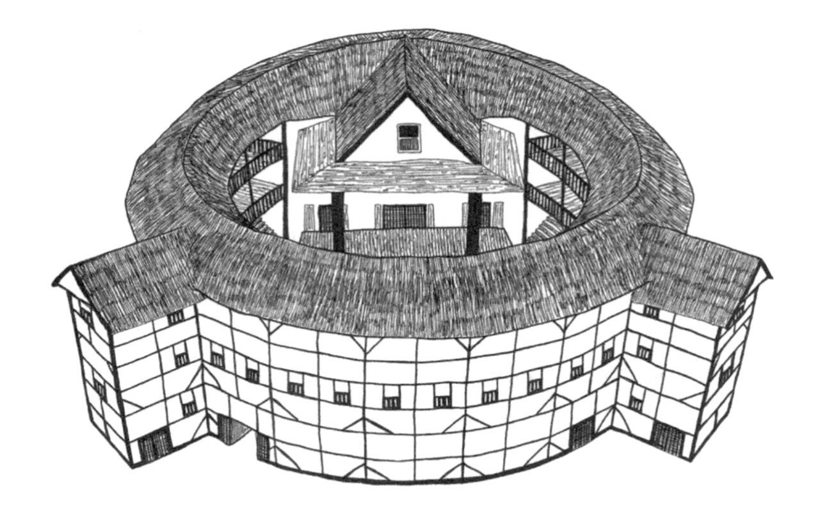

Fig. 8. 6 Shakespear's Globe

1. History of Shakespear's Globe

The original Globe Theatre was built in 1599 by the playing company, Lord Chamberlain's Men, to which Shakespeare belonged. On 29 June 1613 the Globe Theatre went up in flames during a performance of *Henry VIII*. When a theatrical cannon set off during the performance, it misfired and ignited the wooden beams and thatching. According to one of the few surviving documents of the event, no one was hurt except a man whose burning breeches were put out with a bottle of ale. The theatre was rebuilt by June 1614, but was officially closed by pressure of Puritan opinion in 1642 and demolished in 1644.

In 1970, American actor and director Sam Wanamaker founded the Shakespeare Globe Trust and the International Shakespeare Globe Centre, with the objective of building a faithful recreation of Shakespeare's Globe close to its original location at Bankside, Southwark.

Many detractors maintained that a faithful Globe reconstruction was impossible to achieve due to the complications in the 16th century design and modern fire safety requirements; however, Wanamaker persevered in his vision for over twenty years, and a new Globe theatre was eventually built according to a design based on the research of historical adviser John Orrell.

It was Wanamaker's wish that the new building recreate the Globe as it existed during most of Shakespeare's time there; that is, the 1599 building rather than its 1614 replacement. A study was made of what was known of the construction of the Theatre, the building from which the 1599 Globe obtained much of its timber, as a starting point for the modern building's design. To this were added: examinations of other surviving London buildings from the latter part of the 16th century; comparisons with other theatres of the period and contemporary drawings and descriptions of the first Globe. For practical reasons, some features of the 1614 rebuilding were incorporated into the modern design, such as the external staircases. The design team consisted of architect, structural and services engineer, and quantity surveyors. The theatre opened in 1997 under the name "Shakespeare's Globe Theatre", and has staged plays every summer.

2. Description of Shakespear's Globe

The original Globe's actual dimensions are unknown, but its shape and size can be approximated from scholarly inquiry over the last two centuries. The evidence suggests that it was a three-storey, open-air amphitheatre approximately 30m in diameter that could house up to 3,000 spectators. The Globe is shown as round on Wenceslas Hollar's sketch of the building, later incorporated into his etched *Long View of London from Bankside* in 1647. However, in 1988—1989, the uncovering of a small part of the Globe's foundation suggested that it was a polygon of 20 sides.

At the base of the stage, there was an area called the *pit*, where, for a penny, people would stand on the rush-strewn earthen floor to watch the performance. During the excavation of the Globe in 1989 a layer of nutshells was found, pressed into the dirt flooring so as to form a new surface layer. Vertically around the yard were three levels of stadium-style seats, which were more expensive than standing room. A rectangular stage platform, also known as an "apron stage", thrust out into the middle of the open-air yard. The stage measured approximately 13.1m in width, 8.2m in depth and was raised about 1.5m off the ground. On this stage, there was a trap door for use by performers to enter from the "cellarage" area beneath the stage.

The back wall of the stage had two or three doors on the main level, with a curtained inner stage in the centre and a balcony above it. The doors entered into the "tiring house", meaning backstage area, where the actors dressed and awaited their entrances. The floors above may have been used to store costumes and props and as management offices. The balcony housed the musicians and could also be used for scenes requiring an upper space, such as the balcony scene in *Romeo and Juliet*. Rush matting covered the stage, although this may only have been used if the setting of the play demanded it. (See Fig. 8. 7)

Large columns on either side of the stage supported a roof over the rear portion of the stage. The ceiling under this roof was called the "heavens," and was painted with clouds and the sky. A trap door in the heavens enabled performers to descend using some form of rope and harness.

Fig. 8. 7 Interior of Shakespear's Global in the 17th century ①

The theatre is located on Bankside, about 230 meters from the original site—measured from centre to centre. The Thames was much wider in Shakespeare's time and the original Globe was on the riverbank, though that site is now far from the river, and the river-side site for the reconstructed Globe was chosen to recreate the atmosphere of the original theatre. Like the original Globe, the modern theatre has a thrust stage that projects into a large circular yard surrounded by three tiers of raked seating. The only covered parts of the amphitheatre are the stage and the seating areas. Plays are staged during the summer, usually between May and the first week of October; in the winter, the theatre is used for educational purposes. Tours are available all year round. Some productions are filmed and released to cinemas as Globe on Screen productions.

① Walter Hodges. Gallery hip. com.

The reconstruction was carefully researched so that the new building would be as faithful a replica of the original as possible. This was aided by the discovery of the remains of the original Rose Theatre, a nearby neighbor to the Globe, as final plans were being made for the site and structure. Performances are engineered to duplicate the original environment of Shakespeare's Globe; there are no spotlights, plays are staged during daylight hours and in the evenings with the help of interior floodlights, there are no microphones, speakers or amplification. All music is performed live; the actors and the audience can see each other, adding to the feeling of a shared experience and of a community event.

The building itself is constructed entirely of English oak, with mortise and tenon joints and is, in this sense, an "authentic" 16th century timber-framed building, as no structural steel was used. The seats are simple benches and the Globe has the first and only thatched roof permitted in London since the Great Fire of 1666. The modern thatch is well protected by fire retardants, and sprinklers on the roof ensure further protection against fire. The pit has a concrete surface, as opposed to earthen-ground covered with strewn rush from the original theatre. The theatre has extensive backstage support areas for actors and musicians and is attached to a modern lobby, restaurant, gift shop and visitor centre. Seating capacity is 857 with an additional 700 "groundlings" standing in the pit, making up an audience about half the size of a typical audience in Shakespeare's time.

3. Blackfriars Theater[2]

As the modern Globe was under construction, the shell for an indoor theatre was built next door, to house a "simulacrum" of the sixteenth-century Blackfriars Theatre from the opposite side of the Thames. Blackfriars Theatre was the name given to two separate theatres located in the former Blackfriars Dominican priory in the City of London during the Renaissance. The first theatre began as a venue for the Children of the Chapel Royal, child actors associated with the Queen's chapel choirs staged plays in the vast hall in the former monastery. The second theatre dates from the purchase of the upper part of the priory and another building by James Burbage in 1596, which included the Parliament Chamber on the upper floor that was converted into the playhouse. They successfully used it as their winter playhouse until all the theatres were closed in 1642 when the English Civil War began.

As no reliable plans of the Blackfriars are known, the plan for the new theatre was based on drawings found in the 1960s at Worcester College, Oxford, at first thought to date from the early 17th century, and to be the work of Inigo Jones. The shell was built to accommodate a theatre as specified by the drawings,they nevertheless represent the earliest known plan for an English theatre, and are thought to approximate the layout of the

Blackfriars Theatre. Some features believed to be typical of earlier in the 17th century were added to the new theatre's design.

Completed at a cost of £7.5 million, the theatre opened as the Sam Wanamaker Playhouse in January 2014. It is an oak structure built inside the building's brick shell. The thrust stage is surmounted by a musicians' gallery, and the theatre has an ornately painted ceiling. The seating capacity is 340, with benches in a pit and two horse-shoe galleries, placing the audience close to the actors. Shutters around the first gallery admit artificial daylight. When the shutters are closed, lighting is provided by beeswax candles mounted in sconces, as well as on six height-adjustable chandeliers and even held by the actors.

Notes:

1. Shakespeare's Globe：莎士比亚环球剧场，位于英国伦敦。最初的环球剧场由威廉·莎士比亚所在宫内大臣剧团于1599年建造，1613年6月29日毁于火灾，1614年环球剧场重建，并于1642年关闭。1997年，一座现代仿造的环球剧场落成，命名为"莎士比亚环球剧场"，距离原址约205米远。考古证据表明，原环球剧场是一座三层开放式圆形剧场，直径大约为100英尺，能容纳3000名观众。原来的舞台下方，有一个区域庭院。观众只要花一个便士，就可以站着看演出。在庭院的四周是三层体育场风格的座位，比站位的价格要高。舞台两侧的大型柱子撑起了舞台后方的屋顶，屋顶的天花板被称为天空。天空也有一个活板门，使表演者可以利用绳索或其他装置从空中降落。舞台后墙有两到三扇门，上方有一个阳台。门通向后台，演员在那里穿戴服装，等候上场。阳台是演奏者的位置，有时也可以作为一些场景的布景。
2. Blackfriars Theater：黑衣修士剧院(Blackfriars Theatre)，两座伦敦剧院的名字。这两座剧院均坐落于多明我会黑衣修士的修道院原址。16世纪30年代，该修道院被亨利八世没收，在1576至1584年期间成为儿童剧院。第二座剧院位于建筑的另一部分，大约于1600年建立，被另一个儿童剧团使用过一段时间。威廉·莎士比亚在1608年成为它的所有者之一，他的剧团"国王供奉剧团"在此进行了多年的冬季演出。与"国王供奉剧团"夏天使用的环球剧院不同，黑衣修士剧院是封闭式的，1655年该剧院被拆毁。

Palais Garnier[1]

The Palais Garnier is a 1,979-seat opera house, which was built from 1861 to 1875 for the Paris Opera. The theatre is also often referred to as the Opéra de Paris or simply the Opéra, as it was the primary home of the Paris Opera. The Palais Garnier now is mainly used for ballet.

The Palais Garnier is probably the most famous opera house in the world, a symbol of Paris like Notre Dame Cathedral[2], the Louvre, or the Sacré Coeur Basilica. It has been described as the only one that is "unquestionably a masterpiece of the first rank." This opinion is far from unanimous however: the 20th-century French architect Le Corbusier once described it as "a lying art" and contended that the "Garnier movement is a décor of the grave". (See Fig. 8. 8)

Fig. 8. 8 The Palais Garnier ①

1. The History of the Palais Garnier

The Palais Garnier was designed as part of the great reconstruction of Paris during the Second Empire initiated by Emperor Napoleon III, who chose Baron Haussmann[3] to supervise the reconstruction. In 1858 the Emperor authorized Haussmann to clear the required 12,000 square meters of land on which to build a theatre for the world-renowned Parisian Opera and Ballet companies.

The selection of the architect was the subject of an architectural design competition in 1861, a competition which was won by the architect Charles Garnier(1825—1898). Legend has it that the Emperor's wife, the Empress Eugénie, who was likely irritated that her own favored candidate, Viollet-le-Duc, had not been selected, asked the relatively unknown Garnier: "What is this? It's not a style; it's neither Louis Quatorze, nor Louis Quinze, nor Louis Seize!" "Why Ma'am, it's Napoléon Trois" replied Garnier "and you're complaining!" Andrew Ayers has written that Garnier's definition "remains undisputed, so much does the Palais Garnier seem emblematic of its time and of the Second Empire that created it. A giddy mixture of up-to-the-minute technology, rather prescriptive rationalism, exuberant eclecticism and astonishing opulence, Garnier's opera encapsulated the divergent

① Lizstil. Almay. com

tendencies and political and social ambitions of its era." Ayers goes on to say that the judges of the competition in particular admired Garnier's design for "the clarity of his plan, which was a brilliant example of the *beaux-arts*[4] design methods in which both he and they were thoroughly versed."

The site was excavated between 27 August and 31 December. On 13 January 1862 the first concrete foundations were poured, starting at the front and progressing sequentially toward the back, with the laying of the substructure masonry. The opera house needed a much deeper basement in the substage area than other building types, but the level of the groundwater was unexpectedly high. To deal with this problem Garnier designed a double foundation to protect the superstructure from moisture. It incorporated a water course and an enormous concrete cistern which would both relieve the pressure of the external groundwater on the basement walls and serve as a reservoir in case of fire. Soon a persistent legend arose that the opera house was built over a subterranean lake, inspiring Gaston Leroux to incorporate the idea into his novel *The Phantom of the Opera*[5].

The emperor expressed an interest in seeing a model of the building, and a plaster scale model was constructed. After previewing it, the emperor requested several changes to the design of the building, the most important of which was the suppression of a balustraded terrace with corner groups at the top of the facade and its replacement with a massive attic story fronted by a continuous frieze surmounted by imperial *quadrigae* over the end bays.

The emperor's quadrigae were never added, although they can be seen in the model. Instead Charles-Alphonse Gumèry's gilded bronze sculptural groups *Harmony* and *Poetry* were installed in 1869. The linear frieze seen in the model was also redesigned with alternating low-and high-relief decorative medallions bearing the gilded letters from the imperial monogram. After the fall of the empire in 1870, Garnier was relieved to be able to remove them from the medallions. During 1874 Garnier and his construction team worked feverishly to complete the Paris opera house and the theatre was formally inaugurated on 5 January 1875 with a lavish gala performance. During the intermission Garnier stepped out onto the landing of the grand staircase to receive the approving applause of the audience.

2. The Description of the Palais Garnier

The Palais Garnier is a building of exceptional opulence. The style is monumental and considered Second-Empire Beaux-Arts style with axial symmetry in plan and eclectic exterior ornamentation with an abundance of Neo-Baroque[6] decorative elements. These include very elaborate multicolored marble friezes, columns, and lavish statuary, many of which portray deities of Greek mythology. (See Fig. 8. 9)

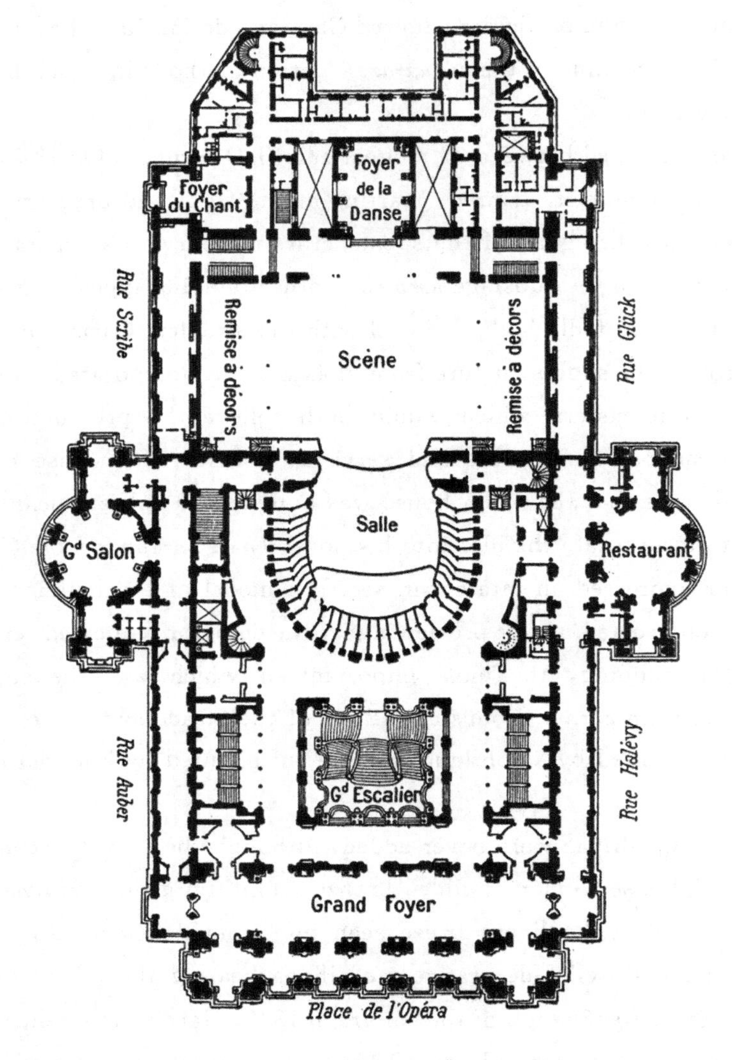

Fig. 8. 9 The plan of Palais Garnier

2. 1 Main Facade

The principal facade is on the south side of the building, overlooking the Place de l'Opéra and terminates the perspective along the Avenue de l'Opéra. Fourteen painters, mosaicists and seventy-three sculptors participated in the creation of its ornamentation.

The two gilded figural groups, Harmony and Poetry, crown the apexes of the principal facade's left and right avant-corps. They are both made of gilt copper electrotype. The bases of the two avant-corps are decorated from left to right with four major multi-figure groups: Harmony, Instrumental Music, The Dance, and Drama.

Gilded galvano plastic bronze busts of many of the great composers are located between the columns of the theatre's front facade and depict from left to right: Rossini, Auber, Beethoven, Mozart, Spontini, Meyerbeer, and Halévy. On the left and right lateral returns of the front facade are busts of the librettists Eugène Scribe and Philippe Quinault, respectively.

The sculptural group Apollo, Poetry, and Music are located at the apex of the south gable of the stage flytower, and the two smaller bronze Pegasus figures at either end of the south gable.

Pavillon de l'Empereur is located on the left side of the building and was designed to allow secure and direct access by the Emperor via a double ramp to the building. When the Empire fell, work stopped, leaving unfinished dressed stonework. It now houses the Paris Opera Library-Museum.

Pavillon des Abonnés is located on the right side of the building as a counterpart to the Pavillon de l'Empereur, this pavilion was designed to allow subscribers direct access from their carriages to the interior of the building. It is covered by a 13.5 meter diameter dome. Two pairs of obelisks marking the entrances of the Rotunda to the north and the south.

2.2 Interior

The interior of Paris Opera consists of interweaving corridors, stairwells, alcoves and landings allowing the movement of large numbers of people and space for socializing during intermission. Rich with velvet, gold leaf, and cherubim and nymphs, the interior is characteristic of Baroque sumptuousness.

The building features a large ceremonial staircase of white marble with a balustrade of red and green marble, which divides into two divergent flights of stairs that lead to the Grand Foyer. The pedestals of the staircase are decorated with female torchères. The ceiling above the staircase was painted by Isidore Pils to depict The Triumph of Apollo.

Grand Foyer is 18 meters high, 154 meters long and 13 meters wide designed to act as a drawing room for Paris society. Its ceiling was painted by Paul-Jacques-Aimé Baudry to represents various moments in the history of music. The foyer opens into an outside loggia at each end of which are the Salon de la Lune and Salon du Soleil.

The auditorium has a traditional Italian horseshoe shape and can seat 1,979. The stage is the largest in Europe and can accommodate as many as 450 artists. The canvas house curtain was painted to represent a draped curtain, complete with tassels and braid. The ceiling area, which surrounds the chandelier, depicts scenes from operas by 14 composers. The 7-ton bronze and crystal chandelier was designed by Garnier. On 20 May 1896, one of the chandelier's counterweights broke free and burst through the ceiling into the auditorium, killing a member of the audience. This incident inspired one of the more famous scenes in Gaston Leroux's classic 1910 gothic novel The Phantom of the Opera.

3. The Influence Paris Opera abroad

The building of Paris Opera inspired many other buildings over the following thirty years. Several buildings in Poland were based on the design of the Palais Garnier. In Ukraine, the influence of the Palais Garnier can be seen at the National Opera House of Ukraine in Kiev, built in 1901. In America, the Thomas Jefferson Building of the Library of Congress in Washington, D. C. is modelled after the Palais Garnier, most notably the facade and Great Hall. The Theatro Municipal do Rio de Janeiro in Brazil was also modeled after Palais Garnier, particularly Great Hall and stairs.

The Hanoi Opera House in Vietnam is considered to be a typical French colonial architectural monument in Vietnam, and it is also a small-scale replica of the Palais Garnier. After the departure of the French the building was used for Vietnamese plays and musicals.

Notes:

1. The Palais Garnier：巴黎歌剧院，又称为加尼叶歌剧院，是一座位于法国巴黎，拥有2200个座位的歌剧院，总面积11237平方米。歌剧院由查尔斯·加尼叶于1861年设计，是折衷主义登峰造极的作品，其建筑将古希腊罗马式柱廊、巴洛克等几种建筑形式完美地结合在一起，规模宏大，精美细致，金碧辉煌，被誉为是一座绘画、大理石和金饰交相辉映的剧院，是拿破仑三世典型的建筑之一。巴黎歌剧院长173米，宽125米，建筑总面积11237平方米，有着全世界最大的舞台，可同时容纳450名演员。演出大厅的悬挂式分枝吊灯重约8吨。富丽堂皇的休息大厅装潢豪华，四壁和廊柱布满巴洛克式的雕塑、挂灯、绘画。巴黎歌剧院具有十分复杂的建筑结构，有6英里长的地下暗道，地下层有一个容量极大的暗湖，湖深6米。
2. Notre Dame Cathedral：巴黎圣母院，是一座位于法国巴黎市中心、西堤岛上的教堂建筑，也是天主教巴黎总教区的主教座堂。圣母院建造于1163年到1250年间，为哥特式建筑形式，始建于1163年，整座教堂历时180多年建成。圣母院平面呈横翼较短的十字形，坐东朝西，正面风格独特，结构严谨。
3. Baron Haussmann：乔治-欧仁·奥斯曼男爵(1809—1891年)，法国政治家、法兰西第二帝国的重要官员。1853年拿破仑三世任命他为塞纳区行政长官，并封为男爵，领导巴黎城市改建工作。以古典式对称中轴线道路和广场为中心，使首都大部地区由陋屋窄巷变为宽街直路，卫生状况改善，交通运输和工商业都有所发展，建立起许多公园、广场、教堂、公共建筑及住宅区，并督建巴黎歌剧院和霍尔斯商场等，美化了首都。
4. *beaux-arts*：“学院派”，是西方现代意义上建筑教育的发端，它始于17世纪后期的法国皇家建筑研究会。该研究会在研讨建筑学说的同时，定期开设建筑方面的学术讲座，是法国皇家的建筑学校雏形。法国大革命开始后，成立了“美术学院”(Ecole des Beaus-Arts)，其学术思想和教学方法作为一种学派，常常被称之为“学院派”。学院有效地促成了建筑学说的规范化整合，完成了早期正规建筑教育成型和发展。巴黎美术学院在学术及教学两方面均得到了长足的发展,至19世纪后期,"学院派"思想和方法开始受到全世界的关注。"学院派"风格的设计具有以下特点:①外观装饰豪华、繁复。有一层或多层的画龙点睛的立面,通常搭配设计一排双柱式的柱廊。②墙面、窗户、窗顶、屋檐等均有精致的雕花装饰,凸显华丽气质。③墙体由石块砌成,立面呈现对称感。④通常有两种外观设计形态:平顶或低坡度屋顶,有双重斜坡的孟莎式屋顶。

5. *The Phantom of the Opera*:《歌剧魅影》,是勒鲁于1911年发表的介于侦探小说和荒诞小说间的作品,从一个记者的角度,讲述了发生在宏伟壮丽的巴黎歌剧院的一个"鬼故事"。问世至今多次被改编成电影和音乐剧,成为悬疑作品当中的经典之作。1986年,韦伯试演音乐剧《剧院魅影》,由韦伯当时的妻子莎拉·布拉曼饰演女主角克里斯蒂娜,使这出音乐剧一夜成名。

6. Neo-Baroque:新巴洛克,也称巴洛克复兴,是19世纪建筑风格,用来表现巴洛克风格的重要方面。新巴洛克风格建筑重点突出巴洛克建筑的基本元素,是巴黎美术学院的核心主张,在19世纪后半叶的法国和欧洲有广泛影响。

Insights: Flying Buttress

The flying buttress is a specific form of buttress composed of an arched structure that extends from the upper portion of a wall to a pier of great mass, in order to convey to the ground the lateral forces that push a wall outwards, which are forces that arise from vaulted ceilings of stone and from wind-loading on roofs. (See Fig. 8.10)

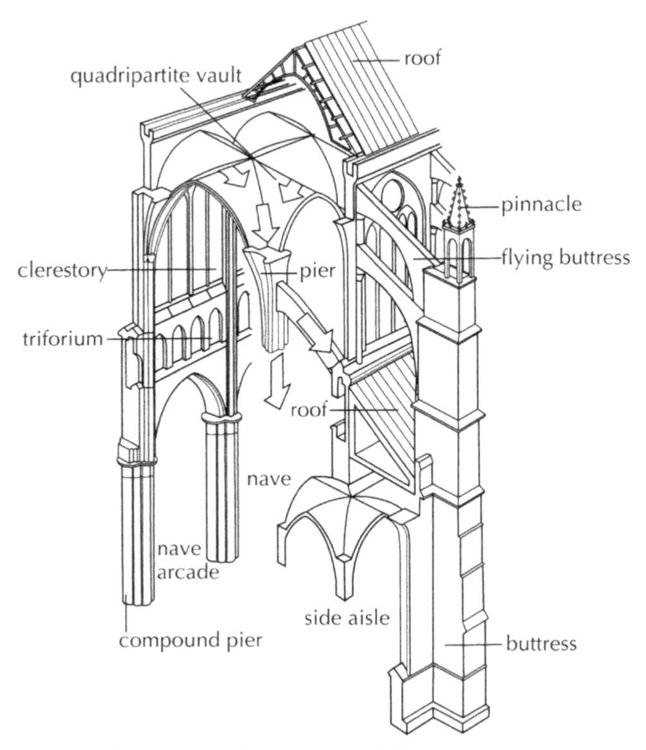

Fig. 8.10 Structure of flying buttress

The flying-buttress systems are composed of two parts: a massive pier, a vertical block of masonry situated away from the building wall, and an arch that bridges the span between the pier and the wall.

The flying buttress was developed during late antiquity and later flourished during the Gothic period of architecture. The advantage of such lateral-support systems is that the outer walls do not have to be massive and heavy in order to resist the lateral-force thrusts of the vault. Instead, the wall surface could be reduced, because the vertical mass is concentrated onto external buttresses.

Moreover, because the flying buttress relieves the load-bearing walls of excess weight and thickness, by way of a smaller area of contact, the wall has a greater surface area in which to install windows.

Another application of the flying-buttress support system is the reinforcement of a leaning wall in danger of collapsing, especially a load-bearing wall; the practical application of a flying buttress to a buckled wall is more practical than dismantling and rebuilding the wall.

Chapter Nine

Mausoleum

A mausoleum is an external free-standing building constructed as a monument enclosing the interment space or burial chamber of a deceased person or people. A mausoleum may be considered a type of tomb, or the tomb may be considered to be within the mausoleum. A tomb is a repository for the remains of the dead. It is generally any structurally enclosed interment space or burial chamber, of varying sizes. As indicated, tombs are generally located in or under religious buildings, such as churches, or in cemeteries or churchyards. However, they may also be found in catacombs, on private land or, in the case of early or pre-historic tombs, in what is today open landscape.

Historically, mausolea were, and still may be, large and impressive constructions for a deceased leader or other person of importance. However, smaller mausolea soon became popular with the gentry and nobility in many countries. In the Roman Empire, these were often ranged in necropoles or along roadsides. However, when Christianity became dominant, mausoleums were out of use.

Later, mausolea became particularly popular in Europe and its colonies during the early modern and modern periods. A single mausoleum may be permanently sealed. A mausoleum encloses a burial chamber either wholly above ground or within a burial vault below the superstructure. This contains the body or bodies, probably within sarcophagi or interment niches.

Egyptian Pyramid[1]

The Egyptian pyramids are ancient pyramid-shaped masonry structures. Most of Egyptian pyramids were built as tombs for the country's pharaohs and their consorts during the Old and Middle Kingdom periods (2600 B. C. —1700 B. C.). The shape of Egyptian pyramids is thought to represent the primordial mound from which the Egyptians believed the earth was created. The shape of a pyramid is also thought to be representative of the descending rays of the sun, and most pyramids were faced with polished, highly reflective white limestone, in order to give them a brilliant appearance when viewed from a distance. (See Fig. 9. 1)

Fig. 9.1 Egyptian Pyramid [1]

1. History of Egyptian Pyramid

Pyramid is a structure whose outer surfaces are triangular and converge to a single point at the top, making the shape roughly a pyramid in the geometric sense. The base of a pyramid can be trilateral, quadrilateral, or any polygon shape, meaning that a pyramid has at least three outer triangular surfaces . The square pyramid, with square base and four triangular outer surfaces, is a common version.

The Egyptians believed the dark area of the night sky around which the stars appear to revolve was the physical gateway into the heavens. One of the narrow shafts that extends from the main burial chamber through the entire body of the Great Pyramid points directly towards the center of this part of the sky. This suggests the pyramid may have been designed to serve as a means to magically launch the deceased pharaoh's soul directly into the abode of the gods.

During the third and fourth dynasties of the Old Kingdom, Egypt enjoyed tremendous economic prosperity and stability. Kings held a unique position in Egyptian society. Somewhere in between human and divine, they were believed to have been chosen by the gods to serve as mediators between them and the people on earth. Because of this, it was in everyone's interest to keep the king's majesty intact even after his death, when he was

---

① sketchit.com

believed to become Osiris, god of the dead. The new pharaoh, in turn, became Horus, the god who served as protector of the sun-god, Ra. Ancient Egyptians also believed that when the king died, part of his spirit, known as "ka" remained with his body. To properly care for his spirit, the corpse was mummified, and everything the king would need in the afterlife was buried with him, including gold vessels, food, furniture and other offerings. The pyramids became the focus of a cult of the dead king that was supposed to continue well after his death. Their riches would provide not only for him, but also for the relatives, officials and priests who were buried near him.

After 2700 B. C. , the Egyptians began building pyramids, until about 1700 B. C. The earliest known Egyptian pyramids are found at Saqqara, northwest of Memphis[2]. Ancient Egyptian pyramids were in most cases placed west of the river Nile because the divine pharaoh's soul was meant to join with the sun during its descent before continuing with the sun in its eternal round. The earliest among these is the Pyramid of Djoser which was built during the third dynasty. This pyramid and its surrounding complex were designed by the architect Imhotep, and are generally considered to be the world's oldest monumental structures constructed of dressed masonry. Imhotep is credited with being the first to conceive the notion of stacking mastabas on top of each other, creating an edifice composed of a number of "steps" that decreased in size towards its apex.

The most prolific pyramid-building phase coincided with the greatest degree of absolutist rule. It was during this time that the most famous pyramids, the Giza pyramid complex, were built. Over time, as authority became less centralized, the ability and willingness to harness the resources required for construction on a massive scale decreased, and later pyramids were smaller, less well-built and often hastily constructed.

2. The Description of the Giza Pyramid Complex

The Giza pyramid complex is an archaeological site on the Giza Plateau, on the outskirts of Cairo, including three pyramid complexes known as the Great Pyramids, the massive sculpture known as the Great Sphinx, several cemeteries, a workers' village and an industrial complex. The pyramids, were popularized in Hellenistic times, when the Great Pyramid was listed as one of the Seven Wonders of the World.

2.1 The Great Pyramid

The Great Pyramid, also known as the Pyramid of Khufu is the oldest and largest of the three pyramids in the Giza pyramid complex. Egyptologists believe that the pyramid was built as a tomb over a 10 to 20-year period. It is thought that the Great Pyramid was originally 146.5 meters, but with erosion and absence of its pyramidion, its present height is 138.8 meters. Each base side was 230.4 meters long. The mass of the pyramid is estimated at 5.9 million tones. The volume, including an internal hillock, is roughly 2,

500,000 cubic meters. The Great Pyramid was the tallest man-made structure in the world for more than 3,800 years.

Originally, the Great Pyramid was covered by casing stones that formed a smooth outer surface; what is seen today is the underlying core structure. Some of the casing stones that once covered the structure can still be seen around the base. In AD 1303, a massive earthquake loosened many of the outer casing stones, which were then carted away by Bahri Sultan in 1356 to build mosques and fortresses in nearby Cairo. Many more casing stones were removed from the great pyramids by Muhammad Ali Pasha in the early 19th century to build the upper portion of his Alabaster Mosque in Cairo not far from Giza. These limestone casings can still be seen as parts of these structures. (See Fig. 9. 2)

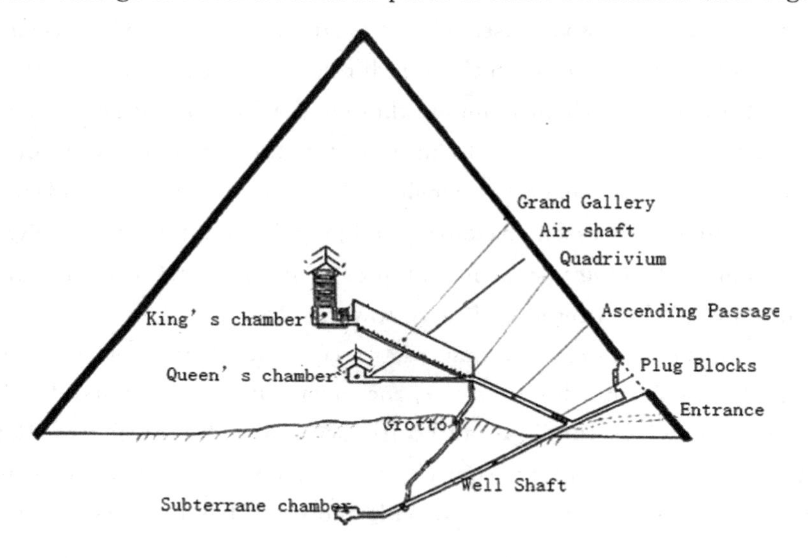

Fig. 9. 2 The diagram of the Great Pyramid

There are three known chambers inside the Great Pyramid. The lowest chamber is cut into the bedrock upon which the pyramid was built and was unfinished. The so-called Queen's Chamber and King's Chamber are higher up within the pyramid structure. The "King's Chamber" has a flat roof 5. 82 meters above the floor, there are two narrow shafts in the north and south walls. The King's Chamber is entirely faced with granite. Above the roof, which is formed of nine slabs of stone weighing in total about 400 tons, are five compartments known as Relieving Chambers. The first four, like the King's Chamber, have flat roofs formed by the floor of the chamber above, but the final chamber has a pointed roof. It is believed that the compartments were intended to safeguard the King's Chamber from the possibility of a roof collapsing under the weight of stone above the Chamber.

The "Queen's Chamber" is exactly halfway between the north and south faces of the pyramid and has a pointed roof with an apex 6. 23 meters above the floor. At the eastern end of the chamber there is a niche. The original depth of the niche was 1.04 meters, but has since been deepened by treasure hunters. In the north and south walls of the Queen's Chamber there are shafts. The shafts in the Queen's Chamber were explored in 1993 by the German engineer Rudolf Gantenbrink using a crawler robot he designed. After a climb of 65 m, he discovered that one of the shafts was blocked by limestone "doors" with two eroded copper "handles". Some years later the National Geographic Society created a similar robot which, in September 2002, drilled a small hole in the southern door, only to find another door behind it.

2.2 The Great Sphinx[3] of Giza

The Great Sphinx of Giza, commonly referred to as the Sphinx of Giza or just the Sphinx, is a limestone statue of a reclining sphinx, a mythical creature with a lion's body and a human head, that stands on the Giza Plateau on the west bank of the Nile in Giza. The face of the Sphinx is generally believed to represent the face of the Pharaoh Khafra. It faces directly West to East.

Cut from the bedrock, the original shape of the Sphinx has been restored with layers of blocks. It measures 73 meters long from paw to tail, 20.21 meters high from the base to top of the head, and 19 meters wide at its rear haunches. It is the oldest known monumental sculpture, and is commonly believed to have been built by ancient Egyptians of the Old Kingdom during the reign of the Pharaoh Khafra.

It is mythicised as treacherous and merciless. Those who cannot answer its riddle suffer a fate typical in such mythological stories, as they are killed and eaten by this ravenous monster. Unlike the Greek sphinx, which was a woman, the Egyptian sphinx is typically shown as a man. In addition, the Egyptian sphinx was viewed as benevolent, but having a ferocious strength similar to the malevolent Greek version and both were thought of as guardians often flanking the entrances to temples.

3. The Construction Techniques

There have been many hypotheses about the Egyptian pyramid construction techniques. Most of the construction hypotheses are based on the idea that huge stones were carved with copper chisels from stone quarries, and these blocks were then dragged and lifted into position. The Greeks believed that the pyramids must have been built by slave labor. Archaeologists now believe that the Great Pyramid of Giza was built by tens of thousands of skilled workers who camped near the pyramids and worked as a form of tax payment until the construction was completed.

One of the major problems faced by the early pyramid builders was the need to move huge quantities of stone. Dr. Parry has suggested a method for rolling the stones, using a cradle-like machine that had been excavated in various new kingdom temples. It is still not known whether the Egyptians used this method but the experiments indicate it could have worked using stones of this size.

There is good information concerning the location of the quarries, some of the tools used to cut stone in the quarries, transportation of the stone to the monument, leveling the foundation, and leveling the subsequent tiers of the developing superstructure. Workmen probably used copper chisels, drills, and saws to cut softer stone, such as most of the limestone. The harder stones, such as granite, cannot be cut with copper tools alone; instead they were worked with time-consuming methods like pounding with dolerite, drilling, and sawing with the aid of an abrasive, such as quartz sand. Blocks were transported by sledge likely lubricated by water.

Notes:

1. Egyptian Pyramid：埃及金字塔是古埃及法老和王后的陵墓。埃及金字塔始建于公元前2600年以前，目前有96座金字塔，大部分位于开罗西南部的吉萨高原的沙漠中。塔内有甬道、石阶、墓室、木乃伊，也就是法老的尸体。最大、最有名的是祖孙三代金字塔——胡夫金字塔、哈夫拉金字塔和门卡乌拉金字塔，其中以胡夫金字塔最为出名。
2. Memphis：孟菲斯，位于埃及尼罗河三角洲南端，今开罗西南23千米处的米特·拉辛纳村，从公元前3100年前起它就是埃及最古老的首都，定都长达800年之久。据传于公元前3000年为法老米那美尼斯所建，名“白城”，后改称孟菲斯，曾作为古王国（公元前27至前22世纪）的都城。公元前2000年代孟菲斯为底比斯所取代，但仍为埃及宗教、文化名城，当时是全世界最壮丽伟大的都市。公元7世纪孟菲斯被毁。
3. Sphinx：斯芬克斯，最初源于古埃及神话，被描述为长有翅膀的怪物，通常为雄性。传说中有三种斯芬克斯——人面狮身的，羊头狮身的，鹰头狮身的。在希腊神话中，赫拉派斯芬克斯坐在忒拜城附近的悬崖上，拦住过往的路人，用缪斯所传授的谜语问他们，猜不中者就会被它吃掉。古埃及神话中的斯芬克斯为人面狮身。著名的斯芬克斯狮身人面像距胡夫金字塔约350米，是世界上最大最著名的一座，身长约73米，高21米，脸宽5米。据说这尊斯芬克斯狮身人面像的头像是按照法老哈夫拉的样子雕成的。

Sanchi Stupa[1]

Sanchi is a small village in Raisen District of the state of Madhya Pradesh, India, it is located 46km north east of Bhopal, the central part of the state. Sanchi is famous for outstanding specimen of Buddhist art and architecture, belonging to the period between the third century B. C. and the twelfth century A. D. The most important of all the Sanchi monuments is the Sanchi Stupa, in which the relics of the Buddha were placed. The Sanchi Stupa is one of the best preserved early stupas in central India and a fine example of the development of the Buddhist architecture and sculpture. (See Fig. 9. 3)

Fig. 9. 3 Sanchi Stupa

1. The History of Sanchi Stupa

The foundation of the great religious establishment at Sanchi as an important centre of Buddhism for many centuries to come, was probably laid by the great Maurya emperor Asoka[2] (273 B. C. —236 B. C.), when he built a stupa and erected a monolithic pillar here. In addition to his marriage with a lady of Vidisa. It is said that the construction work of this stupa was overseen by Ashoka's wife, Devi herself, who was the daughter of a merchant of Vidisha. Sanchi was also her birthplace as well as the venue of her and Ashoka's wedding. The reason for his selection of this particular spot may be due to the fact that the hilltop served as an ideal place for giving a concrete shape to the newly aroused zeal for Buddhism in the emperor, who is said to have opened up seven out of the eight original stupas erected over the body relics of Buddha and to have distributed the relics among innumerable stupas built by himself all over his empire.

After a temporary setback following the break-up of the Maurya empire, when the stupa of Asoka was damaged, the cause of the Buddhist establishment of Sanchi was taken up with a feverish zeal by the monks and the laity alike, not a negligible percentage of the latter being formed by visitors of Vidisa for trade and other purposes. The religious fervor found its expression in vigorous building activity about the middle of the second century B. C. , during which the Sungas[3] were ruling and saw the stone encasing and enlargement of the stupa of Asoka, the erection of balustrades round its ground, berm, stairway and harmika, the reconstruction of Temple 40 and the building of Stupas 2 and 3.

The same intense religious aspiration and creative forces continued unabated in the next century as well, when, during the supremacy of the Satavahanas[4], new embellishments, in the form of elaborately-carved gateways, were added to Stapas 1 and 3.

After a prolonged period of stagnation and lassitude under the Kashtrapas, there was a revival of sculptural activity at Sanchi during the reign of the Guptas[5] who provided peace and prosperity essential for the growth of artistic pursuits. The discovery of few images in Mathura[6], sandstone executed in the early Gupta tradition, proves that Mathura continued, even in the fourth century A. D., to meet the demand of the clientele of Sanchi. But soon afterwards the local art of Sanchi once more came to the fore.

2. The Description of Great Stupa at Sanchi

The Great Stupa' at Sanchi is the oldest stone structure in India, its nucleus was a simple hemispherical brick structure built over the relics of the Buddha. It was crowned by the *chatra*, a parasol-like structure symbolizing high rank, which was intended to honor and shelter the relics. It has four profusely carved ornamental gateways and a balustrade encircling the whole structure.

2.1 The Plan of Great Stupa

The Great stupa is 36. 5m in diameter and 16. 4m high vandalized at one point sometime in the 2nd century B. C., an event some have related to the rise of the Sunga emperor Pusyamitra Sunga who overtook the Mauryan Empire as an army general. It has been suggested that Pushyamitra may have destroyed the original stupa, and his son Agnimitra rebuilt it. During the later rule of the Sunga, the stupa was expanded with stone slabs to almost twice its original size. The dome was flattened near the top and crowned by three superimposed parasols within a square railing. With its many tiers it was a symbol of the dharma, the Wheel of the Law. The dome was set on a high circular drum meant for circumambulation, which could be accessed via a double staircase. (See Fig. 9. 4)

A second stone pathway at ground level was enclosed by a stone balustrade with four monumental gateways facing the cardinal directions. The buildings which seem to have been commissioned during the rule of the Sungas are the Second and Third stupas, and the ground balustrade and stone casing of the Great Stupa. An inscription records the gift of one of the top architraves of the Southern Gateway by the artisans of the Satavahana king Satakarni: "Gift of Ananda, the son of Vasithi, the foreman of the artisans of rajan Siri Satakarni". Although made of stone, they were carved and constructed in the manner of wood and the gateways were covered with narrative sculptures. They showed scenes from the life of the Buddha integrated with everyday events that would be familiar to the onlookers and so make it easier for them to understand the Buddhist creed as relevant to their lives.

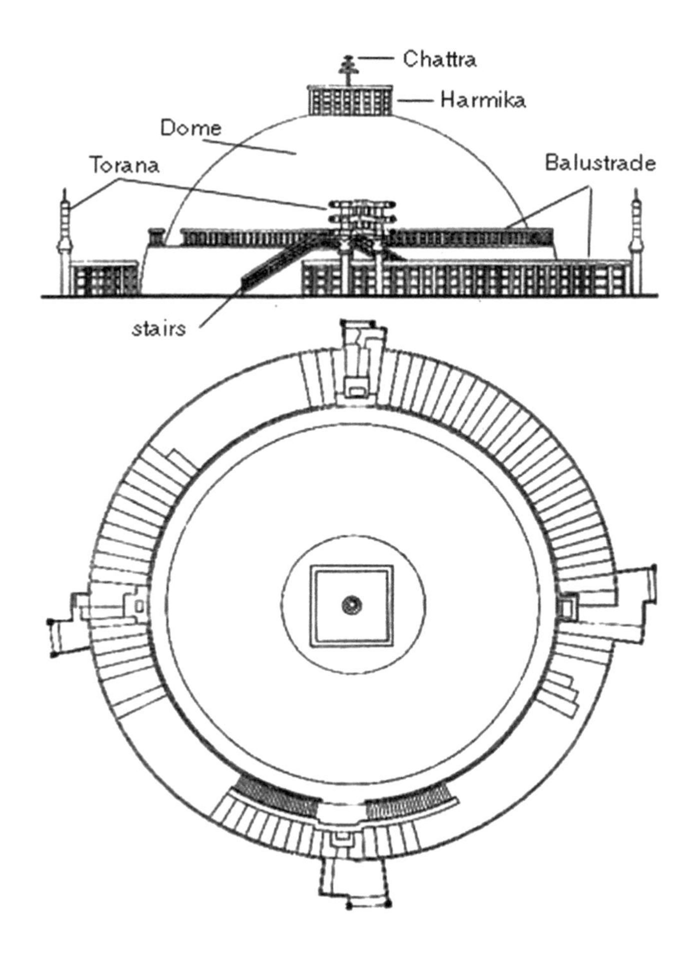

Fig. 9. 4 The plan of Great stupa at Sanchi

2. 2 The Torana

A torana is a type of gateway seen in the Hindu, Buddhist and Jain architecture of the Indian subcontinent, Southeast Asia and East Asia. The Four gateways or Torans constructed in 35 B. C. is the best from of Buddhist expression one can find anywhere in the world. They are covered with explicit carving which depict scenes from the life Buddha and the Jatakas[7], the stories relating to Buddha and his earlier births. At this stage Buddha was not represented directly but symbols were used to portray him—The lotus represents his birth, the tree his enlightenment, the wheel, derived from the title of his first sermon, the footprints and throw symbolizing his presence. The carvings on the Torans are done with inspired imagery, which in harmony with the surrounding figures balance the solidity of massive stupas.

The carving of Western Gate has depictions of the seven incarnations of Buddha. The pillar of East Gate Gate depicts story of the great departure of Prince Gautama. The

Southern Gate is a representation of scenes from the life of Ashoka and Buddha's Birth. The Northern Gate is crowned by the wheel of law and depicts the miracles, which took place during the life of Buddha.

3. Religious Significance of a Stupa

A stupa is a dome-shaped mound that mimics the funerary mounds used to mark the graves of great kings. The first Buddhist stupas enshrined the Buddha's physical relics, and thus gave him royal status. Another sign of this claim is the three-layer stone umbrella visible at the top of the stupa, since the umbrella was also a royal symbol. The Sanchi stupa has a walkway built halfway up the mound; the faithful would use this to circle the stupa to pay homage to the Buddha. Motion was always clockwise, since this kept one's right side toward the relics at the center.

To build a stupa, transmissions and ceremonies from a Buddhist teacher is necessary. Which kind of stupa to be constructed in a certain area is decided together with the teacher assisting in the construction. Sometimes the type of stupa chosen is directly connected with events that have taken place in the area.

A very important element in every stupa is the Tree of Life. It is a wooden pole covered with gems and thousands of mantras, and placed in the central channel of the stupa. It is placed here during a ceremony or initiation, where the participants hold colorful ribbons connected to the Tree of Life. Together the participants make their most positive and powerful wishes, which are stored in the Tree of Life. In this way the stupa is charged up, and will start to function.

Building a stupa is considered extremely beneficial, leaving very positive karmic imprints in the mind. Future benefits from this action will result in fortunate rebirths. Fortunate worldly benefits will be the result, such as being born into a rich family, having a beautiful body, a nice voice, and being attractive and bringing joy to others and having a long and happy life, in which one's wishes are fulfilled quickly. On the absolute level, one will also be able to reach enlightenment, the goal of Buddhism, quickly. Destroying a stupa on the other hand, is considered an extremely negative deed, similar to killing. Such an action is explained to create massive negative karmic imprints, leading to massive future problems. It is said this action will leave the mind in a state of paranoia after death has occurred, leading to totally unfortunate rebirths.

Notes:

1. Sanchi Stupa：窣堵波，是古代佛教特有的建筑类型之一，主要用于供奉和安置佛祖及圣僧的遗骨(舍利)、经文和法物，外形是一座圆冢的样子，也可以称作佛塔。公元前 3 世纪时，窣堵波流行于印度孔雀王朝，是当时重要的建筑。相传公元前 3 世纪时，孔雀王朝

(公元前 324—前 188 年)的第三代君主阿育王斥巨资建起 8 万 4 千座窣堵波,将佛祖释迦牟尼的骨灰分成 8 万 4 千份,分藏于各塔。其中有 8 座佛塔建在今印度中央邦博帕尔附近的桑奇村,经过 2 000 多年岁月,8 座中仅存 3 座,其中的桑奇窣堵波(the Great Stupa of Sanchi)是现存最早、最大而且最完整的佛塔。桑奇窣堵波是一个半球形的、缺乏内部空间但却是十分独特的建筑物,它的中央是覆钵形的半球体坟冢,球体直径 32 米,高 12.8 米,立在一个直径 36.6 米、高 4.3 米的圆形台基上。冢体由砖石砌成,表面镶贴着一层红色砂石。围绕着冢体有一圈高 3.3 米的仿木式石栏杆,栏杆外四面各辟有一座砂石塔门牌坊,这是在巽伽王朝时代(公元前 187—公元前 75 年)增建的。牌坊高约 10 米,包括两根垂直的有柱头的柱子和在立柱之间插榫的 3 根水平的横梁,断面呈橄榄形,造型独特,反映了木结构的传统。牌坊的两面覆满了浮雕,轮廓的外缘则用圆雕装饰。圆冢顶部有一圈石栏杆,正中是一座托名佛邸的小亭,亭上冠戴着 3 层华盖。桑奇窣堵波四周有一圈小径,僧侣们围绕着窣堵坡边走边诵经。小径外侧石栏杆建于公园前 2 世纪。在立柱之间采用插榫的方法横排着三根石料,断面呈橄榄形。立柱顶上用条石连成一个环。

2. Maurya emperor Asoka:阿育王(公元前 273—公元前 232 年在位),是古代印度摩揭陀国孔雀王朝的第三代国王,早年好战杀戮,统一了整个南亚次大陆和今阿富汗的一部分地区,晚年笃信佛教,放下屠刀,又被称为"无忧王"。阿育王在全国各地兴建佛教建筑,据说总共兴建了 84000 座奉祀佛骨的佛舍利塔。为了消除佛教不同教派的争议,阿育王曾邀请著名高僧目犍连子帝须长老召集 1000 比丘,在华氏城举行大结集,驱除了外道,整理了经典,并编撰了《论事》,为佛教在印度的发展做出了巨大的贡献。
3. Sungas:巽伽王朝,是古印度摩揭陀王国的一个王朝,建于公元前 185 年,末于公元前 73 年,在孔雀王朝之后建立。其主要领地包括印度东北部恒河下游地区。巽伽王朝共历 10 帝,首都建于华氏城。
4. Satavahanas:甘婆王朝(Kanva dynasty),又译甘华王朝,是古印度摩揭陀王国的一个取代巽伽王朝的印度王朝。甘婆王朝建于约公元前 75 年(一说约公元前 73 年),末于约公元前 30 年(一说约公元前 26 年),其创建者为婆薮提婆(婆苏提婆)。甘婆王朝共历 4 王,最后为南印安达罗(百乘王朝)所灭。
5. Guptas:笈多王朝(Gupta Dynasty,约公元 320—540 年),中世纪统一印度的第一个封建王朝,疆域包括印度北部、中部及西部部分地区。首都为华氏城。笈多王朝是中世纪印度的黄金时代。大乘佛教盛行,印度教兴起。笈多诸王采取宗教兼容政策,放任各派宗教自由发展。大乘佛教中心那烂陀寺,成为印度中世纪前期的宗教和学术文化中心。
6. Mathura:马图拉,地处中印度和北印度的交界处,自古以来是商业、宗教和艺术名城,也是东西方贸易与文化交汇的地区。贵霜王朝时代,除北方的犍陀罗艺术外,印度中部的马图拉出现了本地的雕刻艺术,它们比犍陀罗更少受希腊艺术的影响,更具有印度本土的民族风格。
7. *Jatakas*:《本生经》,是印度的一部佛教寓言故事集,大约产生于公元前三世纪。它是用古印度的一种方言——巴利语撰写的,主要讲述佛陀释迦牟尼前生的故事。按照佛教的说法,释迦牟尼在成佛以前,只是一个菩萨,还逃不出轮回。他必须经过无数次转生,才能最后成佛。现存的《本生经》共收有 547 个佛本生故事。这些故事绝大部分是长期流传于印度民间的寓言、故事、神话、传奇,佛教徒用以宣传佛教教义。

Ming Tombs in China[1]

The Ming Tombs are a collection of mausoleums built by the emperors of the Ming Dynasty of China. The first Ming emperor's tomb is located near his capital Nanjing. However, the majority of the Ming Tombs are located in a cluster near Beijing and collectively known as the Thirteen Tombs of the Ming Dynasty (明十三陵). They are within the suburban of Beijing. The site, on the southern slope of Tianshou Mountain, was chosen based on the principles of Fengshui (风水) by the third Ming emperor, the Yongle Emperor[2]. After the construction of the Imperial Palace (the Forbidden City) in 1420, the Yongle Emperor selected his burial site and created his own mausoleum. The subsequent emperors placed their tombs in the same valley. From the Yongle Emperor onwards, 13 Ming Dynasty emperors were buried in the same area. But the Xiaoling Tomb (明孝陵) of the first Ming emperor, the Hongwu Emperor[3], is located near his capital Nanjing. (See Fig. 9. 5)

Fig. 9. 5 Ritual Gate of Ming Tomb

1. Xiaoling Tomb in Nanjing

The Ming Xiaoling Mausoleum is the tomb of the Hongwu Emperor, the founder of the Ming Dynasty. It lies at the southern foot of Purple Mountain, located east of the historical centre of Nanjing, China. The construction of the mausoleum began during the Hongwu Emperor's life in 1381 and ended in 1405, during the reign of his son the Yongle Emperor, with a huge expenditure of resources involving 100,000 laborers. The original wall of the mausoleum was more than 22.5 kilometers long.

One enters the site through the monumental Great Golden Gates (大金门), and is soon faced by a giant stone tortoise (赑屃), which resides in the Square city ("四方城") pavilion. The tortoise supports a splendid carved stone stele, crowned by intertwining hornless dragons.

Unlike the similar pavilion at the Thirteen Ming Tombs near Beijing, Nanjing's Square city does not have its roof anymore, as it was destroyed during the Taiping Rebellion. Recently, Chinese engineers have conducted research in regard to the possibility of restoring the roof.

The Sacred Way (神道) is an 1800-metre-long road starting near the Square city pavilion. It includes several sections: the Elephant Road and the Wengzhong(翁仲) Road. The Elephant Road is lined by 12 pairs of 6 kinds of animals such as lions, Xiezhi, camels, elephants, Qilin, and horses, guarding the tomb. Beyond them is a column called huabiao (华表) in Chinese. One then continues along the Wengzhong Road. Four pairs of ministers and generals of stone have been standing there for centuries to guard the journey to the after life.

One enters the central area of the mausoleum complex through the Gate of the Civil and the Military (文武方门). On an inscribed stone tablet outside of the gate is an official notification of the local government in the Qing Dynasty (1644—1911) to protect the tomb. Inside the gate, there the Tablet Hall (碑殿) in which 5 steles stand.

Behind the pavilion, there used to be other annexes; The emperor and his queen were buried in a clay tumulus, 400 meters in diameter, known as the Lone Dragon Hill . A stone wall with a terrace on top, known as Ming Mansion (明楼) or the Soul Tower is half-embedded into the front face of the tumulus.

2. Ming Tombs in Beijing

The sitting of the Ming Dynasty imperial tombs was carefully chosen according to Fengshui principles. According to these, bad spirits and evil winds descending from the North must be deflected; therefore, an arc-shaped valley area at the foot of the Jundu

Mountains, north of Beijing, was selected.

2.1 The Layout of Ming Tombs Complex

A 7-kilometer road named the "Spirit Way" (神道) leads into the complex, lined with statues of guardian animals and officials, with a front gate consisting of a three-arches, painted red, and called the "Great Red Gate". The Spirit Way, or Sacred Way, starts with a huge stone memorial archway lying at the front of the area. Constructed in 1540, during the Ming Dynasty, this archway is one of the biggest stone archways in China today. (See Fig. 9.6)

Fig. 9.6 The Ming Tombs complex plan

Further in, a Stele Pavilion can be seen in which is a 50-ton stone statue of tortoise carrying a memorial tablet. Four white marble Huabiao are positioned at each corner of the stele pavilion. At the top of each pillar is a mythical beast. Then come two pillars on each side of the road, whose surfaces are carved with the cloud design, and tops are shaped like a rounded cylinder. They are of a traditional design and were originally beacons to guide the soul of the deceased, The road leads to 18 pairs of stone statues of mythical animals, which are all sculpted from whole stones and larger than life size, leading to a three-arched gate known as the Dragon and Phoenix Gate(龙凤门).

2.2 Dingling Tomb (定陵)

Dingling Tomb is the mausoleum of Emperor Wanli (万历 1563—1620) of Ming Dynasty (1368—1644) and his two empresses. Wanli was the thirteenth emperor and occupied the throne for 48 years, the longest among all of the emperors of the Ming Dynasty. Built over six years between 1584 and 1590, the tomb, which covers an area of 180,000 square meters, is of great historical value.

The above ground buildings of Dingling Tomb is rectangular in shape, square in front and round in rear. In order on the axis are the similar buildings like mausoleum gate, Stele Pavilion, Gate of Prominent Favor, Mausoleum Tower, and Treasure City. Yet the craft is more meticulous everywhere, such as the three additional single arched stone bridges in front of the mausoleum gate, the Stele Pavilion behind the bridges in the style of double-eave between four ridges, and the stele carrier tortoise was carved with lively sea waves and coast cliffs, etc.

The above ground part of Dingling Tomb symbolized the ancient Chinese philosophical concept of "heaven is round and the earth is square". Three white marble stone bridges lead you to the entrance of Dingling Tomb, where you will see a high tablet pavilion. Further back, there is an enclosing wall named Wailuo Wall (外罗城) around the mausoleum. At the axis position of the wall a palace gate was set, which is the first gate. The yellow glazed tiles, eaves, archway, rafters and columns are all sculptured from stone, and colorfully painted. Inside the Wailuo Wall, there are three courtyards in the square front part, and the Treasure City in the circular rear part. The first courtyard has no buildings and facilities, but three Divine Kitchens on the left side outside the courtyard, and three Divine Storerooms on the right side. The gate of the second courtyard is named Blessing and Grace Gate (祾恩门). There is a base with railings, and the top of the railings are decorated with stone dragon heads and phoenix heads. The Blessing and Grace Palace (祾恩殿) is in the third courtyard. It is the place for making sacrifice to Emperor Zhu Yijun and his two empresses. The stone road in the middle of the courtyard is engraved with a dragon and a phoenix playing with a pearl. The third courtyard has a two-column archway door called Lingxing Gate (棂星门) and a few stone tables on which sacrificial items are placed. The circular rear part has the Treasure City, where Emperor and his two empresses were buried. It is covered with earth and the middle part stands out, looking like a round castle.

The underground part is the Underground Palace, which was unearthed between 1956 and 1958. It is the only unearthed palace of the Thirteen Imperial Tombs of Ming Dynasty. Starting from the ground, after more than 40 meters of the underground tunnel, you can access to the hidden palace. The stone structure of the palace is a representative style of the Ming Dynasty. The entire palace is divided into five communicant vaulted halls: the front, the middle, the rear, the left and the right halls, among which the rear hall is the main and largest. The entrance of each hall is made of sculptured jade, and the floors are covered with gilded bricks. In the middle of each hall is a white marble coffin. On each coffin there is a square hole called Gold Well filled with loess. A paved path leads to the central hall where there are three white marble thrones, in front of which incense, candles and flowers were set. Before each of them, there are the glazed Five Offerings and a blue china jar that would have been filled with sesame oil to be used for lamps. The coffins of Emperor Zhu Yijun and his two empresses are in the rear hall. There are also some precious items displayed with these coffins, such as jades, vases, red lacquer boxes, golden crown, silver, silk and so on.

The Underground Palace unearthed a total of over 3,000 pieces of cultural relics, including four national treasures: the gold imperial crown, the gold empress crown, glowing pearl and tri-colored glazed pottery of the Ming Dynasty. These relics are all stored in the Dingling Tomb Museum.

3. The Excavation of Dingling

The excavation of Dingling began in 1956, after a group of prominent scholars led by Guo Moruo(郭沫若) and Wu Han(吴晗) began advocating the excavation of Changling(长陵), the tomb of the Yongle Emperor, the largest and oldest of the Ming tombs near Beijing. Despite winning approval from premier Zhou Enlai, this plan was vetoed by archaeologists because of the importance and public profile of Changling. Instead, Dingling, the third largest of the Ming Tombs, was selected as a trial site in preparation for the excavation of Changling. Excavation completed in 1957, and a museum was established in 1959.

The excavation revealed an intact tomb, with thousands of items of silk, textiles, wood, and porcelain, and the skeletons of the Wanli Emperor and his two empresses. However, there was neither the technology nor the resources to adequately preserve the excavated artifacts. After several disastrous experiments, the large amount of silk and other textiles were simply piled into a storage room that leaked water and wind. As a result, most of the surviving artifacts today have severely deteriorated, and many replicas are instead displayed in the museum.

The lessons learned from the Dingling excavation has led to a new policy of the People's Republic of China government not to excavate any historical site except for rescue purposes. In particular, no proposal to open an imperial tomb has been approved since Dingling, even when the entrance has been accidentally revealed, as was the case of the Qianling Mausoleum. The original plan, to use Dingling as a trial site for the excavation of Changling, was abandoned.

Notes:

1. Ming Tombs in China：明十三陵，是明朝迁都北京后13位皇帝陵墓的皇家陵寝的总称。长陵是永乐皇帝和皇后的合葬陵寝，在十三陵中建筑规模最大，营建时间最早，地面建筑也保存得最为完好。长陵的陵宫建筑，占地约12万平方米。其平面布局呈前方后圆形状。其前面的方形部分，由前后相连的三进院落组成。
2. Yongle Emperor：明成祖朱棣(1360—1424年)，明太祖第四子，明朝第三位皇帝，1402年登基，年号永乐。在他统治期间，明朝经济繁荣、国力强盛，文治武功都有了很大提升，史称永乐盛世。
3. Hongwu Emperor：大明太祖高皇帝朱元璋(1328—1398年)，明朝开国皇帝。朱元璋在位期间，社会生产逐渐恢复和发展，史称洪武之治。

Taj Mahal[1]

Taj Mahalis a white Marble mausoleum located in Agra, India. It was built by Mughal emperor Shah Jahan[2] in memory of his third wife, Mumtaz Mahal. It is the finest example of Mughal architecture[3], a style that combines elements from Persian, Turkish and Indian architectural styles. Taj Mahal is widely recognized as the jewel of Muslim art in India and one of the universally admired masterpieces of the world's heritage. (See Fig. 9.7)

Fig. 9.7 Mausoleum of Taj Maha①

① F. Kohil

1. The History of Taj Mahal

In 1631, Shah Jahan, the emperor during the Mughal empire's period of greatest prosperity, was grief-stricken when his third wife, Mumtaz Mahal, died during the birth of their 14th child. The construction of Taj Mahal began around 1632 and the principal mausoleum was completed in 1648 and the surrounding buildings and garden were finished five years later. The construction was entrusted to a board of architects under imperial supervision. Lahauri is generally considered to be the principal designer.

In the tomb area, wells were dug and filled with stone and rubble to form the footings of the tomb. Instead of lashed bamboo, workmen constructed a colossal brick scaffold that mirrored the tomb. The scaffold was so enormous that foremen estimated it would take years to dismantle. According to the legend, Shah Jahan decreed that anyone could keep the bricks taken from the scaffold, and thus it was dismantled by peasants overnight.

A fifteen kilometer tamped-earth ramp was built to transport marble and materials to the construction site and teams of twenty or thirty oxen pulled the blocks on specially constructed wagons. An elaborate post-and-beam pulley system was used to raise the blocks into desired position. Water was drawn from the river by a series of *purs*, an animal-powered rope and bucket mechanism, into a large storage tank and raised to a large distribution tank. It was passed into three subsidiary tanks, from which it was piped to the complex.

The Taj Mahal was constructed using materials from all over India and Asia and over 1,000 elephants were used to transport building materials. The translucent white marble was brought from Makrana, Rajasthan, the jasper from Punjab, jade and crystal from China. The turquoise was from Tibet and the Lapis lazuli from Afghanistan, while the sapphire came from Sri Lanka and the carnelian from Arabia. In all, twenty eight types of precious and semi-precious stones were inlaid into the white marble.

The plinth and tomb took roughly 12 years to complete. The remaining parts of the complex took an additional 10 years and were completed in order of minarets, mosque, and gateway. Estimates of the cost of construction vary due to difficulties in estimating costs across time. The total cost has been estimated to be about 32 million Rupees at that time.

A labor force of twenty thousand workers was recruited across northern India. Sculptors from Bukhara, calligraphers from Syria and Persia, inlayers from southern India, stonecutters from Baluchistan, a specialist in building turrets, another who carved only marble flowers were part of the thirty-seven men who formed the creative unit.

2. The Description of Taj Mahal

The Taj Mahal incorporates and expands on design traditions ofPersian architecture and earlier Mughal architecture. Specific inspiration came from successful Timurid and Mughal buildings. While earlier Mughal buildings were primarily constructed of red sandstone, Shah Jahan promoted the use of white marble inlaid with semi-precious stones, and buildings under his patronage reached new levels of refinement.

2.1 The Layout of Taj Mahal Complex

The Taj Mahal complex is bordered on three sides by crenellated red sandstone walls; the side facing the river is open. Outside the walls are several additional mausoleums, including those of Shah Jahan's other wives, and a larger tomb for Mumtaz's favourite servant.

The main gateway is a monumental structure built primarily of marble, and reminiscent of the Mughal architecture of earlier emperors. Its archways mirror the shape of the tomb's archways, and its pishtaq[4] arches incorporate the calligraphy that decorates the tomb. The vaulted ceilings and walls have elaborate geometric designs like those found in the other sandstone buildings in the complex.

At the far end of the complex are two grand red sandstone buildings that mirror each other, and face the sides of the tomb. The backs of the buildings parallel the western and eastern walls. The western building is a mosque and the other is the jawab, thought to have been constructed for architectural balance although it may have been used as a guesthouse. Distinctions between the two buildings include the jawab's lack of a mihrab, and its floors of geometric design whereas the floor of the mosque is laid with outlines of 569 prayer rugs in black marble. The mosque's basic design of a long hall surmounted by three domes is similar to others built by Shah Jahan. (See Fig. 9.8)

2.2 The Tomb

The tomb is the central focus of the entire complex of the Taj Mahal. This large, white marble structure stands on a square plinth and consists of a symmetrical building with an iwan topped by a large dome and finial. Like most Mughal tombs, the basic elements are Persian in origin.

The base structure is essentially a large, multi-chambered cube with chambered corners, forming an unequal octagon that is approximately 55 metres on each of the four long sides. On each of these sides, a huge vaulted archway, frames the iwan with two similarly shaped, arched balconies stacked on either side. This motif of stacked pishtaqs is replicated on the chambered corner areas, making the design completely symmetrical on all sides of the building. Four minarets frame the tomb, one at each corner of the plinth facing the chamfered corners. The main chamber houses the false sarcophagi of Mumtaz Mahal and Shah Jahan; the actual graves are at a lower level.

The marble dome that surmounts the tomb is the most spectacular feature. Its height of around 35 metres is about the same as the length of the base, and is accentuated as it sits

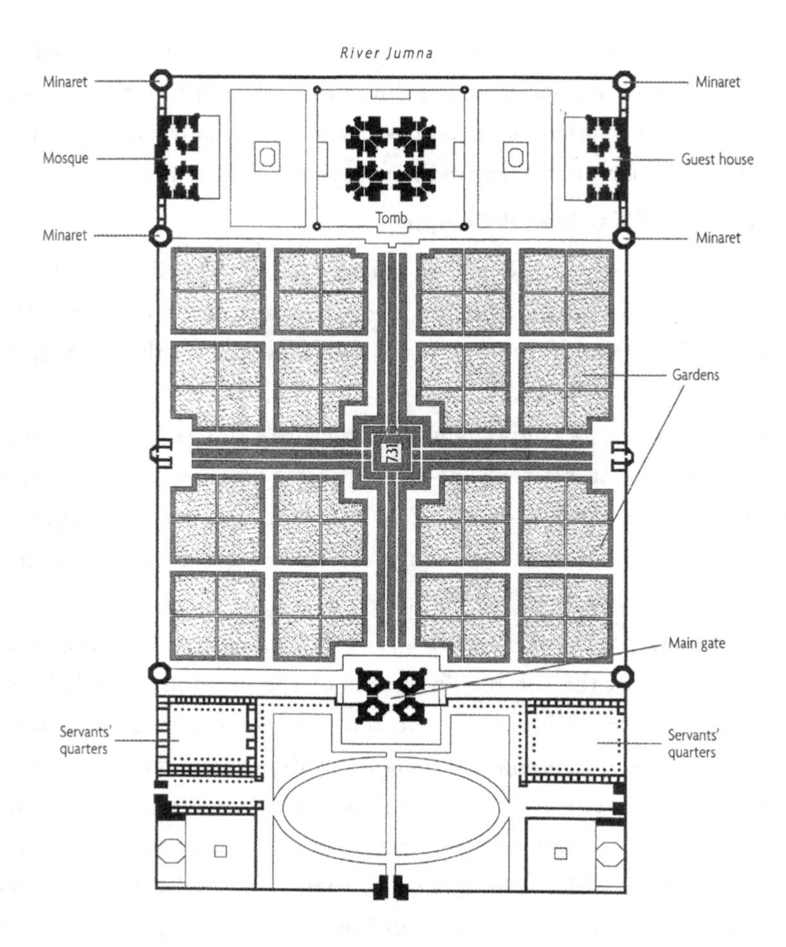

Fig. 9. 8 The site plan of Taj Mahal complex

on a cylindrical "drum" which is roughly 7 meters high. Because of its shape, the dome is often called an onion dome . The top is decorated with a lotus design, which also serves to accentuate its height. The shape of the dome is emphasized by four smaller domed kiosks placed at its corners, which replicate the onion shape of the main dome. Their columned bases open through the roof of the tomb and provide light to the interior. Tall decorative spires extend from edges of base walls, and provide visual emphasis to the height of the dome. The dome is topped by a gilded finial, which mixes traditional Persian and Hindustani decorative elements.

The minarets, which are each more than 40 meters tall, display the designer's penchant for symmetry. They were designed as working minarets—a traditional element of mosques, used by the muezzin to call the Islamic faithful to prayer. Each minaret is effectively divided into three equal parts by two working balconies that ring the tower. At the top of the tower is a final balcony surmounted by a chattri that mirrors the design of those on the tomb. The chattris all share the same decorative elements of a lotus design topped by a gilded finial. The minarets were constructed slightly outside of the plinth so that, in the event of collapse, the material from the towers would tend to fall away from the tomb.

2.3 The Decoration

The exterior decorations of the Taj Mahal are among the finest in Mughal architecture. As the surface area changes the decorations are refined proportionally. The decorative elements were created by applying paint, stucco, stone inlays, or carvings. In line with the Islamic prohibition against the use of anthropomorphic forms, the decorative elements can be grouped into either calligraphy, abstract forms or vegetative motifs.

Near the lines from the Qur'an at the base of the interior dome is the inscription, much of the calligraphy is composed of florid thuluth script, made of jasper or black marble, inlaid in white marble panels. Higher panels are written in slightly larger script to reduce the skewing effect when viewed from below. The calligraphy found on the marble cenotaphs in the tomb is particularly detailed and delicate.

Abstract forms are used throughout, especially in the plinth, minarets, gateway, mosque, to a lesser extent, on the surfaces of the tomb. The domes and vaults of the sandstone buildings are worked with tracery of incised painting to create elaborate geometric forms. Herringbone inlays define the space between many of the adjoining elements. White inlays are used in sandstone buildings, and dark or black inlays on the white marbles. Mortared areas of the marble buildings have been stained or painted in a contrasting color, creating geometric patterns of considerable complexity. Floors and walkways use contrasting tiles or blocks in tessellation patterns.

On the lower walls of the tomb there are white marble dados that have been sculpted with realistic bas-relief depictions of flowers and vines. The marble has been polished to emphasize the exquisite detailing of the carvings and the dado frames and archway spandrels have been decorated with pietra dura[5] inlays of highly stylised, almost geometric vines, flowers and fruits. The inlay stones are of yellow marble, jasper and jade, polished and leveled to the surface of the walls.

3. Taj Mahal Garden

The Taj Mahal complex is set around a large 300-metre square Mughal garden. The garden uses raised pathways that divide each of the four quarters of the garden into 16 sunken flower beds. A raised marble water tank at the center of the garden, halfway between the tomb and gateway with a reflecting pool on a north-south axis, reflects the image of the mausoleum. The raised marble water tank is called the "Tank of Abundance" promised to Muhammad. Elsewhere, the garden is laid out with avenues of trees and fountains.

Most Mughal charbaghs are rectangular with a tomb or pavilion in the center. The Taj Mahal garden is unusual in that the main element, the tomb, is located at the end of the garden. Early accounts of the garden describe its profusion of vegetation, including abundant roses, daffodils, and fruit trees. As the Mughal Empire declined, the tending of the garden also declined, and when the British took over the management of Taj Mahal during the time of the British Empire, they changed the landscaping to resemble that of lawns of London.

Notes:

1. Taj Mahal：泰姬陵全称为泰姬·玛哈尔陵，是一座白色大理石建成的巨大陵墓清真寺，是莫卧儿皇帝沙·贾汗为纪念他心爱的妃子于1631年至1653年在阿格拉而建的。泰姬陵位于今印度距新德里200多千米外的北方邦的阿格拉(Agra)城内，亚穆纳河右侧。陵园由殿堂、钟楼、尖塔、水池等构成，全部用纯白色大理石建筑，用玻璃、玛瑙镶嵌，具有极高的艺术价值。
2. Shah Jahan：沙·贾汗，是印度莫卧儿帝国的皇帝，1628年到1658年在位。"沙·贾汉"在波斯语中的意思是"世界的统治者"。在沙·贾汉的统治期间，莫卧儿帝国的艺术和建筑成就达到顶峰。他为他所钟爱的妻子慕塔芝·玛哈尔在阿格拉修建了著名的泰姬陵，在德里修建了红堡。
3. Mughal architecture：莫卧儿建筑，是印度莫卧儿王朝时代(1526—1858年)的伊斯兰建筑，集中分布于莫卧儿帝国都城德里、阿格拉与陪都拉合尔(今属巴基斯坦)等地。建筑形式主要包括城堡、宫殿、清真寺、陵墓等，一般以尖拱门、尖塔、大圆顶穹窿、小圆顶凉亭等建筑构件的组合为特征。建筑材料主要采用印度特产的红砂石和白色大理石，这使莫卧儿建筑堂皇的外观与坚固的程度都超过中亚的彩釉瓷砖建筑。建筑风格主要吸收了波斯伊斯兰建筑的影响，同时融合了印度传统建筑的因素，形成了一种既简洁明快又装饰富丽的莫卧儿风格，代表着印度伊斯兰教美术的最大成就。
4. pishtaq：一种以旺式拱门，有长方形空间，上有拱顶，三面有墙，一端辟入口，是典型的波斯大门。大门通常有伊斯兰书法装饰条带、釉面砖以及几何纹饰，通常与伊斯兰教建筑有密切联系。
5. pietra dura：佛罗伦萨的一种马赛克饰面。

Insights: Fresco

Fresco is a technique of mural painting executed upon freshly-laid, or wet lime plaster. Water is used as the vehicle for the pigment to merge with the plaster, and with the setting of the plaster, the painting becomes an integral part of the wall. The word fresco is derived from the Italian adjective fresco meaning "fresh", and may thus be contrasted with secco mural painting techniques, which are applied to dried plaster, to supplement painting in fresco. The fresco technique has been employed since antiquity and is closely associated with Italian Renaissance painting.

The earliest known examples of frescoes date from the Fourth Dynasty of Egypt of the Old Kingdom of Egypt. The oldest frescoes done in the Buon Fresco method date from the first half of the second millennium BCE during the Bronze Age and are to be found among Aegean civilizations. Some art historians believe that fresco artists from Crete may have been sent to various locations as part of a trade exchange, a possibility which raises to the fore the importance of this art form within the society of the times.

The most common form of fresco was Egyptian wall paintings in tombs, usually using the a secco technique. Thanks to large number of ancient rock-cut cave temples, valuable ancient and early medieval frescoes have been preserved in more than 20 locations of India. The frescoes on the ceilings and walls of the Ajanta Caves were painted between 200 B. C. and 600 A. D. and are the oldest known frescoes in India.

Chapter Ten

Museum

A museum is an institution that cares for a collection of artifacts and other objects of artistic, cultural, historical, or scientific importance. Some public museums makes them available for public viewing through exhibits that may be permanent or temporary. Museums have varying aims, ranging from serving researchers and specialists to serving the general public.

While there is an ongoing debate about the purposes of interpretation of a museum's collection, there has been a consistent mission to protect and preserve artifacts for future generations. Much care, expertise, and expense is invested in preservation efforts to retard decomposition in aging documents, artifacts, artworks, and buildings. All museums display objects that are important to a culture.

Museums are, above all, storehouses of knowledge. Some seek to reach a wide audience, such as a national or state museum, while some museums have specific audiences. Although most museums do not allow physical contact with the associated artifacts, there are some that are interactive and encourage a more hands-on approach.

The design of museum involves planning the actual mission of the museum along with planning the space that the collection of the museum will be housed in. The way that museums are planned and designed vary according to what collections they house, but overall, they adhere to planning a space that is easily accessed by the public and easily displays the chosen artifacts.

Louvre Museum[1]

Louvre Museum is one of the world's largest museums and a historic monument. A central landmark of Paris, France, it is located on the Right Bank of the Seine. The museum is housed in the Louvre Palace, originally built as a fortress in the late 12th century under Philip II. The building was extended many times to form the present Louvre Palace. In 1682, Louis XIV chose the Palace of Versailles for his household, leaving the Louvre primarily as a place to display the royal collection. During the French Revolution, the National Assembly decreed that the Louvre should be used as a museum to display the nation's masterpieces. (See Fig. 10. 1)

Fig. 10. 1 The main entrance of Louvre Museum ①

1. History of Louvre Musuem

The Louvre Palace was begun as a fortress in the 12th century, with remnants of this building still visible in the crypt. The Louvre Palace was altered frequently throughout the Middle Ages. In the 14th century, Charles V converted the building into a residence and in 1546, Francis I renovated the site in French Renaissance style. After Louis XIV chose Versailles as his residence in 1682, constructions slowed; however, the move permitted the Louvre to be used as a residence for artists.

By the mid-18th century there was an increasing number of proposals to create a public gallery, a hall was opened for public viewing on Wednesdays and Saturdays. Under Louis XVI, the royal museum idea became policy, many proposals were offered for the Louvre's renovation into a museum.

During the French Revolution[2] the Louvre was transformed into a public museum. In May 1791, the Assembly declared that the Louvre would be "a place for bringing together monuments of all the sciences and arts". On 10 August 1792, Louis XVI was imprisoned and the royal collection in the Louvre became national property. The museum opened on 10 August 1793, the first anniversary of the monarchy's demise. The public was given free access on three days per week, which was "perceived as a major accomplishment and was generally appreciated".

By 1874, the Louvre Palace had achieved its present form of an almost rectangular structure with the Sully Wing to the east containing the square Cour Carrée and the oldest parts of the Louvre; and two wings which wrap the Cour Napoléon, the Richelieu Wing to the north and the Denon Wing, which borders the Seine to the south. In 1983, French President Francois Mitterrand proposed the Grand Louvre plan to renovate the building and

---

① keyword suggestion. com

relocate the Finance Ministry, allowing displays throughout the building. Architect I. M. Pei(贝聿铭) was awarded the project and proposed a glass pyramid to stand over a new entrance in the main court, the Cour Napoléon. The pyramid was completed in 1989 and the Inverted Pyramid was completed in 1993.

2. The Description of Louvre Museum

The present-day Louvre Palace is a vast complex of wings and pavilions on four main levels which, although it looks to be unified, is the result of many phases of building, modification, destruction and restoration. The Louvre complex may be divided into the "Old Louvre": the medieval and Renaissance pavilions and wings surrounding the Cour Carrée, as well as the Grande Galerie extending west along the bank of the Seine; and the "New Louvre": those 19th-century pavilions and wings extending along the north and south sides of the Cour Napoléon along with their extensions to the west which were originally part of the Tuileries Palace, burned during the Paris Commune in 1871. (See Fig. 10. 2)

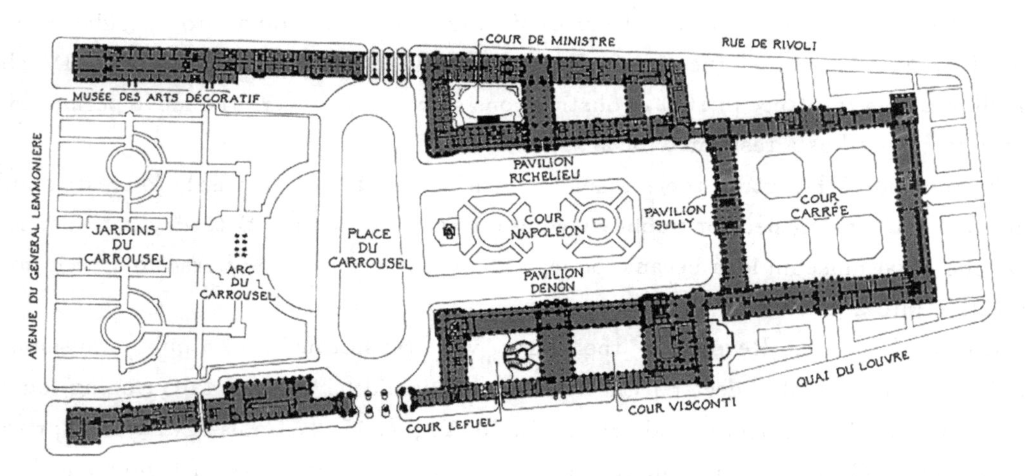

Fig. 10. 2 The ground floor plan of Louvre Museum

2.1 Old Louvre

The Old Louvre occupies the site of the 12th-century fortress of King Philip Augustus, also called the Louvre. Its foundations are viewable in the basement level as the "Medieval Louvre" department. The Old Louvre is a quadrilateral approximately 160 meters on a side consisting of 8 wings which are articulated by 8 pavilions. Between the Pavilion du Roi and the Pavilion Sully is the Lescot Wing built between 1546 and 1551, which is the oldest part of the visible external elevations and was important in setting the mould for later French architectural classicism. Between the Pavilion Sully and the Pavilion de Beauvais is the Lemercier Wing built in 1639 by Louis XIII and Cardinal Richelieu, which is a symmetrical extension of Lescot's wing in the same Renaissance style[3].

Pavilion Sully is a prominent pavilion located in the center of the west wing of the Square Court . The two names Pavilion de l'Horloge and Pavilion Sully are now often reserved for the central pavilion's eastern and western facade respectively.

2.2 New Louvre

The New Louvre is the name often given to the wings and pavilions extending the Palace for about 500 meters westwards on the north and on the south sides of the Cour Napoléon and Cour du Carrousel. It was Napoléon III who finally connected the Tuileries Palace with the Louvre in the 1850s.

The northern limb of the new Louvre consists of three great pavilions, these pavilions and their wings define three subsidiary Courts, Cour Khorsabad, Cour Puget and Cour Marly. For simplicity, the New Louvre north limb, the New Louvre south limb, and the Old Louvre are designated as the "Richelieu Wing", the "Denon Wing" and the "Sully Wing" respectively in order to allow the casual visitor to avoid becoming totally mystified at the bewildering array of named wings and pavilions.

The Pavilion de Flore and the Pavilion de Marsan, at the western most extremity of the Palace were destroyed when the Third Republic razed the ruined Tuileries, but were subsequently restored beginning in 1874. The Flore then served as the model for the renovation of the Marsan by architect Gaston Redon. The Pavilion de Flore was originally constructed in 1607—1610, during the reign of Henry IV, as the corner pavilion between the Tuileries Palace to the north and the Louvre's Grande Galerie to the east. The pavilion was entirely redesigned and rebuilt by Hector Lefuel in 1864—1868 in a highly decorated Second Empire Neo-Baroque style. The most famous sculpture on the exterior of the Louvre, Jean-Baptiste Carpeaux's The Triumph of Flora, was added below the central pediment of the south facade at this time. The Tuileries Palace was destroyed by fire in 1871, and a north facade, similar to the south facade, was added to the pavilion in 1874—1879. Currently.

2.3 The Louvre Pyramid

The Louvre Pyramid is a large glass and metal pyramid designed by Chinese-American architect I. M. Pei, surrounded by three smaller pyramids, in the main courtyard of the Louvre . The large pyramid serves as the main entrance to the Louvre Museum. Completed in 1989, it has become a landmark of the city of Paris. The structure, which was constructed entirely with glass segments, reaches a height of 21.6 meters . Its square base has sides of 34 meters and a base surface area of 1,000 square meters . It consists of 603 rhombus-shaped and 70 triangular glass segments.

The pyramid and the underground lobby beneath it were created because of a series of problems with the Louvre's original main entrance, which could no longer handle the enormous number of visitors on an everyday basis. Visitors entering through the pyramid descend into the spacious lobby then re-ascend into the main Louvre buildings.

3. The Tuileries Garden

The Tuileries Garden is a public garden located between the Louvre Museum and the Place de la Concorde. Created by Catherine de Medici as the garden of the Tuileries Palace in 1564, it was eventually opened to the public in 1667, and became a public park after the French Revolution. In the 19th and 20th century, it was the place where Parisians celebrated, met, promenaded, and relaxed.

In July 1559, Catherine commissioned a landscape architect from Florence, Bernard de Carnesse, to build an Italian Renaissance garden, with fountains, a labyrinth, and a grotto, decorated with faience images of plants and animals. The garden of Catherine de Medicis was an enclosed space five hundred meters long and three hundred meters wide, separated from the new chateau by a lane. It was divided into rectangular compartments by six alleys, and the sections were planted with lawns, flower beds, and small clusters of five trees. The Tuileries was the largest and most beautiful garden in Paris at the time.

In 1664, Colbert, the superintendent of buildings of the King, commissioned the landscape architect André Le Nâtre, to redesign the entire garden. He began transforming the Tuileries into a formal garden *ǎ* la francaise, a style he had first developed at Vaux-le-Vicomte[4] and perfected at Versailles, based on symmetry, order and long perspectives.

He eliminated the street which separated the palace and the garden, and replaced it with a terrace looking down upon parterres bordered by low boxwood hedges and filled with designs of flowers. In the centre of the parterres he placed three basins with fountains. In front of the center first fountain he laid out the grand allée, which extended 350 meters. He built two other alleys, lined with chestnut trees, on either side. He crossed these three main alleys with small lanes, to create compartments planted with diverse trees, shrubs and flowers.

On the west side of the garden, beside the present-day Place de la Concorde, he built two ramps in a horseshoe shape and two terraces overlooking a octagonal water basin sixty meters in diameter with a fountain in the centre. These terraces frame the western entrance of the garden, and provide another viewpoint to see the garden from above. After the French Revolution, the Tuileries became the National Garden of the new French Republic.

Notes:

1. Louvre Museum：卢浮宫博物馆，位于法国巴黎市中心的塞纳河北岸，位居世界四大博物馆之首。卢浮宫始建于1204年，原是法国的王宫，居住过50位法国国王和王后，是法国文艺复兴时期最珍贵的建筑物之一。拿破仑时期，从许多国家掠夺过来大量名贵艺术品，其艺术品收藏之丰堪称世界之冠。1973年，法国国民议会决定将其改为国立美术馆。卢浮宫整个建筑群和广场及草坪总共占地45万平方米，展厅面积大约为138000平方米，为新古典主义风格建筑。
2. French Revolution：法国大革命，是1789年7月14日在法国爆发的革命，统治法国多个世纪的波旁王朝统治下的君主制在三年内土崩瓦解。过往的贵族和宗教特权不断受到自由主义政治组织及上街抗议的民众的冲击，旧的观念逐渐被全新的天赋人权、三权分立等的民主思想所取代。1794年7月27日，热月政变推翻了雅各宾派的激进集权统治，宣告了法国大革命中市民革命的结束。1830年7月，巴黎人民发动七月革命，建立了以路易·菲利浦为首的七月王朝，至此法国大革命才彻底结束。
3. Renaissance style：文艺复兴建筑风格，是欧洲建筑史上继哥特式建筑之后出现的一种建筑风格，15世纪产生于意大利，后传播到欧洲其它地区，形成了带有各自特点的各国文艺复兴建筑。意大利文艺复兴建筑在文艺复兴建筑中占有最重要的位置。文艺复兴建筑最明显的特征是扬弃了中世纪时期的哥特式建筑风格，而在宗教和世俗建筑上重新采用古希腊罗马时期的柱式构图要素。他们一方面采用古典柱式，一方面又灵活变通，大胆创新，甚至将各个地区的建筑风格同古典柱式融合一起。他们还将文艺复兴时期的许多科学技术上的成果，如力学上的成就、绘画中的透视规律、新的施工机具等等，运用到建筑创作实践中去。
4. Vaux-le-Vicomte：子爵城堡，是一座巴洛克风格的法国城堡，位于法国法兰西岛大区塞纳—马恩省，在巴黎东南方55千米处。它兴建于1658年到1661年，主人是路易十四的财政大臣——尼古拉斯·富凯。子爵城堡是法国众多雄伟壮观的城堡之一，也是目前被法国列入历史性建筑保护名单中面积最大的私有财产。子爵城堡位于花园中心，不论从园中的任何一个地方看过来，城堡都耸立在中心，显得威严、气派。离城堡500米远的水池起到镜子的效果。城堡内部的房间和客厅，走廊都十分宽广，主人的卧室里装饰得富丽堂皇，女主人的房间里布满了镜子，这可能也是凡尔赛宫里镜廊的前身。

British Museum[1]

The British Museum, located in the Bloomsbury area of London, is a museum dedicated to human history, art, and culture. Its permanent collection, numbering some 8 million works, is among the largest and most comprehensive in existence and originates from all continents, illustrating and documenting the story of human culture from its beginnings to the present. Until 1997, when the British Library, previously centered on the Round Reading Room moved to a new site, the British Museum housed both a national museum of antiquities and a national library in the same building. (See Fig. 10. 3)

Fig. 10. 3 British Museum

1. The History of British Museum

The British Museum was established in 1753, largely based on the collections of the physician and scientist Sir Hans Sloane[2]. The museum first opened to the public on 15 January 1759, in Montagu House in Bloomsbury, on the site of the current museum building. Its expansion over the following two and a half centuries was largely a result of an expanding British colonial footprint and has resulted in the creation of several branch institutions.

The British Museum was founded as a "universal museum". Its foundations lie in the will of the Irish-born British physician and naturalist Sir Hans Sloane (1660—1753). During the course of his lifetime Sloane gathered an enviable collection of curiosities and, not wishing to see his collection broken up after death, he bequeathed it to King George II, for the nation, for a sum of £20,000. At that time, Sloane's collection consisted of around 71,000 objects of all kinds including some 40,000 printed books, 7,000 manuscripts, extensive natural history specimens including 337 volumes of dried plants, prints and drawings.

On 7 June 1753, King George II gave his formal assent to establish the British Museum. The body of trustees decided on a converted 17th-century mansion, Montagu House, as a location for the museum, which it bought from the Montagu family for £20,000. With the acquisition of Montagu House the first exhibition galleries and reading room for scholars opened on 15 January 1759.

From 1778, a display of objects from the South Seas brought back from the round-the-world voyages of Captain James Cook and the travels of other explorers fascinated visitors with a glimpse of previously unknown lands. Montagu House became increasingly crowded and decrepit and it was apparent that it would be unable to cope with further expansion. The neoclassical architect, Sir Robert Smirke, was asked to draw up plans for an eastern extension to the Museum. The dilapidated Old Montagu House was demolished.

The Museum became a construction site as Sir Robert Smirke's grand neo-classical building gradually arose. The King's Library, on the ground floor of the East Wing, was handed over in 1827, and was described as one of the finest rooms in London. Although it was not fully open to the general public until 1857, special openings were arranged during The Great Exhibition of 1851.

With the departure and the completion of the new White Wing in 1884, more space was available for antiquities and ethnography and the library could further expand. This was a time of innovation as electric lighting was introduced in the Reading Room and exhibition galleries. In 1895 the trustees purchased the 69 houses surrounding the Museum with the intention of demolishing them and building around the West, North and East sides of the Museum.

In 1931, the art dealer Sir Joseph Duveen offered funds to build a gallery for the Parthenon sculptures. Designed by the American architect John Russell Pope, it was completed in 1938. The appearance of the exhibition galleries began to change as dark Victorian reds gave way to modern pastel shades.

By the 1970s the Museum was again expanding. The Government suggested a site at St Pancras for the new British Library but the books did not leave the museum until 1997. The departure of the British Library created the opportunity to redevelop the vacant space in Robert Smirke's 19th-century central quadrangle into the Queen Elizabeth II Great Court—the largest covered square in Europe—which opened in 2000.

2. The Description of British Museum

2.1 The Layout of British Museum

The core of today's British Museum building was designed by the architect Sir Robert Smirke (1780—1867) in 1823. It was a quadrangle with four wings: the north, east, south and west wings. The building was completed in 1852. It included galleries for classical sculpture and Assyrian[3] antiquities as well as residences for staff.

Smirke designed the building in the Greek Revival style[4], with 44 columns in the Ionic order 14 meters high, closely based on those of the temple of Athena Polias at Priene in Asia Minor[5], which emulated classical Greek architecture. Greek features on the building also include the pediment at the South entrance, which is decorated by sculptures by Sir Richard Westmacott depicting "The Progress of Civilization", consisting of fifteen allegorical figures, installed in 1852. Since the Greek Revival style had become increasingly popular since the 1750s when Greece and its ancient sites were "rediscovered" by western Europeans, the design of the columns has been borrowed from ancient Greek temples, and the pediment at the top of the building is a common feature of classical Greek architecture. (See Fig. 10. 4)

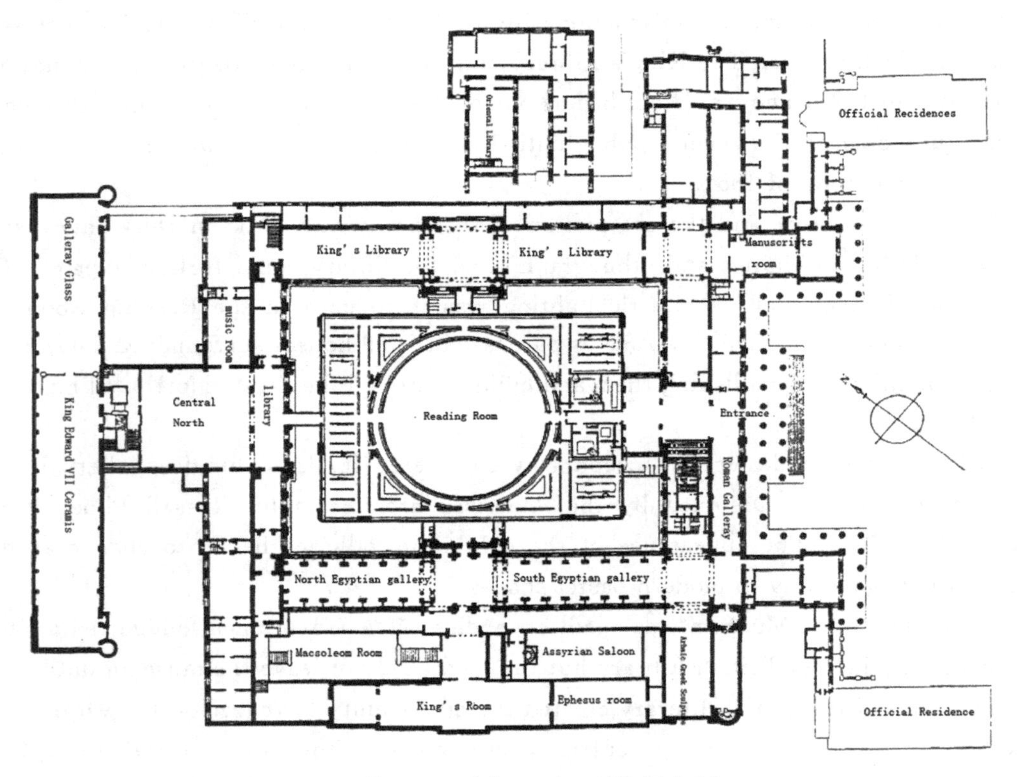

Fig. 10. 4 The ground floor plan of British Museum

The east and west residences of British Museum, which are to the left and right of the entrance have a more modest exterior. This is an example of mid-nineteenth century domestic architecture and reflects the domestic purpose of these wings. They housed the Museum's employees, who originally lived on site.

The White Wing, facing Montague Street, was designed by the architect Sir John Taylor (1833—1912) and constructed 1882—1885. It was designed in the same style as the quadrangle building.

The Museum had again been looking to expand and a bequest made by William White. He had two requests about the design of the building: that it had a monumental entrance steps which run up to the entrance and an inscription which is above the doorway. These can both be seen from Montague Street.

The building was constructed using up-to-the-minute 1820s technology. Built on a concrete floor, the frame of the building was made from cast iron and filled in with London stock brick. The public facing sections of the building were covered in a layer of Portland stone. In 1853, the quadrangle building won the Royal Institute of British Architects' Gold Medal.

2.2 The Interior Design of British Museum

The Weston Hall was designed by Sydney Smirke, who took over from his brother, Sir Robert Smirke, in 1845. The patterns and colors on the ceiling of the Weston Hall were borrowed from classical Greek buildings, which would have been brightly decorated. The electric lamps in the entrance hall are replicas of the original lighting lamps in the Museum. The Museum was the first public building to be electrically lit.

The King Edward VII galleries, designed by Sir John Burnet (1859—1939), were originally intended to be part of a larger development at the north side of the Museum. The design of these galleries and north entrance are predominantly marked by imperialistic features and draw on Roman rather than Greek characteristics. Imperial features include the royal coat of arms, above the entrance to the gallery, and sculptures of crowns, lions' heads and coats of arms of Edward VII on the stonework above the north entrance.

The north entrance was never originally intended to be a public entrance. Instead this entrance and gallery were meant to face a long avenue which would be part of a victory parade route. The saluting gallery, a reminder of this grand scheme, can be seen above the north entrance.

2.3 The British Museum Reading Room

The British Museum Reading Room, situated in the centre of the Great Court of the British Museum, used to be the main reading room of the British Library. In 1997, this function moved to the new British Library building at St Pancras, London, There were forty kilometers of shelving in the stacks prior to the library's relocation to the new site, but the Reading Room remains in its original form at the British Museum.

In the early 1850s the museum library was in need of a larger reading room and the then-Keeper of Printed Books, Antonio Panizzi came up with the thought of a round room in the central courtyard. The building was designed by Sydney Smirke and was constructed between 1854 and 1857. The building used cast iron, concrete, glass and the latest technology in ventilation and heating. The dome, inspired by the Pantheon in Rome, has a diameter of 42. 6 meters but is not technically free standing: constructed in segments on cast iron, the ceiling is suspended and made out of papier-mâché. Book stacks built around the reading room were made of iron to take the huge weight and add fire protection.

3. The Restoration of the Great Court

In the original Robert Smirke's design the courtyard was meant to be a garden. However, in 1852 the Reading Room and a number of bookstacks were built in the courtyard to house the library department of the Museum and the space was lost.

In 1997, the Museum's library department was relocated to the new British Library building in St Pancras and there was an opportunity to re-open the space to public. An architectural competition was launched to re-design the courtyard space. The competition had three aims: Revealing hidden spaces, Revising old spaces and Creating new spaces. There were over 130 entries and it was eventually won by Lord Foster.

Revealing hidden spaces: the courtyard. The courtyard had been a lost space since 1857. The re-design of the Great Court meant that this hidden space could be seen again. The design of the Great Court was loosely based on Foster's concept for the roof of the Reichstag in Berlin, Germany. A key aspect of the design was that with every step in the Great Court the vista changed and allowed the visitor a new view on their surroundings.

Work on the Great Court's magnificent glass and steel roof began in September 1999. The canopy was designed and installed by computer. It was constructed out of 3,312 panes of glass, no two of which are the same. At two acres, the Great Court increased public space in the Museum by forty per cent, allowing visitors to move freely around the main floor for the first time in 150 years.

Revising old spaces: the Ford Centre. Previously a storage room for Egyptian sculpture, the Ford Centre for Young Visitors provides dedicated facilities and a range of tailor-made educational programs for the hundreds of thousands of young people who visit every year. At weekends and during school holidays these areas are used for family and community events.

Creating new spaces: galleries, education facilities, visitor facilities and forecourt Gallery space. The design of the Great Court provided two new gallery spaces. The Sainsbury Gallerieshouse a display of objects from the Museum's Africa collection. The Wellcome Trust Gallery is home to a series of long term, cross-cultural, thematic exhibitions, currently based around Living and Dying. In addition to this a new space for temporary exhibitions, Room 35, was also built.

Notes:

1. British Museum：英国国家博物馆，位于英国伦敦新牛津大街北面的大罗素广场，成立于1753年，1759年1月15日起正式对公众开放，是世界上历史最悠久、规模最宏伟的综合性博物馆，也是世界上规模最大、最著名的博物馆之一。1823年，乔治四世将其父亲的图书馆作为礼物捐赠给国家，促使建造了由罗伯特·斯默克爵士(Sir Robert Smirke，1780—1867年)设计的四方院。1857年，大楼和圆形阅览室(Reading Room)建成。女王伊丽莎白二世大中庭建在了原图书馆的位置，是博物馆中1997年来公开的扩建项目。它占地两英亩，是欧洲最大的有顶广场。
2. Sir Hans Sloane：斯隆，是一名内科医生、博物学家和收藏家，一生中共收藏71000多件物品，他希望自己在去世后它们还可以完好地保存。他将所有的收藏遗赠给了国王乔治二世，回报是给他的继承人20000英镑。国家接受了他的赠品。1753年6月7日，国会法案批准建立英国国家博物馆。
3. Assyrian：亚述帝国(公元前935年—公元前612年)，是兴起于美索不达米亚，即两河流域，今伊拉克境内幼发拉底河和底格里斯河一带的国家。公元前8世纪末，亚述逐步强大，先后征服了小亚细亚东部、叙利亚、腓尼基、巴勒斯坦、巴比伦尼亚和埃及等地。国都定于尼尼微，今伊拉克摩苏尔附近。亚述人在两河流域古代历史上频繁活动时间前后约有2000年。后来亚述人失去了霸主地位，不再有独立的国家。
4. Greek Revival style：希腊复兴风格。18世纪中叶，欧洲人对古希腊建筑的知识逐渐丰富，19世纪初，在英国兴起了希腊复兴建筑。希腊复兴的主要特点是使用希腊式的多立克和爱奥尼柱，并且追求体形的单纯。大英博物馆正立面便采用单层的爱奥尼柱廊。
5. Asia Minor：又名安纳托利亚(Anatolia)，或西亚美尼亚，是亚洲西南部的一个半岛。半岛北临黑海，西临爱琴海，南濒地中海，东接亚美尼亚高原，主要由安纳托利亚高原和土耳其西部低矮山地组成，大部分属土耳其领土。公元前1200年，著名的特洛伊战争在土耳其的爱琴海沿岸发生，其后，强大的波斯大流士大帝、马其顿亚历山大大帝、古罗马帝国、拜占廷帝国、塞尔柱帝国、奥斯曼帝国等先后统治这块大地。

The Victoria and Albert Museum[1]

The Victoria and Albert Museum, named after Queen Victoria and Prince Albert, often abbreviated as the V&A, located in the Brompton district of the Royal Borough of Kensington and Chelsea, London, is the world's largest museum of decorative arts and design, housing a permanent collection of over 4.5 million objects. The area has become known as "Albertopolis" because of its association with the Albert Memorial and the major cultural institutions with which he was associated. (See Fig. 10.5)

Fig. 10.5 The Victoria and Albert Museum

1. The History of V& A

The V&A has its origins in the Great Exhibition[2] of 1851 as the Museum of Manufactures initially, at this stage the collections covered both applied art and science. The official opening by Queen Victoria was on 22 June 1857. In the following year, late night openings were introduced, made possible by the use of gas lighting. This was to enable "to ascertain practically what hours are most convenient to the working classes"—this was linked to the use of the collections of both applied art and science as educational resources to help boost productive industry. In these early years the practical use of the collection was very much emphasized as opposed to that of "High Art" at the National Gallery and scholarship at the British Museum.

George Wallis (1811—1891), the first Keeper of Fine Art Collection, passionately promoted the idea of wide art education through the museum collections. This led to the transfer to the museum of the School of Design that had been founded in 1837 at Somerset House to as the Art School or Art Training School, later to become the Royal College of Art which finally achieved full independence in 1949. From the 1860s to the 1880s the scientific collections had been moved from the main museum site to various improvised galleries to the west of Exhibition Road.

The laying of the foundation stone of the Aston Webb building to the left of the main entrance on 17 May 1899 was the last official public appearance by Queen Victoria. It was during this ceremony that the change of name from the South Kensington Museum to the Victoria and Albert Museum was made public.

In 2001, "Future Plan" was launched, which involves redesigning all the galleries and public facilities in the museum that have yet to be remodeled. This is to ensure that the exhibits are better displayed, more information is available and the museum meets modern expectations for museum facilities.

2. The Description of V&A

2.1 The Layout of V&A

The Victorian parts of the building have a complex history, with piecemeal additions by different architects. In 1890 the government launched a competition to design new buildings for the museum, this would give the museum a new imposing front entrance. The main facade, built from red brick and Portland stone, stretches 220 meters along Cromwell Gardens and was designed by Aston Webb. Stylistically it is a strange hybrid, although much of the detail belongs to the Renaissance there are medieval influences at work. The main entrance consisting of a series of shallow arches supported by slender columns and niches with twin doors separated by pier is Romanesque in form but Classical in detail. Likewise the tower above the main entrance has an open work crown surmounted by a statue of fame, a feature of late Gothic architecture and a feature common in Scotland, but the detail is Classical. The main windows to the galleries are also mullioned and transomed, again a Gothic feature, the top row of windows are interspersed with statues of many of the British artists whose work is displayed in the museum. (See Fig. 10.6)

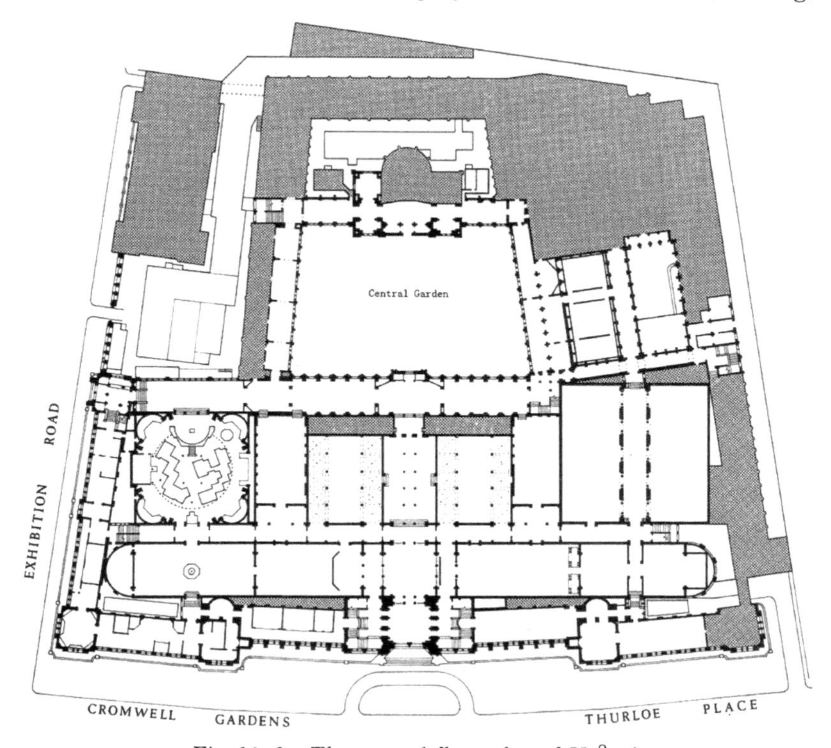

Fig. 10.6 The ground floor plan of V & A

Prince Albert appears within the main arch above the twin entrances, Queen Victoria above the frame around the arches and entrance, sculpted by Alfred Drury. These facades surround four levels of galleries. Other areas designed by Webb include the Entrance Hall and Rotunda, the East and West Halls, the areas occupied by the shop and Asian Galleries as well as the Costume Gallery. The interior makes much use of marble in the entrance hall and flanking staircases, although the galleries as originally designed were white with restrained classical detail and mouldings, very much in contrast to the elaborate decoration of the Victorian galleries, although much of this decoration was removed in the early 20th century.

The lower ground-floor galleries in the south-west part of the museum were redesigned, opening in 1978 to form the new galleries covering Continental art 1600—1800, which is in the style of late Renaissance, Baroque through Rococo and neo-Classical. As part of the 2006 renovation the mosaic floors in the sculpture gallery were restored—most of the Victorian floors were covered in linoleum after the Second World War.

2.2 The Interiors of the Three Refreshment Rooms

The interiors of the three refreshment rooms were assigned to different designers. The Green Dining Room 1866—1868 was the work of Philip Webb and William Morris, and displays Elizabethan influences. The lower part of the walls are paneled in wood with a band of paintings depicting fruit and the occasional figure, with molded plaster foliage on the main part of the wall and a plaster frieze around the decorated ceiling and stained-glass windows by Edward Burne-Jones.

The Centre Refreshment Room 1865—1877 was designed in a Renaissance style by James Gamble, the walls and even the Ionic columns are covered in decorative and molded ceramic tile, the ceiling consists of elaborate designs on enameled metal sheets and matching stained-glass windows, the marble fireplace was designed and sculpted by Alfred Stevens and was removed from Dorchester House prior to that building's demolition in 1929.

The Grill Room 1876—1881 was designed by Sir Edward Poynter, the lower part of the walls consist of blue and white tiles with various figures and foliage enclosed by wood paneling, above there are large tiled scenes with figures depicting the four seasons and the twelve months, these were painted by ladies from the Art School then based in the museum, the windows are also stained glass, there is an elaborate cast-iron grill still in place.

2.3 The Central Garden

The central garden was redesigned by Kim Wilkie and opened on 5 July 2005. The design is a subtle blend of the traditional and modern, the layout is formal; there is an elliptical water feature lined in stone with steps around the edge which may be drained to use the area for receptions, gatherings or exhibition purposes. This is in front of the bronze doors leading to the refreshment rooms, a central path flanked by lawns leads to the sculpture gallery; the north, east and west sides have herbaceous borders along the museum walls with paths in front which continues along the south facade; in the two corners by the north facade there is planted an American Sweet gum tree; the southern, eastern and western edges of the lawns have glass planters which contain orange and lemon trees in summer, these are replaced by bay trees in winter.

At night both the planters and water feature may be illuminated, and the surrounding facades lit to reveal details normally in shadow, especially noticeable are the mosaics in the loggia of the north facade. The garden is also used for temporary exhibits of sculpture, it has also played host to the museum's annual contemporary design showcase.

3. The Architecture Exhibition in V&A

In 2004, the V&A alongside Royal Institute of British Architects opened the first permanent architecture display in the UK covering the history of architecture with models, photographs, elements from buildings and original drawings. With the opening of the new gallery, the RIBA Drawings and Archives Collection has been transferred to the museum, joining the already extensive collection held by the V&A to form the world's most comprehensive architectural resource.

Not only are all the major British architects of the last four hundred years represented, but many European, especially Italian and American architects' drawings are held in the collection. As well as period rooms, the collection includes parts of buildings, for example the two top stories of the facade of Sir Paul Pindar's house dated 1600 A. D. from Bishopsgate with elaborately carved wood work and leaded windows, a rare survivor of the Great Fire of London, there is a brick portal from a London house of the English Restoration period and a fireplace from the gallery of Northumberland house. European examples include a dormer window dated 1523—1535 from the chateau of Montal. There are several examples from Italian Renaissance buildings including, portals, fireplaces, balconies and a stone buffet that used to have a built in fountain.

The main architecture gallery has a series of pillars from various buildings and different periods, for example a column from the Alhambra. Examples covering Asia are in those galleries concerned with those countries, as well as models and photographs in the main architecture gallery.

Notes:

1. The Victoria and Albert Museum：维多利亚和阿尔伯特博物馆，简称 V&A，是位于英国伦敦的一间装置及应用艺术的博物馆。该博物馆是为 1851 年在伦敦召开的第一届万国博览会而建，是规模仅次于大英博物馆的第二大国立博物馆。博物馆的中心是面向花园的一栋红砖结构建筑物，是按照文艺复兴风格设计的，用现代工业生产的材料——红砖和赤陶土建造。

2. Great Exhibition：万国工业博览会，是1851年英国维多利亚时期的一次真正意义上的第一次世界性的博览会，时间从1851年5月1日至1851年10月15日，普遍被认为是维多利亚中期的象征，并确立了大英帝国世界工厂的主导地位。这次博览会的组织者包括阿尔伯特亲王。展地为水晶宫，是维多利亚时代最重要的里程碑。

Museum Island in Berlin[1]

Museum Island is the name of the northern half of an island in the Spree river in the central Mitte district of Berlin, Germany, the site of the old city of Cölln. It is so called for the complex of five internationally significant museums, all part of the Berlin State Museums, that occupy the island's northern part: The Old Museum, the New Museum, the Old National Gallery, the Bode Museum and the Pergamon Museum. The island, originally a residential area, was dedicated to "art and science" by King Frederick William IV of Prussia in 1841. Further extended under succeeding Prussian kings, the museum's collections of art and archaeology were turned into a public foundation after 1918. (See Fig. 10. 7)

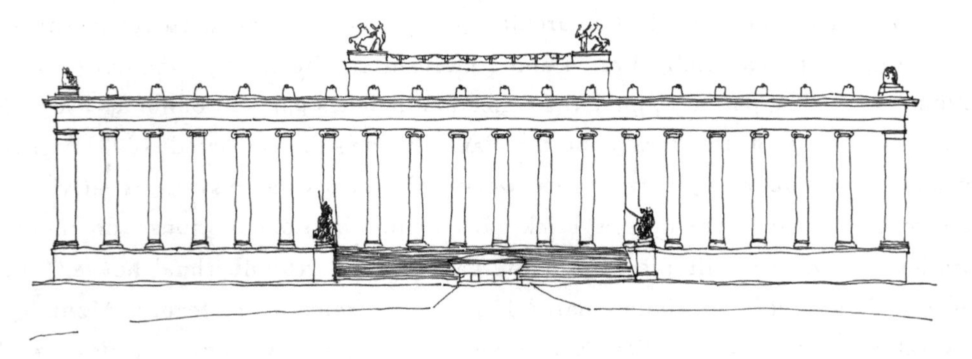

Fig. 10. 7 Berlin Museum

1. History of Museum Island

A first exhibition hall was erected in 1797 at the suggestion of the archaeologist Aloys Hirt. In 1822, Schinkel designed the plans for the Altes(old) Museum to house the royal Antikensammlung, the arrangement of the collection was overseen by Wilhelm von Humboldt[2]. The island, originally a residential area, was dedicated to "art and science" by King Frederick William IV of Prussia in 1841. Further extended under succeeding Prussian kings, the museum's collections of art and archeology were turned into a public foundation after 1918. They are today maintained by the Berlin State Museums branch of the Prussian Cultural Heritage Foundation.

The Prussian collections became separated during the Cold War during the division of the city, but were reunited after German reunification, except for some art and artefacts removed after World War II by Allied troops. These include the Priam's Treasure, also called the gold of Troy[3], excavated by Heinrich Schliemann[4] in 1873, then smuggled out of Turkey to Berlin and today kept at the Pushkin Museum in Moscow.

Before World War II, these museums were connected by bridge passages above ground; they were destroyed due to the effects of the war. There have never been plans to rebuild them; instead, the central courts of individual museums will be lowered, which has already been done in the Bode Museum and in the New Museum. They will be connected by subterranean galleries. In a way, this archaeological promenade can be regarded as the sixth museum in the Island, because it is devised not only as a connecting corridor but also as a strung-out exhibition room for interdisciplinary presentations.

The Archaeological Promenade may be characterized as a cross-total of the collections that are shown separately in the individual museums of the Island. Once the Museum Island Master Plan is completed, the so-called Archaeological Promenade will connect four of the five museums in the Museum Island. The Promenade will begin at the Old Museum in the south, lead through the New Museum and the Pergamon Museum and end at the Bode Museum, located at the northern tip of the Island. (See Fig. 10. 8)

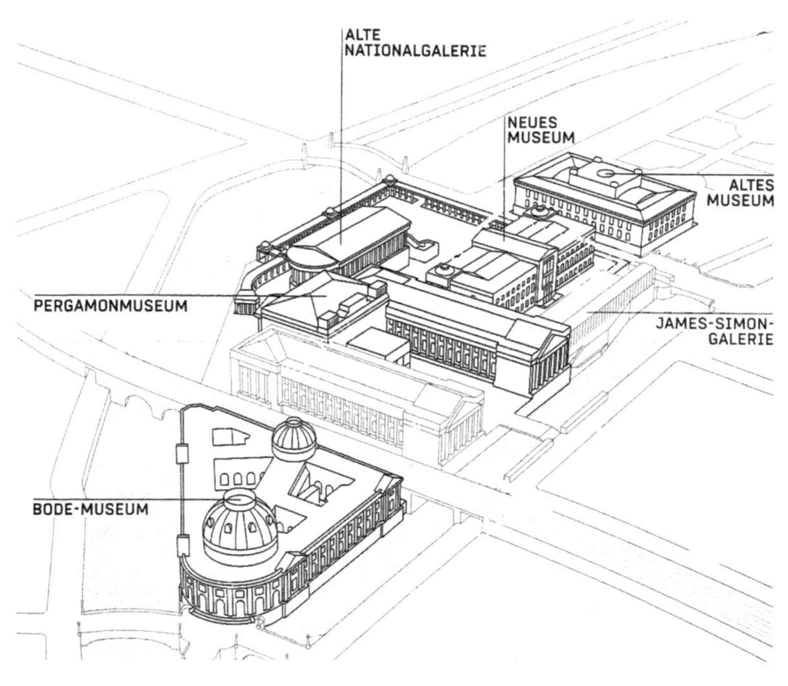

Fig. 10. 8 Plan of Museum island

2. The Description of the Altes Museum

The Altes Museum was built between 1823 and 1830 by the architect Karl Friedrich Schinkel in the neoclassical style[5] to house the Prussian royal family's art collection. In the early nineteenth century, Germany's bourgeoisie had become increasingly self-aware and self-confident. This growing class began to embrace new ideas regarding the relationship between itself and art, and the concepts that art should be open to the public and that citizens should be able to have access to a comprehensive cultural education began to pervade society. King Friedrich Wilhelm III of Prussia was a strong proponent of this humboldtian ideal for education and charged Karl Friedrich Schinkel with planning a public museum for the royal art collection.

The historic, protected building counts among the most distinguished in neoclassicism and is a high point of Schinkel's career. Schinkel's plans for the Museum were also influenced by drafts of the crown prince, later King Friedrich Wilhelm IV, who desired a building that was heavily influenced by antiquity. The crown prince even sent Schinkel a pencil sketch of a large hall adorned with a classical portico.

Schinkel had developed plans for the Altes Museum as early as 1822, but construction did not begin until 1825. Construction was completed in 1828. Schinkel was also responsible for the renovation of the Berliner Dom in the neo-classical style, which was originally a baroque cathedral, and he exercised considerable influence on Peter Joseph Lenné's renovation of the Lustgarten, which coincided with the construction of the museum, resulting in a harmonized and integrated ensemble.

In 1841, King Friedrich Wilhelm IV announced in a royal decree, that the entire northern part of the Spree Island be transformed into a sanctuary for art and science. In 1845, theRoyal Museum was renamed the Altes Museum, the name it holds to this day.

Schinkel's plans incorporated the Museum into an ensemble of buildings which surround the Berliner pleasure garden. TheBerlin City Palace in the south was a symbol of worldly power, it was a royal and imperial palace in the centre of Berlin, the historical capital of Prussia, and subsequently Germany. It was also the winter residence of the Kings of Prussia and the German Emperors. The palace was originally built in the 15th century and bore features of the Baroque style, its shape, finalized by the middle 18th century; The old Arsenal in the west represented military might, it is the oldest structure and was built by the Brandenburg Elector Frederick III between 1695 and 1730 in the baroque style, to be used as an artillery arsenal for the display of cannons from Brandenburg and Prussia; The Berliner Dom in the east was the embodiment of divine authority, it is the parish church of the congregation which was also in baroque style originally. The museum to the north of the garden, which was to provide for the education of the people, stood as a symbol for science and art—and not least for their torchbearer: the self-aware bourgeoisie.

3. The Description of Neues Museum

The Neues museum was built between 1843 and 1855 according to plans by Friedrich August Stüler, a student of Karl Friedrich Schinkel. The museum was heavily damaged during the bombing of Berlin in 1939 and was overseen by the English architect David Chipperfield. The museum officially reopened in October 2009 and received a 2010 RIBA European Award and the 2011 European Union Prize for Contemporary Architecture.

The New Museum was the second museum to be built on Museum Island and was intended as an extension to house collections which could not be accommodated in the Old Museum. It is an important monument in the history of construction and technology. With its various iron constructions, it is the first monumental building of Prussia to consistently apply new techniques made possible by industrialization. As a further innovation, a steam engine was used for the first time in construction in Berlin. It was used to ram pilings into the building ground. The soft, spongy soil around the River Spree means that buildings in the central area of Berlin require deep foundations. Therefore, a pile structure was necessary under the whole building, consisting of 2344 wooden foundation piles between 6. 9 and 18. 2 meters long. To ram the piles in, a 5-horsepower steam engine was used, whose power could be increased if necessary to 10 hp . It drove the pumps that drained the building site, the elevators, and the mortar mixing machines.

As originally built, the New Museum was nearly rectangular, with the 105 meters long axis of the building oriented north to south, parallel to the street to the west, across the River Spree, and a width of 40 meters . The building is nearly perpendicular to the Old Museum, with the Bodestraße between them. The bridge connecting the two museums, which was destroyed during World War II was 6. 9 meters wide, 24. 5 meters long, and supported by three arches. The main stairway was located in the center of the building, which was the highest section 31 meters.

The three main wings surround two interior courtyards, the Greek courtyard and the Egyptian courtyard. The northern Egyptian courtyard was covered with a glass ceiling from the beginning, but the southern Greek courtyard was first covered with a glass ceiling between 1919 and 1923.

When the New Museum opened, there were the Egyptian, patriotic and ethnographic collections in the ground floor, while the collections of the plaster casts of Greek and Roman sculptures from antiquity and Byzantine, Romanesque, Gothic, Renaissance and Classic art works occupied the first floor. A collection of architectural models, furniture, clay, pottery and glass containers, and church articles shared the second floor, along with smaller works of art from the Middle Ages and modern times.

In the Greek courtyard, the courtyard covered with a glass roof, and a new floor at the same height as the ground floor was added. Thus several rooms and cabinets for the display of the Amarna collection were created. In the adjacent areas of the ground floor, suspended ceilings were added to produce modern, neutral display rooms by covering the original decorations.

In the bombardments on 23 November 1943, the central stairway and its frescos were burned, along with other great treasures of human history. In February 1945, Allied bombs destroyed the northwest wing as well as the connection to the Old Museum and damaged the southwest wing as well as the south-east facade.

In the post-war period, the ruin of the New Museum was left decaying for a long time. Reconstruction work was started in 1986 by the East German government, but it was halted after the fall of the Berlin Wall and German reunification. In the process historical parts of the building were lost, the last remnants of the Egyptian courtyard were eliminated.

In 1997, planning for the reconstruction project was resumed and English architect David Chipperfield was officially appointed for the project. Sections and fragments of the building were taken out and put in storage. A new reception building for visitors to Museum island, the "Cube", was also planned to be completed . The northwest wing and the south-east facade, which were completely destroyed in the war, have been reconstructed according to Chipperfield's plan, in a manner close to their original layout in the museum building. At the reopening ceremony, the Chancellor of Germany Angela Merkel described Chipperfield's work as "impressive and extraordinary" and the museum as "one of the most important museum buildings in European cultural history". However, Chipperfield's construction design has been a subject of debates by those who preferred a more traditional reconstruction of Friedrich August Stüler's original 19th-century design.

Notes:

1. Museum Island in Berlin：柏林博物馆岛，是一组独特的建筑文化遗产。岛上的五座博物馆形态各异，却又和谐统一。岛的最南端，紧邻宫殿大桥和柏林大教堂的是老博物馆，最北端是新博物馆和老国家艺术画廊。面向西侧的是佩加蒙博，最外侧的是博德博物馆。这组经过百年历史建造完善起来的博物馆群体建筑在二次大战中70%以上被毁，二次大战之后，东西柏林分裂，博物馆中的藏品也被分散在东西柏林多处。博物馆岛位于前东柏林地段，当时由于经济困难，岛上各大博物馆年久失修，破烂不堪。两德统一后德国政府投入大量资金对岛上的所有建筑进行重新维修。博物馆岛上的建筑群被联合国教科文组织列入世界文化遗产名录 。
2. Wilhelm von Humboldt：威廉·冯·洪堡，是柏林洪堡大学的创始者，也是著名的教育改革者、语言学者及外交官。1819年后，他辞去公职，专门从事学术研究。威廉·冯·洪堡是德国著名博物学家、自然地理学家，19世纪科学界中最杰出的人物亚历山大·冯·洪堡的哥哥。

3. Troy：特洛伊，为公元前16世纪前后由古希腊人所建城市，位于小亚细亚半岛，今土耳其的希萨利克。公元前12世纪初，迈锡尼联合希腊各城邦组成联军，渡海远征特洛伊，战争延续十年之久，史称"特洛伊战争"。
4. Heinrich Schliemann：海因里希·施里曼，德国传奇式的考古学家。他自幼陶醉于荷马诗篇，决心有朝一日探挖古迹，寻找出化为废墟的特洛伊故城。1870年，施里曼在达达尼尔海峡附近土耳其境内的希沙里克山丘开始考古发掘，发现了多层城墙遗址，并终于在一座地下古建筑物的围墙附近掘出了大量珍贵的金银制作器皿，仅一顶金冕就由16353个金片和金箔组成。施里曼兴奋地宣布，他发现了特洛伊国王普里阿姆的宝藏。
5. neoclassical style：新古典主义，最早出现于17世纪中叶欧洲的建筑装饰设计界以及与之密切相关的家具设计界。新古典主义，一方面起源于对巴洛克(Baroque)和洛可可(Rococo)艺术的反对，另一方面则是希望以重振古希腊、古罗马的艺术为信念。新古典主义是在古典美学规范下，采用现代先进的工艺技术和新材质，重新诠释传统文化的精神内涵，具有端庄、雅致、明显的时代特征。采用这种建筑风格的主要是法院、银行、交易所、博物馆、剧院等公共建筑和一些纪念性建筑，而对一般的住宅、教堂、学校影响不大。法国在18世纪末、19世纪初是欧洲新古典建筑活动的中心。英国以复兴希腊建筑形式为主，如伦敦的不列颠博物馆。德国柏林的勃兰登堡门、柏林宫廷剧院都是复兴希腊建筑形式的。

Insights: Mossaic

A mosaic is a piece of art or image made from the assemblage of small pieces of colored glass, stone, or other materials. It is often used in decorative art or as interior decoration. Most mosaics are made of small, flat, roughly square, pieces of stone or glass of different colors. The floor mosaics are made of small rounded pieces of stone called "pebble mosaics".

Mosaics have a long history, starting in Mesopotamia in the 3rd millennium B. C. Mosaics with patterns and pictures became widespread in classical times, both in Ancient Greece and Ancient Rome. Early Christian basilicas from the 4th century onwards were decorated with wall and ceiling mosaics. Mosaic art flourished in the Byzantine Empire from the 6th to the 15th centuries; Mosaic fell out of fashion in the Renaissance.

Mosaic was also widely used on religious buildings and palaces in early Islamic art, including Islam's first great religious building, the Dome of the Rock in Jerusalem, and the Umayyad Mosque in Damascus. Mosaic went out of fashion in the Islamic world after the 8th century.

Glossary

词汇表

A

abacus [ˈæbəkəs] *n.* 算盘；柱头的顶板

abandon [əˈbændən] *vt.* 放弃，抛弃；使屈从；终止 *n.* 放任，放纵；完全屈从

abbreviate [əˈbriːvieit] *vt.* 缩略；缩简；使用缩写词 *n.* 缩短，缩写

abdication [ˈæbdiˈkeiʃn] *n.* 退位，辞职；放弃

abode [əˈbəʊd] *n.* 住所；公寓；逗留 *vt.* 预兆，预示 *v.* 容忍；等候；逗留；停留

aboriginal [ˈæbəˈridʒənl] *adj.* 土著人的；原产的，原始的

abrasive [əˈbreisiv] *adj.* 有磨蚀作用的；粗糙的 *n.* 研磨料；琢料；金刚砂

abundance [əˈbʌndəns] *n.* 丰度；丰富，充裕；大量，极多；盈余

acanthus [əˈkænθəs] *n.* 叶形的装饰板；柱头上的莨苕叶装饰

accentuate [əkˈsentʃueit] *vt.* 重读；强调，着重指出

accession [ækˈseʃn] *n.* 增加；就职；增加物；正式加入条约或同盟

accountancy [əˈkaʊntənsi] *n.* 会计学；会计工作

acoustics [əˈkuːstiks] *n.* 声学；音响效果；听觉的；原声的

acropolis [əˈkrɔpəlis] *n.* (古希腊城市的)卫城

acute [əˈkjuːt] *adj.* 尖的，锐的；敏锐的，敏感的；严重的，剧烈的

adjacent [əˈdʒeisnt] *adj.* 相邻；邻近的，毗邻的

adjoin [əˈdʒɔin] *vt.* 邻近，毗连；附加，使结合

aeration [eəreiʃn] *n.* 通风；增氧；换气

affix [əˈfiks] *vt.* 粘贴；附加 *n.* 附加物；附件；词缀

agora [ˈægɔrə] *n.* 集会，市场

aisle [ail] *n.* 过道，通道；侧廊

alabaster [ˈæləbɑːstə(r)] *n.* 雪花石膏；条纹大理石，蜡石 *adj.* 雪花石膏的；白色的，光滑的

alcove [ˈælkəʊv] *n.* 凹室，壁龛；花园凉亭

algae [ˈældʒiː] *n.* 水藻；藻类

alien [ˈeiliən] *adj.* 外国的；相异的 *n.* 外星人；外国人 *vt.* 让渡，转让；疏远

align [əˈlain] *vt.* 使成一线，使结盟；排整齐 *vi.* 排列；成一条线

alleged [əˈledʒd] *adj.* 声称的，所谓的；可疑的，靠不住的 *v.* 宣称，断言

allegorical [ˈæləˈgɔrikl] *adj.* 寓言的，讽喻的
alleviate [əˈliːvieit] *vt.* 减轻，缓和
altar [ˈɔːltə(r)] *n.* 祭坛，圣坛；圣餐台
alteration [ˈɔːltəˈreiʃn] *n.* 变更；变化，改变
altitude [ˈæltitjuːd] *n.* 高度，海拔高度；地平纬度
amalgamation [əˈmælgəˈmeiʃn] *n.* 合并，混合；混一
ambassador [æmˈbæsədə(r)] *n.* 大使，使节；代表；特使
ambitious [æmˈbiʃəs] *adj.* 有雄心的；有野心的；费力的
ambulatory [ˈæmbjələtəri] *adj.* 走动的，流动的，非固定的回廊
ammunition [ˈæmjuˈniʃn] *n.* 子弹；弹药，军火
anatomy [əˈnætəmi] *n.* 解剖，分解，分析；剖析；解剖结构；骨骼
analogous [əˈnæləgəs] *adj.* 相似的，可比拟的
animation [ˈæniˈmeiʃn] *n.* 生气，活泼；动画片制作，动画片
annex [əˈneks] *vt.* 附加，追加；兼并；获得，得到 *n.* 附录；附属物(品)，附属建筑
antechamber [ˈæntitʃeimbə(r)] *n.* 前厅
anthropomorphic [ˈænθrəpəˈmɔːfik] *adj.* 拟人的，赋予人性的
antiquated [ˈæntikweitid] *adj.* 过时的，陈旧的；古色古香的；有古风的 *v.* 使古旧，废弃
antiquity [ænˈtikwəti] *n.* 古人；古老，古代；古迹，古物；古代的风俗习惯
apex [ˈeipeks] *n.* 顶；顶峰；尖端
apogee [ˈæpədʒiː] *n.* 远地点；最高点；最远点.
apostle [əˈpɔsl] *n.* 基督教的使徒；早期基督教的传教士
apricot [ˈeiprikɔt] *n.* 杏；杏树；杏仁；杏黄色 *adj.* 杏黄色的
appropriate [əˈprəʊpriət] *adj.* 适当的；合适的 *v.* 盗用；侵吞
approximately [əˈprɔksimətli] *adv.* 近似地，大约
approximation [əˈprɔksiˈmeiʃn] *n.* 接近；近似值；近似额；粗略估计
apse [æps] *n.* 教堂半圆形的后殿
aquatic [əˈkwætik] *adj.* 水生的；水产的 *n.* 水生动植物；水上运动
aqueduct [ˈækwidʌkt] *n.* 导水管；沟渠；引水渠；高架渠
aquifer [ˈækwifə(r)] *n.* 地下蓄水层，砂石含水层
arabesque [ˈærəˈbesk] *n.* 阿拉伯图案；阿拉伯式花饰；错综图饰 *adj.* 阿拉伯式图案的
arbor [ˈɑːbə] *n.* 凉亭；树木；心轴
arcade [ɑːˈkeid] *n.* 商场；拱廊，连拱廊；有拱廊的街道
arch [ɑːtʃ] *n.* 拱门；弓形；拱形物 *vt.* (使)弯成拱形；用拱连接 *vi.* 拱起
archaic [ɑːˈkeiik] *adj.* 古代的；古体的；过时的，陈旧的；古色古香的
archaeological [ˈɑːkiəˈlɔdʒikl] *adj.* 考古学的，考古学上的
archbishop [ˈɑːtʃˈbiʃəp] *n.* 大主教；主教长
archer [ˈɑːtʃə(r)] *n.* 弓箭手，射箭运动员
architrave [ˈɑːkitreiv] *n.* 额枋；柱顶过梁；框缘；门窗的线角

archway [ˈɑːtʃwei] *n.* 拱门，拱道
arena [əˈriːnə] *n.* 竞技场；表演场地，舞台；古罗马圆形剧场中央的竞技场
aristocrat [ˈæristəkræt] *n.* 贵族；贵族政治论者；有贵族派头的人
armourer [ˈɑːmərə(r)] *n.* 军械官；枪械或盔甲制造者
aroma [əˈrəʊmə] *n.* 芳香，香味；气派，风格
array [əˈrei] *n.* 数组；队列，阵列　*vt.* 排列；部署兵力
artesian [aːˈtiːzjən] *adj.* 喷水井的，自流井的
articulate [ɑːˈtikjuleit] *adj.* 善于表达的；有关节的　*vt.* 关节连接　*vi.* 清楚地讲
artifact [ˈɑːtəˈfækt] *n.* 人工制品，手工艺品，加工品；石器
ascent [əˈsent] *n.* 上升；上坡；登高
assemblage [əˈsemblidʒ] *n.* 装配；聚集；聚集在一起的一堆东西；组合成的艺术品
assent [əˈsent] *vi.* 赞同；赞成　*n.* 同意，赞同
athletic [æθˈletik] *adj.* 运动的；运动员的；体格健壮的；行动敏捷的
atrium [ˈeitriəm] *n.* 心房；中庭，天井
attic [ˈætik] *n.* 阁楼；顶楼　*adj.* 文雅的，古雅的
audience [ˈɔːdiəns] *n.* 观众；听众；读者
auspicious [ɔːˈspiʃəs] *adj.* 吉利的；有前途的；有希望的；有利的
austere [ɔˈstiə(r)] *adj.* 朴素的，简朴的；严峻的，严厉的；苦行的；一丝不苟的
avant-corps [əˌvɑːntˈkɔː] *n.* 前亭
avert [əˈvəːt] *vt.* 转移；防止，避免
awning [ˈɔːniŋ] *n.* 雨篷；遮篷；布做的凉篷
axis [ˈæksis] *n.* 轴，轴线；轴心；轴心国

B

backdrop [ˈbækdrɔp] *n.* 背景；背景幕布
bailey [ˈbeili] *n.* 城堡的外墙；堡场
balcony [ˈbælkəni] *n.* 阳台；包厢；剧场的楼厅，楼座
baldachin [ˈbɔːldəkin] *n.* 祭坛华盖；鲜艳的锦缎
ballet [ˈbælei] *n.* 芭蕾舞；芭蕾舞团
banisters [ˈbænistə(r)] *n.* 楼梯的栏杆，扶手
banquet [ˈbæŋkwit] *n.* 筵席；宴会，盛宴；宴请，款待　*vt.* 宴请，设宴招待　*vi.* 参加宴会
barbican [ˈbɑːbikən] *n.* 外堡，碉楼，瓮城
barley-sugar [ˈbɑːliːʃʊgər] *n.* 麦芽糖
baroque [bəˈrəʊk] *adj.* 巴罗克式的；巴罗克风格的　*n.* 巴罗克风格；新奇作品
barrack [ˈbærək] *n.* 营房，兵营；棚屋　*vt.* 向……提供营房；使(如军人)住在营房内
basilica [bəˈzilikə] *n.* 长方形廊柱大厅；古罗马长方形大会堂；长方形基督教堂
bas-relief [ˈbæs riˈliːf] *n.* 浅浮雕
battlement [ˈbætlmənt] *n.* 城垛
bay [bei] *n.* 湾，海湾；月桂树；凸窗；开间

bazaar [bəˈzɑː(r)] *n.* 街市，市场；百货商店

beacon [ˈbiːkən] *n.* 烽火；灯塔；警标，界标 *vt.* 指引；为……设置信标 *vi.* 像灯塔般照耀

bedrock [ˈbedrɔk] *n.* 基岩；牢固基础；基本事实，基本原则；最低点

beeswax [ˈbiːzwæks] *n.* 蜂蜡 *vt.* 给……上蜡

belfry [ˈbelfri] *n.* 钟塔；钟室

benefactor [ˈbenifæktə(r)] *n.* 恩人；捐助者；施主

benign [biˈnain] *adj.* 温和的，仁慈的；善良的；有利于健康的

bequeath [biˈkwiːð] *vt.* 将财物等遗赠给；将知识等传给

bequest [biˈkwest] *n.* 遗赠；遗产，遗赠物

berm [bəːm] *n.* 护道；阶地；城墙与外濠间的狭道，崖径；栈道

besiege [biˈsiːdʒ] *vt.* 包围，为敌对势力包围；使感到丧气或焦虑

bespeak [biˈspiːk] *vt.* 证明；预定；预先请求

beverage [ˈbevəridʒ] *n.* 饮料

bewilder [biˈwildə(r)] *vt.* 使迷惑；使为难；使手足无措

bishop [ˈbiʃəp] *n.* 基督教管辖大教区的主教；国际象棋的象；香甜葡萄酒

blend [blend] *vt.* 混合；把……掺在一起；（使）调和；协调 *vi.* 掺杂；结合 *n.* 混合；混合物

blessing [ˈblesiŋ] *n.* 上帝的祝福；福分 *v.* 称颂上帝；求神赐福于

booth [buːð] *n.* 售货棚，摊位；公用电话亭；选举投票站

bouleuterion [bəʊˈluːʃraiən] *n.* 古希腊的议事厅，会议室

bourgeoisie [bʊəʒwɑːˈziː] *n.* 资产阶级；中产阶级

boutique [buːˈtiːk] *n.* 精品店；时装店

bracket [ˈbrækit] *n.* 支架，悬臂，斗拱，雀替 *vt.* 为……装托架

bracket set [ˈbrækit set] 斗拱

bravura [brəˈvjʊərə] *n.* 大胆的尝试；华美乐段 *adj.* 壮勇华丽

breadth [bredθ] *n.* 宽度；宽容

breeze [briːz] *n.* 微风；轻而易举的事 *vi.* 吹微风；逃走

brocade [brəˈkeid] *n.* 凸花纹织物，锦缎 *vt.* 织成浮花织锦

Broderies *n.* （法语）装饰图案

Buddhism [ˈbʊdizəm] *n.* 佛教；释；佛门

bucket [ˈbʌkit] *n.* 水桶；一桶；大量 *v.* 用桶装，用桶运

buckle [ˈbʌkl] *vt.* 用搭扣扣紧；变形，弯曲 *n.* 搭扣，扣环

bulge [bʌldʒ] *n.* 膨胀；凸出部分；暴涨，突增 *v.* 凸出；膨胀；急增

bungalow [ˈbʌŋgələʊ] *n.* （英）平房；（美）单层小屋，多于一层的小屋

bureaucracy [bjʊəˈrɔkrəsi] *n.* 官僚主义；官僚机构；官僚政治

bust [bʌst] *vt.* 打破，打碎 *n.* 胸围；半身雕塑像 *vt.* 突击搜查

buttress [ˈbʌtrəs] *n.* 扶壁；支撑物 *vt.* 支持，鼓励；用扶壁支撑，加固

C

caisson ['keisən] *n*. 藻井；沉箱
canopy ['kænəpi] *n*. 天篷，华盖；顶篷，顶盖；苍穹；树荫
cantonment [kæn'tɔnmənt] *n*. 宿营地；兵营
capital ['kæpitl] *n*. 柱头
capitulation [kəˌpitʃʊ'leiʃn] *n*. 有条件的投降；概要；协议；一系列条款
caravanserai ['kærə'vænsərai] *n*. 商队旅馆，大旅舍
cart [kɑːt] *n*. 运货马车，手推车　*vt*. 用推车或卡车运送；抓走
casing ['keisiŋ] *n*. 壳；套；罩；框
catacomb ['kætəkəʊm] *n*. 地下墓穴，茔窟
catapult ['kætəpʌlt] *n*. 弹弓；石弩；弹射座椅　*vt*. 用弩炮发射，猛投　*vi*. 用弹射器弹射
clad [klæd] *adj*. 覆盖的；镀金属的　*v*. 穿衣；在外面包上金属
cavern ['kævən] *n*. 大山洞；凹处　*v*. 挖空；置……于山洞中
cascade [kæ'skeid] *n*. 串联；倾泻；小瀑布　*vi*. 流注；大量落下
castle ['kɑːsl] *n*. 城堡，堡垒；象棋中的车　*vt*. 把……置于城堡中
cella ['selə] *n*. 内殿，内堂
ceremony ['serəməni] *n*. 典礼，仪式；礼仪，礼节
chancel ['tʃɑːnsl] *n*. 教堂的高坛
chandelier ['ʃændə'liə(r)] *n*. 枝形吊灯
chant [tʃɑːnt] *vt*. 吟颂，咏唱　*n*. 咏唱
chapel ['tʃæpl] *n*. 小教堂，附属教堂；礼拜仪式；特殊小房间
chateau [ʃæ'təʊ] *n*. 法国封建时代的城堡；法国的别墅，庄园
cherubim ['tʃerəbim] *n*. 智天使，小天使
chisel ['tʃizl] *n*. 凿子，錾子　*vt*. 凿，雕；錾，镌
chorus ['kɔːrəs] *n*. 合唱；合唱队；副歌；合唱歌曲　*vt*. 合唱；齐声背诵
chronicle ['krɔnikl] *n*. 编年史；记录；年代记　*vt*. 记录；将(某事物)载入编年史
chryselephantine [kriˈselə'fæntiːn] *adj*. 用金和象牙做成的
chute [ʃuːt] *n*. 降落伞；斜槽，滑道　*vt*. 用斜槽或斜道运送　*vi*. 顺斜道而下
ciborium [si'bɔːriəm] *n*. 祭坛上的天盖，圣礼容器
cedar ['siːdə(r)] *n*. 雪松；香柏；香椿；西洋杉
cistern ['sistən] *n*. 蓄水池，储水箱；地下储水池
citadel ['sitədəl] *n*. 城堡，要塞；大本营，根据地，避难所
civic ['sivik] *adj*. 城市的；公民的，市民的
clerestory ['kliəstɔːri] *n*. 天窗，通风窗
clergy ['kləːdʒi] *n*. 尤指基督教堂内的 牧师，教士
clientele ['kliːən'tel] *n*. 诉讼委托人；顾客，客户；被保护者，追随者
cobble ['kɔbl] *vt*. 粗劣地制作；匆匆制作；胡乱拼凑　*n*. 鹅卵石，圆石
cohesion [kəʊ'hiːʒn] *n*. 凝聚，内聚；连着

coil [kɔil] *n.* 线圈；盘；纷乱，纠缠不清　*vt.* 盘绕；卷成一圈　*vi.* 绕成盘状
coincide [ˈkəʊinˈsaid] *vi.* 相符；与……一致；想法、意见等相同
colonial [kəˈləʊniəl] *adj.* 殖民地的；殖民地化的　*n.* 殖民地居民；殖民时代建筑
colony [ˈkɔləni] *n.* 殖民地；群体；聚居地；聚居人群
combat [ˈkɔmbæt] *n.* 格斗，搏斗，战斗　*vt.* 与……战斗；与……斗争
commemorate [kəˈmeməreit] *vt.* 纪念，庆祝；成为……的纪念
commission [kəˈmiʃn] *n.* 委员会，委员；佣金，手续费；委任　*vt.* 委任，授予；使服役
complex [ˈkɔmpleks] *adj.* 复杂的；复合的　*n.* 建筑群；相关联的一组事物
compliance [kəmˈplaiəns] *n.* 服从，听从；承诺
conceive [kənˈsiːv] *vt.* 构思；想象，设想；持有　*vi.* 设想；考虑
concerted [kənˈsəːtid] *adj.* 协调的；协定的；协商好的
concrete [ˈkɔŋkriːt] *adj.* 具体的，有形的；混凝土制的　*n.* 混凝土　*vt.* 混凝土修筑
concubine [ˈkɔŋkjubain] *n.* 妾，妃子
confession [kənˈfeʃn] *n.* 忏悔；承认，自首；信条，教义
configuration [kənˈfigəˈreiʃn] *n.* 配置；布局，构造
congregation [ˈkɑːŋgriˈgeiʃn] *n.* 集会；集合；教堂里的会众；人群
conglomeration [kənˈglɔməˈreiʃn] *n.* 团块，聚集，混合物
consecutive [kənˈsekjətiv] *adj.* 连续的，连贯的
consecrate [ˈkɔnsikreit] *vt.* 把……奉为神圣；奉献，献祭　*adj.* 神圣的；被献给神的
constance [ˈkɔnstəns] *n.* 恒定性
constrict [kənˈstrikt] *vt.* 收缩；压缩，压紧；妨害，阻碍；挤压
contour [ˈkɔntʊə(r)] *n.* 外形，轮廓；等高线；概要；电路　*vt.* 画轮廓　*adj.* 显示轮廓的
contributor [kənˈtribjətə(r)] *n.* 贡献者；捐助者；投稿者
controversy [ˈkɔntrəvəːsi] *n.* 论战；公开辩论
conveniently [kənˈviːnjəntli] *adv.* 随手；方便地，便利地，合宜地；顺便
converge [kənˈvəːdʒ] *vi.* 聚集；会于一点；人或车辆汇集　*vt.* 使聚集
convert [kənˈvəːt] *vt.*（使）转变；使皈依；兑换，换算；侵占
convex [ˈkɔnveks] *adj.* 凸形；凸的，凸面的
corbel [ˈkɔːbl] *n.* 承材，枕梁　*vt.* 给……装上托臂；支撑
cordial [ˈkɔːdiəl] *adj.* 热诚的；诚恳的；兴奋的
coronation [ˈkɔrəˈneiʃn] *n.* 加冕礼
corporal [ˈkɔːpərəl] *adj.* 人体的
corrosion [kəˈrəʊʒn] *n.* 腐蚀，侵蚀，锈蚀；衰败
console [kənˈsəʊl] *vt.* 安慰，慰问　*n.* 控制台，操纵台；演奏台；悬臂；肘托
consort [ˈkɔnsɔːt] *n.* 配偶；伙伴　*vi.* 结交；陪伴；符合　*vt.* 使陪伴；使联系
coulisse [kuːˈliːs] *n.* 有槽木料，侧面布景
counteract [ˈkaʊntərˈækt] *vt.* 抵消；中和；阻碍
counterpart [ˈkaʊntəpɑːt] *n.* 副本；配对物；相对物；极相似的人或物

courtier [ˈkɔːtiə(r)] *n.* 侍臣，廷臣；谄媚者
cradle [ˈkreidl] *n.* 摇篮；发源地，发祥地 *vt.* 将……置于摇篮中；轻轻地抱或捧
cramp [kræmp] *n.* 痛性痉挛 *vt.* 使痉挛；限制，束缚 *adj.* 难懂的，难认的；狭窄的
crawler [ˈkrɔːlə(r)] *n.* 爬行者，爬行动物
creed [kriːd] *n.* 宗教信条，教义；主义，纲领；宗派
creek [kriːk] *n.* 小湾；小河，小溪
crenellate [ˈkrenileit] *vt.* 使成钝锯齿形
crescent [ˈkresnt] *n.* 新月，月牙；伊斯兰教的标记 *adj.* 新月形的；渐圆的，渐强的
crossbeam [ˈkrɔːsbiːm] *n.* 大梁，横梁
cross-section [ˈkrɔsˈsekʃən] *n.* 横断面，剖面；横截面图
cruciform [ˈkruːsifɔːm] *adj.* 十字形的
crucifixion [ˈkruːsəˈfikʃn] *n.* 受难；刑罚；被钉死在十字架；苦痛的考验
crypt [kript] *n.* 地窖，教堂地下室
cuisine [kwiˈziːn] *n.* 菜肴；烹饪，烹调法
culminate [ˈkʌlmineit] *vt.* 达到极点
cultivation [ˈkʌltiˈveiʃn] *n.* 栽培；耕作；教养；培植
cupola [ˈkjuːpələ] *n.* 炮塔
curio [ˈkjʊəriəʊ] *n.* 小件珍奇物品
curvature [ˈkəːvətʃə(r)] *n.* 曲率；弯曲；曲度
cusped [kʌspt] *adj.* 有尖的，似尖头的
cylindrical [səˈlindrikl] *adj.* 圆筒形；圆柱形的，圆筒状的
cypress [ˈsaiprəs] *n.* 柏属植物，柏树；柏树枝

D

dado [ˈdeidəʊ] *n.* 护墙板，台座
dais [ˈdeiis] *n.* 台，讲台
decadence [ˈdekədəns] *n.* 衰落，堕落，颓废
deceased [diˈsiːst] *adj.* 已故的，已死的 *n.* 死者 *v.* 死，死亡
decomposition [ˈdiːkɔmpəˈziʃn] *n.* 分解；腐烂
decree [diˈkriː] *n.* 法令，命令；法院的判决，裁定；教会的教令 *vt.* 命令；颁布……为法令
decrepit [diˈkrepit] *adj.* 衰朽；衰老的，破旧的
dedicate [ˈdedikeit] *vt.* 奉献，献身；献给；题献辞；以……供奉
definitive [diˈfinətiv] *adj.* 最后的；确定的，决定性的 *n.* 限定词
degeneration [diˈdʒenəˈreiʃn] *n.* 变性；退化；恶化；变质
deification [ˈdiːifiˈkeiʃn] *n.* 祀为神，奉若神明，被奉为神者
deify [ˈdeiifai] *vt.* 神化，把……奉若神明
deity [ˈdeiəti] *n.* 上帝；神祇
deliberately [diˈlibərətli] *adv.* 故意地；深思熟虑地；从容不迫地

delimit [di'limit] *vt.* 限制，定……的界
demise [di'maiz] *n.* 死亡；让位；转让 *v.* 让位；遗赠
democracy [di'mɔkrəsi] *n.* 民主政治；民主主义；民主国家
depict [di'pikt] *vt.* 描述；描绘，描画
depot ['depəʊ] *n.* 仓库；补给站 *v.* 把……存放在储藏处
destination ['desti'neiʃn] *n.* 目的，目标；目的地，终点
detractor [di'træktə(r)] *n.* 贬低者
diagonal [dai'ægənl] *n.* 斜线；对角线；斜列；斜纹布 *adj.* 对角线的；斜的；斜线的
diagonal beams 斜梁
diagram ['daiəgræm] *n.* 图表；示意图；图解
dilapidate [di'læpideit] *v.* (使)荒废，毁坏，浪费
diameter [dai'æmitə(r)] *n.* 直径，直径长；放大率
dignitary ['dignitəri] *n.* 高僧；高官；显要人物 *adj.* 权贵的；高官的
dismantle [dis'mæntl] *vt.* 拆卸；拆开；废除；取消
dispense [di'spens] *vt.* 分配，分给；实施，施行；免除，豁免 *vi.* 特许，豁免
disrepute ['disri'pju:t] *n.* 声名狼藉；坏名声；不光彩
dissimulate [di'simjuleit] *vt.* 掩饰，假装
distinguish [di'stiŋgwiʃ] *vi.* 区分，辨别，分清 *vt.* 区分，辨别，识别；引人注目，使著名
divan [di'væn] *n.* 矮沙发
diverge [dai'və:dʒ] *vi.* 分歧；分开，叉开；偏离，背离 *vt.* 使发散；使偏离
diverse [dai'və:s] *adj.* 不同的，多种多样的；变化多的；形形色色的
doctrine ['dɔktrin] *n.* 教条，教义；法律原则；声明
dolerite ['dɔlərait] *n.* 粗玄岩；粗粒玄武岩
dome [dəʊm] *n.* 圆屋顶；穹顶；圆顶体育场 *vt.* 加圆屋顶于……上 *vi.* 成圆顶状
dominate ['dɔmineit] *v.* 支配，影响；占有优势；在……中具有最重要的位置
donor ['dəʊnə(r)] *n.* 捐赠者；施主
dormer ['dɔ:mə] *n.* 屋顶窗
duomo ['dwɔ:mɔ:] *n.* 大教堂，中央寺院
drainage ['dreinidʒ] *n.* 排水系统；排水，放水；排走物，废水；排水区域
dwelling ['dweliŋ] *n.* 住处，处所
dynamic [dai'næmik] *adj.* 动态的；动力的，动力学的；充满活力的 *n.* 动态；动力学

E

eaves [i:vz] *n.* 屋檐
echelon ['eʃəlɔn] *n.* 等级，阶层
echinus [e'kainəs] *n.* 海胆，钟形圆饰
eclectic [i'klektik] *adj.* 折中的；选择的 *n.* 折中主义者
edifice ['edifis] *n.* 大建筑物；知识的结构
egg-and-dart ['eg'ændd'ɑ:t] *n.* 卵镖形装饰

elastic [i'læstik] *adj*. 有弹力的；可伸缩的；灵活的　*n*. 松紧带，橡皮圈
elevation ['eli'veiʃn] *n*. 高处，高度，海拔；升级，上进；正视图，立视图
eliminate [i'limineit] *vt*. 淘汰；排除，消除；除掉
elliptical [i'liptikl] *adj*. 椭圆的；省略的；像椭圆形的
elm [elm] *n*. 榆树，榆木
elongate ['iːlɔŋgeit] *vt*. 延长，加长
embankment [im'bæŋkmənt] *n*. 路堤；筑堤
embodiment [im'bɔdimənt] *n*. 体现；化身；具体化
emblem ['embləm] *n*. 象征，标记；纹章，徽章；标记，典型　*vt*. 象征；用象征表示
emerald ['emərəld] *n*. 祖母绿；翠绿色；绿宝石，翡翠　*adj*. 翡翠的；祖母绿的；翠绿色的
emir [e'miə] *n*. 埃米尔(对穆斯林统治者的尊称)；穆罕默德后裔的尊称
emulate ['emjuleit] *vt*. 仿真；竞争；努力赶上
enact [i'nækt] *vt*. 制定法律；规定；颁布，颁布；担任……角色
enamel [i'næml] *n*. 搪瓷；珐琅；指甲油　*vt*. 给……上珐琅；在……涂瓷漆
encapsulate [in'kæpsjuleit] *vt*. 封装；概述　*vi*. 形成囊状物
encompass [in'kʌmpəs] *vt*. 围绕，包围；包含或包括某事物；完成
encrust [in'krʌst] *vt*. 用一层硬壳覆盖，包裹；用贵重物品在表面装饰　*vi*. 结壳，形成硬壳
endeavor [in'devə] *vt*. 尝试，试图　*n*. 努力，尽力
endure [in'djʊə(r)] *vt*. 忍耐；容忍　*vi*. 忍耐；持续，持久
enfilade ['enfi'leid] *n*. 相对成行排列　*vt*. 向……纵射
enlightenment [in'laitnmənt] *n*. 启蒙运动；启迪，启发；教化，觉悟
enliven [in'laivn] *vt*. 使活泼；使生动；使有生气
enrichment [in'ritʃmənt] *n*. 浓缩；丰富；改进；肥沃
ensemble [ɔn'sɔmbl] *n*. 全体；全套服装；总效果
enshrine [in'ʃrain] *vt*. 珍藏，铭记；把……奉为神圣；秘藏；把……置于神龛内
entablature [en'tæblətʃə] *n*. 古典建筑的柱上楣构，檐部
entasis ['entəsis] *n*. 凸肚状，微凸，卷杀
enthronement [in'θrəʊnmənt] *n*. 即位，就任主教的仪式，登基典礼
entirety [in'taiərəti] *n*. 完全；整体，全面；作为一个整体
enviable ['enviəbl] *adj*. 值得羡慕的；引起妒忌的
erosion [i'rəʊʒn] *n*. 糜烂；烧蚀；腐蚀，侵蚀
eponymous [i'pɔniməs] *adj*. 齐名的；使得名的
equestrian [i'kwestriən] *adj*. 骑马的，骑术的；骑士团的　*n*. 骑手；骑马者；马戏演员
equilibrium ['iːkwi'libriəm] *n*. 平衡，均势；平静
equivalent [i'kwivələnt] *adj*. 相等的，相当的，等效的　*n*. 对等物；当量
essence ['esns] *n*. 香精；本质，实质；精华，精髓
esthetics [es'θetiks] *n*. 美学

Ethnic ['eθnik] *adj*. 种族的，部落的；某文化群体的

ethnography [eθ'nɔgrəfi] *n*. 人种论，民族志

evaporation [i'væpə'reiʃn] *n*. 汽化；蒸发法；蒸发，发散；消失

evict [i'vikt] *vt*. 依法驱逐；依法收回

excursion [ik'skəːʃn] *n*. 集体远足；短途旅行，游览

execution ['eksi'kjuːʃn] *n*. 依法处决；实行，执行

exotic [ig'zɔtik] *adj*. 异国的；外来的；奇异的；吸引人的 *n*. 舶来品，外来物

expansive [ik'spænsiv] *adj*. 易扩张的，膨胀性的；辽阔的

exquisite [ik'skwizit] *adj*. 精致的；细腻的；优美的

external [ik'stəːnl] *adj*. 外面的，外部的 *n*. 外部，外面；外观

extremity [ik'stremәti] *n*. 端点；尽头；手和足；极窘迫的境地

exuberant [ig'zjuːbərənt] *adj*. 生气勃勃的；茂盛的，繁茂的 *adv*. 生气勃勃地；充沛地

F

facade [fə'sɑːd] *n*. 外表；建筑物的正面；虚伪，假象

faceted ['fæsitid] *adj*. 有小面的，有切割面的

facilely ['fæsail] *adv*. 轻易地，轻快地

faience [fai'ɑːns] *n*. 彩瓷；精巧彩色陶器

feat [fiːt] *n*. 功绩，卓绝的手艺，技术，本领；武艺 *adj*. 合适的；灵巧的；整洁的

festive ['festiv] *adj*. 喜庆的；欢乐的；节日的

festoon [fe'stuːn] *n*. 垂花饰；花彩装饰物，彩旗

feverishly ['fiːvəriʃli] *adv*. 发热地；狂热地，兴奋地

feudal ['fjuːdl] *adj*. 封建的；领地的

filigree ['filigriː] *n*. 用金丝等制成的精工制品；精致的物品；华而不实的东西

fireplace ['faiәpleis] *n*. 壁炉

fiscal ['fiskl] *adj*. 财政上的，会计的；国库的 *n*. 财政年度，会计年度

flank [flæŋk] *n*. 侧翼；侧面 *vt*. 位于……之侧面；侧面有 *vi*. 侧面与……相接

fleur-de-lys (法)鸢尾花

floodlighting ['flʌdlaitiŋ] *n*. 泛光灯照明 *v*. 用泛光灯照明

flourish ['flʌriʃ] *vi*. 挥舞；茂盛，繁荣 *vt*. 挥动，挥舞 *n*. 挥舞，挥动

fluid ['fluːid] *n*. 液体，流体 *adj*. 流体的，流体的，不固定的

flute [fluːt] *n*. 长笛；柱身上的凹槽；女服的裙褶 *v*. 吹长笛；在……上刻凹槽

fold-over gable roof 折叠式人字屋顶

foliage ['fəʊliidʒ] *n*. 树叶；植物的叶子

footing ['fʊtiŋ] *n*. 立足点；立场，基础

foremen ['fɔːmən] *n*. 工头，领班

foreshadow [fɔː'ʃædəʊ] *vt*. 预示，是……的先兆

forestall [fɔː'stɔːl] *vt*. 先发制人，预先阻止；垄断；领先

fornix ['fɔːniks] *n*. 穹窿

fortification [ˌfɔːtifiˈkeiʃn] *n.* 防御工事；筑垒，设防
fortress [ˈfɔːtrəs] *n.* 堡垒，要塞 *vt.* 设置要塞
forum [ˈfɔːrəm] *n.* 论坛，讨论会；集会的公共场所；法庭
fountain [ˈfaʊntən] *n.* 泉源；喷泉水；源头，来源
fragile [ˈfrædʒail] *adj.* 易碎的，脆的；虚弱的
fragrant [ˈfreigrənt] *adj.* 芳香的，香的；愉快的；喷香；芳菲
fresco [ˈfreskəʊ] *n.* 壁画；湿壁画 *v.* 在……作壁画；用壁画法画出
frieze [friːz] *n.* 柱顶过梁和挑檐间的雕带，墙顶的饰带
fringe [frindʒ] *n.* 边缘；穗；次要 *vt.* 作为……的边缘，围绕着 *adj.* 边缘的，外围的
full-fledged [fʊl fledʒd] *adj.* 羽毛生齐的；经过充分训练的，成熟的
funerary [ˈfjuːnərəri] *adj.* 葬礼的，埋葬的
furnishing [ˈfəːniʃiŋ] *v.* 提供；陈设，布置
fusion [ˈfjuːʒn] *n.* 融合；熔解，熔化；融合物；核聚变

G

gabled roof 有山墙的三角形，人字形屋顶
gala [ˈgɑːlə] *n.* 庆祝；节日 *adj.* 节日的；欢乐的
gallery [ˈgæləri] *n.* 画廊，走廊；教堂，议院等的边座；旁听席
gauge [geidʒ] *n.* 评估；尺度，标准 *vt.* 测量；评估，判断；采用
gentry [ˈdʒentri] *n.* 绅士们；贵族们；上层阶级
geomagnetism [ˈdʒiːəʊˈmægnətizəm] *n.* 地磁学
geometric [ˈdʒiːəˈmetrik] *adj.* 几何学的；几何装饰的
giddy [ˈgidi] *adj.* 头晕的；眼花的；轻浮的 *vt.* 使眩晕；使眼花
gilding [ˈgildiŋ] *n.* 贴金箔，镀金
girder [ˈgəːdə(r)] *n.* 大梁
gladiatorial [ˈglædiəˈtɔːriəl] *adj.* 角斗士，斗剑者的，争论的
glimpse [glimps] *n.* 一瞥，一看；隐约的闪现；模糊的感觉 *vi.* 瞥见
gothic [ˈgɔθik] *adj.* 哥特式的 *n.* 哥特体
grandeur [ˈgrændʒə(r)] *n.* 伟大；富丽堂皇；宏伟，壮观
granite [ˈgrænit] *n.* 花岗岩，花岗石；坚毅，坚韧不拔
grief-stricken [ˈgriːfˌstrikən] *adj.* 极度忧伤的；黯然销魂的
grievance [ˈgriːvəns] *n.* 委屈；不满；苦衷；牢骚
grotto [ˈgrɔtəʊ] *n.* 岩洞，洞穴，人造石窟
groove [gruːv] *n.* 沟，槽 *vt.* 刻沟，刻槽
grove [grəʊv] *n.* 小树林；果树林，果园
guillotine [ˈgilətiːn] *n.* 断头台；切纸机；铡刀 *vt.* 在断头台上处死
gutter [ˈgʌtə(r)] *n.* 排水沟；屋顶的天沟 *vt.* 形成沟或槽于……

H

haberdashery [ˈhæbəˈdæʃəri] *n.* 男子服饰用品，男子服饰用品店

harem [ˈhɑːriːm] *n.* 伊斯兰教徒的女眷；女眷居室，后宫
Harmika 塔身
harmony [ˈhɑːməni] *n.* 和谐；协调；融洽，一致
harsh [hɑːʃ] 刺耳的；残酷的；粗糙的；严厉的，严格的
haunch [hɔːntʃ] *n.* 腰腿
hawk [hɔːk] *n.* 鹰 *vt.* 兜售货物 *vi.* 叫卖，用训练好的鹰狩猎
headpiece [ˈhedpiːs] *n.* 顶梁；头盔，帽子；顶部
hedge [hedʒ] *n.* 树篱；保护手段 *vt.* 用树篱围起；受……的束缚 *vt.* 回避，避免
heliocentric [ˈhiːliəˈsentrik] *adj.* 以太阳为中心的
hemispherical [ˈhemiˈsferikl] *adj.* 半球的，半球状的
hemp [hemp] *n.* 大麻；长纤维的植物；绞索 *adj.* 大麻类植物的
herbaceous [həːˈbeiʃəs] *adj.* 草本的；叶状的
hexagonal [heksˈægənl] *adj.* 六方；六角形的，六边形的
hieroglyphic [ˈhaiərəˈglifik] *n.* 象形文字；难以辨认或理解的文字
Hinduism [ˈhinduːizəm] *n.* 印度教
hinge [hindʒ] *n.* 铰链，折叶；转折点；枢要 *vt.* 用铰链连接 *vi.* 依……而转移
hipped roof [hipt ruːf] 四坡屋顶
hoist [hɔist] *vt.* 升起，提起 *vi.* 被举起或抬高 *n.* 升起；起重机
hollow [ˈhɔləʊ] *adj.* 空洞的；空的；虚伪的 *n.* 洞，坑；山谷 *vt.* 挖空(某物)
homage [ˈhɔmidʒ] *n.* 敬意；效忠；顺从
honeysuckle [ˈhʌnisʌkl] *n.* 忍冬，金银花
horizontal [ˈhɔriˈzɔntl] *adj.* 水平的，卧式的；地平线的 *n.* 水平线；水平面；水平的物体
horseshoe [ˈhɔːsʃuː] *n.* 马蹄铁；马蹄形的东西 *vt.* 装蹄铁于
horticulture [ˈhɔːtikʌltʃə(r)] *n.* 园艺学；园艺学家 *adj.* 园艺学的
hybrid [ˈhaibrid] *n.* 杂种；杂交生成的生物体；混合物；混合词 *adj.* 混合的；杂种的
hydraulic [haiˈdrɔːlik] *adj.* 水力的，水压的；用水发动的；水力学的
hygiene [ˈhaidʒiːn] *n.* 卫生，卫生学；保健法
hypostyle [ˈhaipəʊstail] *n.* 多柱式建筑

I

iconic [aiˈkɔnik] *adj.* 图标的；符号的；图符的；偶像的
ignite [igˈnait] *vt.* 点燃；使燃烧 *vi.* 点火；燃烧
ignore [igˈnɔː(r)] *vt.* 忽视，不顾；驳回诉讼
ill-judged [ilˈdʒʌdʒd] *adj.* 欠思考的，判断失当的
illumination [iˈluːmiˈneiʃn] *n.* 照明；阐明，解释清楚；彩饰，图案花饰
imaret [iˈmɑːret] *n.* 土耳其的小客店，朝圣者的旅舍
imbrice [ˈimbrais] *n.* 接缝用的瓦片，装饰用鳞瓦
imitation [ˈimiˈteiʃn] *n.* 仿制品；模仿，仿效 *adj.* 仿制的；人造的
impart [imˈpɑːt] *vt.* 传授；给予；告知，透露

imperialistic [imˈpiəriəˈlistik] *adj*. 帝国主义的，帝制的
implant [imˈplɑːnt] *n*. 移植物 *vt*. 移植；植入，插入 *vi*. 被移植
imprint [imˈprint] *n*. 痕迹；盖印；特征 *v*. 盖印；刻上记号；使铭记
inaugurate [iˈnɔːgjəreit] *vt*. 开创；创始；举行开幕典礼；举行就职典礼
incarnation [ˈinkɑːˈneiʃn] *n*. 化身；前身；转世
inception [inˈsepʃn] *n*. 开始，开端，初期
incise [inˈsaiz] *vt*. 在表面雕刻
indigenous [inˈdidʒənəs] *adj*. 土生土长的；生来的，固有的；本地的；根生土长
indispensable [ˈindiˈspensəbl] *adj*. 不可缺少的；绝对必要的 *n*. 不可缺少的人或物
ingenious [inˈdʒiːniəs] *adj*. 精巧的；灵巧的；设计独特的；有天才的，聪明的
inherit [inˈherit] *vt*. 继承 *vt*. 经遗传获得品质、身体特征等，继任
inlay [ˈinˈlei] *v*. 嵌入，镶嵌，插入 *n*. 镶嵌物，镶补
in situ [ˈin ˈsitjuː] 在原位置，在原处
inspiration [ˈinspəˈreiʃn] *n*. 灵感；吸气；鼓舞人心的人或事；启发灵感
intact [inˈtækt] *adj*. 完好无损；原封不动的；完好无缺
integrate [ˈintigreit] *vt*. 使一体化；使整合 *vi*. 成为一体；结合在一起 *adj*. 整体的；完整的
integral [ˈintigrəl] *adj*. 完整的；必须的 *n*. 整体
inter [inˈtəː(r)] *vt*. 埋，葬
interlaced [ˈintəˈleist] *adj*. 交织的，交错的
interlock [ˈintəˈlɔk] *n*. 互锁设备；双螺纹针织品 *v*. 连锁；互锁
interpretation [inˈtəːpriˈteiʃn] *n*. 理解；解释，说明；翻译；表演，演绎
intersperse [ˈintəˈspəːs] *vt*. 点缀；散布，散置
interstice [inˈtəːstis] *n*. 空隙，裂缝，缺口 *v*. 使交错，使交织
intimidate [inˈtimideit] *vt*. 恐吓，威胁
intricate [ˈintrikət] *adj*. 错综复杂的；难理解的；曲折
inturned [ˈintəːnd] *n*. 内弯的，性格内向的
inverted [inˈvəːtid] *adj*. 反向的，倒转的 *v*. 使倒置，使反转
investiture [inˈvestitʃə(r)] *n*. 授权，授职仪式
irritate [ˈiriteit] *vt*. 刺激，使兴奋 *vi*. 引起恼怒；引起不愉快

J

Jainism [ˈdʒeinizəm] *n*. 耆那教(印度非婆罗门教的宗教派别)
jasper [ˈdʒæspə] *n*. 碧玉
jawab [dˈʒəwəb] 对称房屋；配称建筑
judicial [dʒuˈdiʃl] *adj*. 司法的；法庭的；明断的；公正的
juniper [ˈdʒuːnipə(r)] *n*. 杜松；刺柏；刺柏属丛木或树木
juxtaposition [ˈdʒʌkstəpəˈziʃn] *n*. 并列；并置；毗邻

K

karmic [kɑmik] *n.* 命运，因果报应
kiosk ['kiːɔsk] *n.* 亭子，凉亭；书报摊，公共电话亭；小摊棚，售货亭

L

labyrinthine ['læbə'rinθain] *adj.* 迷宫(似)的，曲折的
lacquer ['lækə(r)] *n.* 漆器；漆，天然漆 *vt.* 涂漆于；使……表面或外观光滑
lagoon [lə'guːn] *n.* 泻湖；环礁湖；咸水湖
landslide ['lændslaid] *n.* 滑坡，山崩；大胜利
lapis lazuli ['læpis 'læzjʊli] *n.* 天青石，青金石，天青石色
lash [læʃ] *n.* 鞭挞，鞭子 *vi.* 抽打；猛击 *vt.* 鞭打；摆动；激励，捆绑
lassitude ['læsitjuːd] *n.* 无精打采；不努力，懒怠
latrine [lə'triːn] *n.* 公共厕所
lattice ['lætis] *n.* 格子框架，平棊 *vt.* 把……制成格子状；用格子覆盖或装饰
lavishly ['læviʃli] *adv.* 非常大方地，浪费地，极其丰富地
laity ['leiəti] *n.* 俗人；普通信徒；外行
legislation ['ledʒis'leiʃn] *n.* 立法，制定法律；法律，法规
limestone ['laimstəʊn] *n.* 石灰岩，石灰石
lintel ['lintl] *n.* 楣，过梁
liturgical [li'təːdʒikl] *adj.* 礼拜仪式的
livestock ['laivstɔk] *n.* 家畜，牲畜
load-bearing ['ləʊdb'eəriŋ] *n.* 承重；受荷，承载
lodge [lɔdʒ] *vi.* 存放；暂住 *vt.* 提出报告；容纳；寄存 *n.* 小屋，草屋；帐篷
loggia ['ləʊdʒə] *n.* 凉廊，房屋敞向花园的部分
logistics [lə'dʒistiks] *n.* 物流；后勤；逻辑学；组织工作
loophole ['luːphəʊl] *n.* 漏洞；枪眼；观察孔
loot [luːt] *n.* 抢劫；掠夺物，战利品；赃物，外快 *vt.* 抢劫，掠夺；贪污
lotus petal 莲花瓣
lubricate ['luːbrikeit] *vt.* 加油润滑；使润滑 *vi.* 润滑
lunettes [luː'nets] *n.* 潜水面罩；半月形拱顶
luxurious [lʌg'ʒʊəriəs] *adj.* 豪华的；奢侈的；放纵的

M

magnetic [mæg'netik] *adj.* 磁性的；有磁性的，有吸引力的；有吸引力的；有魅力的
malevolent [mə'levələnt] *adj.* 恶毒的；凶狠；心肠坏的；幸灾乐祸的
mandate ['mændeit] *n.* 授权；命令；委任 *vt.* 托管；批准 *vi.* 强制执行
mantra ['mæntrə] *n.* 咒语；颂歌，圣歌
marble ['mɑːbl] *n.* 大理石 *adj.* 大理石的；冷酷无情的；有大理石花纹的
martyr ['mɑːtə(r)] *n.* 烈士；殉道者 *vt.* 牺牲；折磨

marvel [ˈmɑːvl] *n.* 奇迹；令人惊奇的事物；成就　*vt.* 惊奇，对……感到惊奇
mastaba [ˈmæstəbə] *n.* 石室坟墓
mausoleum [ˈmɔːsəliːəm] *n.* 陵墓
maximum [ˈmæksiməm] *adj.* 最大值的，最大量的　*n.* 最大的量、体积、强度
maze [meiz] *n.* 迷宫；迷惑；错综复杂；迷宫图　*vt.* 使困惑；使混乱；迷失
meander [miˈændə(r)] *vi.* 漫谈；河流蜿蜒而流；无目的地走动　*n.* 曲径
medallion [məˈdæliən] *n.* 大奖章，大勋章；圆形图案饰物
meditation [ˈmediˈteiʃn] *n.* 冥想；沉思；默想；默念
medrese 伊斯兰学校
meld [meld] *vt.* 使融合，合并，结合　*n.* 混合，合并
metaphor [ˈmetəfə(r)] *n.* 象征；隐喻，暗喻
metaphysics [ˈmetəˈfiziks] *n.* 形而上学，玄学
metope [ˈmetəʊp] *n.* 墙面，柱间壁
metropolis [məˈtrɔpəlis] *n.* 大都市，大都会；首府，首都
mihrab [ˈmiːræb] *n.* 米哈拉布清真寺中用以指示麦加方向的壁龛
mimic [ˈmimik] *n.* 巧于模仿的人；复写品或仿制品　*vt.* 摹拟；模仿　*adj.* 模仿的，摹拟的
minaret [ˈminəˈret] *n.* 清真寺旁召唤祈祷时刻的光塔，宣礼塔
minbar [ˈminbɑː] *n.* 敏拜尔，清真寺殿内的宣教台，形如楼梯
mingle [ˈmiŋgl] *vt.* 混合，混淆　*vi.* 混进，与……交往
miniature [ˈminətʃə(r)] *adj.* 小型的　*n.* 微型复制品；微型画　*vt.* 使成小型；把……画成纤细画
minimize [ˈminimaiz] *vt.* 把……减至最低数量；对(某事物)作最低估计，极度轻视
mnemonic [niˈmɔnik] *adj.* 记忆的，记忆术的
moat [məʊt] *n.* 壕沟，护城河　*v.* 挖壕围绕
monarchy [ˈmɔnəki] *n.* 君主政体；君主国；君主政治
monograph [ˈmɔnəgrɑːf] *n.* 专著，专论
monolith [ˈmɔnəliθ] *n.* 独块巨石
monolithic [ˈmɔnəˈliθik] *adj.* 整体的；独块巨石的；庞大的
monsoon [ˈmɔnˈsuːn] *n.* 印度洋的季风；季风雨；夏季季风
monument [ˈmɔnjumənt] *n.* 纪念碑；丰碑；遗迹；遗址
mortal [ˈmɔːtl] *adj.* 致命的；不共戴天的；终有一死的；极度的　*n.* 凡人，人类
mortar [ˈmɔːtə(r)] *n.* 砂浆　*vt.* 用灰泥涂抹
mosaic [məʊˈzeiik] *n.* 镶嵌；镶嵌图案；镶嵌工艺
mosque [mɔsk] *n.* 清真寺，伊斯兰教寺院
moss [mɔs] *n.* 苔藓；藓沼　*vt.* 以苔藓覆盖；使长满苔藓
mortise [ˈmɔːtis] *n.* 榫眼，卯眼，接合　*vt.* 榫接，给……开榫眼；使牢固相接
mound [maʊnd] *n.* 堆；土堆，土丘

muezzin [mu:'ezin] *n.* 报告祷告时刻的人
mullion ['mʌliən] *n.* 竖框，直棂，放射状框；门窗扇中梃
mummify ['mʌmifai] *v.* 干瘪；使干瘪；成木乃伊状
municipality [mju:ˌnisi'pæləti] *n.* 自治市；市政当局
mural ['mjʊərəl] *n.* 大型壁画 *adj.* 墙壁的，墙的；画在或者挂在墙上的
myrtle ['mə:tl] *n.* 桃金娘科植物
mythical ['miθikl] *adj.* 神话的；虚构的
mythicise ['miθisaiz] *vt.* 视为神话，解释为神话
mythology [mi'θɔlədʒi] *n.* 神话学；神话；虚构的事实；错误的观点

N

naiad ['naiæd] *n.* 水中的仙女
naos ['neiɔs] *n.* 古寺院，内殿
narthex ['nɑ:θeks] *n.* 古代礼拜堂的前廊，教堂前厅
nave [neiv] *n.* 教堂正厅；本堂；中仓；车轮的中心部
necropolis [nə'krɔpəlis] *n.* 古代的大墓地
negligible ['neglidʒəbl] *adj.* 微不足道的；可以忽略的
nestle ['nesl] *vi.* 安居，舒适地居住；偎依，贴靠 *vt.* 满意地依偎或紧贴；安置
niche [nitʃ] *n.* 壁龛；合适的位置
niece [ni:s] *n.* 侄女；外甥女
nomadic [nəʊ'mædik] *adj.* 游牧的；流浪的
nucleus ['nju:kliəs] *n.* 中心，核心；原子核；细胞核
nymph [nimf] *n.* 仙女

O

obelisk ['ɔbəlisk] *n.* 方尖石塔，短剑号，疑问记号
obscured [əb'skjʊəd] *v.* 使……模糊不清，掩盖
octastyle ['ɔktəstail] *n.* 八柱式建筑 *adj.* 八柱式的
offensive [ə'fensiv] *adj.* 无礼的，冒犯的；进攻(性)的，攻势的 *n.* 进攻，攻势
offstage ['ɔf'steidʒ] *n.* 舞台内部，舞台后面
omen ['əʊmən] *n.* 预兆；征兆；前兆；兆头 *vt.* 预示；预告；有……的前兆
oppression [ə'preʃn] *n.* 压迫；被压迫的状态；压迫物；沉闷，苦恼
opulent ['ɔpjələnt] *adj.* 豪华的；富裕的
oratorio ['ɔrə'tɔ:riəʊ] *n.* 以宗教为主题的清唱剧
orchestra ['ɔ:kistrə] *n.* 管弦乐队；管弦乐队的全部乐器
orchid ['ɔ:kid] *n.* 兰花
order ['ɔ:də(r)] *n.* 秩序；命令；次序；规则，制度；柱式 *vt.* 命令；订购
ornamentation ['ɔ:nəmen'teiʃn] *n.* 装饰，装饰品
outstretch [aʊt'stretʃ] *v.* 伸出，伸展

overlapping [əuvəˈlæpiŋ] *n.* 重叠，搭接 *v.* 交叠；部分重叠

overscale [əuvəsˈkeil] *adj.* 过大的，超大型的

P

pagoda [pəˈgəudə] *n.* 塔，佛塔

palatial [pəˈleiʃl] *adj.* 宫殿的，宏伟的，壮丽的

panel [ˈpænl] *n.* 镶板；面；嵌板；控制板 *vt.* 把……分格；把……镶入框架内

panoramic [ˈpænəˈræmik] *adj.* 全景的；全貌的

parabolic [ˈpærəˈbɔlik] *adj.* 抛物线的；比喻的

parallel [ˈpærəlel] *adj.* 平行的；相同的，类似的 *adv.* 平行地，并列地 *n.* 平行线；相似物

paranoia [ˈpærəˈnɔiə] *n.* 偏执狂；瞎猜疑；疑神疑鬼

parasol [ˈpærəsɔl] *n.* 太阳伞

parish [ˈpæriʃ] *n.* 教区以下的地方行政区

parliament [ˈpɑːləmənt] *n.* 议会，国会

parlour [ˈpɑːlə(r)] *n.* 客厅，会客室；美容院，冷食店，殡仪馆

parterre [pɑːˈteə(r)] *n.* 宅地；花坛，花圃

pastel [ˈpæstl] *n.* 彩色粉笔，蜡笔画；轻淡柔和的色彩 *adj.* 彩色粉笔画的；色彩柔和的

patina [ˈpætinə] *n.* 绿锈；铜绿；年久而产生的光泽

patriotic [ˈpeitriˈɔtik] *adj.* 爱国主义的；爱国的，有爱国心的

patron [ˈpeitrən] *n.* 赞助人，资助人；保护人；守护神

pavilion [pəˈviliən] *n.* 亭，阁楼；临时建筑物；大型文体馆；看台

peculiar [piˈkjuːliə(r)] *adj.* 奇怪的，古怪的；特有的 *n.* 专有特权，专有财产

peculiarity [piˈkjuːliˈærəti] *n.* 特性；特质；怪癖；奇形怪状

pedestrian [pəˈdestriən] *n.* 行人；步行者 *adj.* 徒步的；平淡无奇的

pediment [ˈpedimənt] *n.* 山花；山墙；山形墙，三角墙

pepper [ˈpepə(r)] *n.* 辣椒；胡椒；胡椒粉 *vt.* 在……上撒胡椒粉

peplos [ˈpepləs] *n.* 古希腊的女式长外衣，女式大披肩

pent [pent] *adj.* 被关闭的

penthouse [ˈpenthaʊs] *n.* 大楼平顶上的楼顶房屋，遮篷

peony [ˈpiːəni] *n.* 牡丹，芍药

perch [pəːtʃ] *n.* 鲈鱼；栖息处 *vi.* 暂栖，停留 *vt.* 栖息；停留

perimeter [pəˈrimitə(r)] *n.* 周长；周围，边界

perpendicular [ˈpəːpənˈdikjələ(r)] *adj.* 垂直的，成直角的；险陡的 *n.* 垂直线，垂直面；直立

prescriptive [priˈskriptiv] *adj.* 规定的；约定俗成的；指定的；惯例的

persecution [ˈpəːsiˈkjuːʃn] *n.* 迫害或受迫害，烦扰；苛求，困扰

perspective [pəˈspektiv] *n.* 透镜；观点，看法；远景；洞察力 *adj.* 透视画法的

periphery [pəˈrifəri] *n.* 外围；边缘；圆周；边缘地带

peripteral [pəˈriptərəl] *adj*. 绕柱式的，周围列柱的；围柱殿
peristyle [ˈperistail] *n*. 列柱廊；列柱走廊，以柱围绕的内院
pervade [pəˈveid] *vt*. 遍及，弥漫；渗透，充满
petit [ˈpetiː] *adj*. 次要的，没价值的
petition [pəˈtiʃn] *n*. 请愿书；请愿，上诉状 *vi*. 请愿；祈求，请求 *vt*. 向法庭申诉
phenomena [fəˈnɔminə] *n*. 现象(复数)
physiognomy [ˈfiziˈɔnəmi] *n*. 人相学；脸；相面术
piazza [piˈætsə] *n*. 广场，市场；走廊，游廊
pictorial [pikˈtɔːriəl] *adj*. 绘画的；图画似的；形象化的 *n*. 画报；画刊
picturesque [ˈpiktʃəresk] *adj*. 别致的；美丽的；生动的；奇特的，独创的
piecemeal [ˈpiːsmiːl] *adv*. 逐渐地；零碎地 *adj*. 零碎的；逐个完成的 *n*. 片，块
pier [piə(r)] *n*. 桥墩；码头，防波堤；窗间壁；柱墩
pierce [piəs] *vt*. 刺破；刺穿，戳穿；洞察 *vi*. 进入；透入
pigment [ˈpigmənt] *n*. 颜料，色料 *vt*. 给……着色 *vi*. 呈现颜色
pilaster [piˈlæstə(r)] *n*. 壁柱，半露柱
pilgrimage [ˈpilgrimidʒ] *n*. 朝圣之旅；参拜圣地 *vi*. 去朝圣，去参拜圣地
pinnacle [ˈpinəkl] *n*. 顶峰；顶点；尖顶；哥德式建筑的小尖塔 *vt*. 为……加尖塔
pious [ˈpaiəs] *adj*. 虔诚的，信神的；孝敬的；好心的
plank [plæŋk] *n*. 厚木板；支持物 *vt*. 在……上铺板；重重放下
plaque [plæk] *n*. 匾，饰板；胸章，徽章
plague [pleig] *n*. 瘟疫；灾害，折磨 *vt*. 使染瘟疫；使痛苦，造成麻烦
plunder [ˈplʌndə(r)] *vt*. 掠夺；偷；私吞 *n*. 抢劫；掠夺物
plateau [ˈplætəʊ] *n*. 高原；平稳时期；停滞期 *v*. 进入停滞期；达到平稳状态
plinth [plinθ] *n*. 柱子的底座，基座；花瓶、塑像等的底座
polychromatic [pɔlikrəʊˈmætik] *adj*. 多色的
polygon [ˈpɔligən] *n*. 多边形，多角形；龟裂状
polygonal [pəˈligənl] *adj*. 多角形的，多边形的
polymorphic [ˈpɔliˈmɔːfik] *adj*. 多形的，多态的，多形态的；多晶形
pontiff [ˈpɔntif] *n*. 罗马教皇，主教，大祭司
populace [ˈpɔpjələs] *n*. 平民；百姓；人口
porch [pɔːtʃ] *n*. 门廊；游廊，走廊
porphyry [ˈpɔːfəri] *n*. 斑岩
portico [ˈpɔːtikəʊ] *n*. 柱廊，有圆柱的门廊
portray [pɔːˈtrei] *vt*. 描绘；描述；画像；描画
posterior [pɔˈstiəriə(r)] *adj*. 后面的；时间上较晚的；在后的；尾部的 *n*. 身体后部，臀部
postern [ˈpəʊstəːn] *n*. 后门；后门的
precinct [ˈpriːsiŋkt] *n*. 管辖区；选区；界限，范围
precursor [priːˈkəːsə(r)] *n*. 前辈，前驱，先锋，前任；预兆，先兆，初期形式

predators ['predətə(r)] *n.* 以掠夺为生的人；捕食其他动物的动物，食肉动物
predominantly [pri'dəminəntli] *adv.* 占主导地位地；显著地；占优势地
prefab ['pri:fæb] *adj.* 预制的，活动房
prevention [pri'venʃn] *n.* 预防；阻止，制止，妨碍
premise ['premis] *n.* 逻辑学中的前提；厨房 *vt.* 预述；提出……为前提；假设
primordial [prai'mɔ:diəl] *adj.* 初生的，初发的，原始的
principal ['prinsəpl] *adj.* 主要的；最重要的 *n.* 首长，负责人；主要演员，当事人
privilege ['privəlidʒ] *n.* 特权；免责特权；特殊荣幸 *vt.* 给与……特权，特免
proclaim [prə'kleim] *vt.* 表明；宣告，公布；赞扬，称颂
procurator ['prɔkjʊəreitə] *n.* 代理人
profile ['prəʊfail] *n.* 侧面，半面；外形，轮廓；人物简介 *vt.* 描……的轮廓
profusely [prə'fju:sli] *adv.* 十分慷慨地；毫不吝惜地；丰富地；繁茂地
prolific [prə'lifik] *adj.* 富饶的；多产的；众多的；丰硕的
prompt [prɔmpt] *adj.* 迅速的；敏捷的；立刻的；准时的 *v.* 提示；促使 *n.* 提示
protract [prə'trækt] *vt.* 拖延某事物
protrude [prə'tru:d] *vt.* 使突出；使伸出 *vi.* 突出；伸出
prominence ['prɔminəns] *n.* 突出；声望；卓越
prominent ['prɔminənt] *adj.* 著名的；突出的，杰出的
prune [pru:n] *vi.* 删除；减少 *vt.* 修剪树木；剪去 *n.* 西梅脯，西梅干；深紫红色
propagandistic [prɔpəgæn'distik] *adj.* 宣传的，宣传家的
proscenium [prə'si:niəm] *n.* 古希腊，古罗马的舞台，包括幕布，拱形墙等
prosperity [prɔ'sperəti] *n.* 繁荣；成功；兴旺，昌盛
prototype ['prəʊtətaip] *n.* 原型，雏形，蓝本
provision [prə'viʒn] *n.* 规定，条款；预备，准备，设备；供应；生活物质 *vt.* 为……提供所需物品
provoke [prə'vəʊk] *vt.* 煽动；激起，挑起；招致；触怒，使愤怒
prow [praʊ] *n.* 船首
Prussia ['prʌʃə] *n.* 普鲁士，1871 年建立统一的德意志帝国
punctuate ['pʌŋktʃueit] *vt.* 加标点符号；加强，强调 *vt.* 使用标点符号；不时打断
pyramidion [pirə'midiɔn] *n.* 小金字塔，方尖塔的顶角锥

Q

qibla wall 清真寺中信徒礼拜面向的墙，通常米哈拉布位于正向墙中央
quarry ['kwɔri] *n.* 采石场；菱形 *vt.* 挖出；努力挖掘 *vi.* 费力地找
quadrangle ['kwɔdræŋgl] *n.* 四边形；四方院子，尤指大学学院的四方院
quadrangular [kwɔ'dræŋgjələ(r)] *adj.* 四边形的；四棱柱
quadrigae [kwəd'rigi:] *n.* 四马二轮战车；四马战车雕饰
quadripartite ['kwɔdri'pɑ:tait] *adj.* 分成四组的，由四部组成的
quadrilateral ['kwɔdri'lætərəl] *adj.* 四边(形)的 *n.* 四边形

R

rafter [ˈrɑːftə(r)] *n.* 椽子 *vt.* 装椽于

rail [reil] *n.* 围栏；轨道，钢轨；扶手 *vi.* 责备；抱怨 *vt.* 将……围起来；用围栏围

rake [reik] *n.* 耙子；倾斜度 *v.* 耙；梳理；扫视；搜寻

rammed [ˈræmd] *v.* 猛撞；夯实(土等)；反复灌输

rampart [ˈræmpɑːt] *n.* 城堡等周围宽阔的防御土墙；防御，保护

ravage [ˈrævidʒ] *vt.* 毁坏；劫掠 *n.* 蹂躏；破坏

ravenous [ˈrævənəs] *adj.* 极饿的；极度的；强取豪夺

ravine [rəˈviːn] *n.* 沟壑，深谷

rebellion [riˈbeljən] *n.* 反抗；造反，叛乱

recessed [riˈsest] *v.* 把某物放在墙壁的凹处；将(墙)做成凹形，在(墙)上做壁龛

reclaim [riˈkleim] *vt.* 取回；开拓，开垦 *n.* 改造，感化；教化；回收再利用

reclusive [riˈkluːsiv] *adj.* 隐遁的，隐居的；孤独

reconciliation [ˈrekənsiliˈeiʃn] *n.* 和解，调停；一致；服从，顺从；和谐

recruit [riˈkruːt] *n.* 新兵；新成员；新学生 *vt.* 招聘，征募；吸收某人为新成员；雇用

redundant [riˈdʌndənt] *adj.* 多余的，累赘的；重沓

reference [ˈrefrəns] *n.* 参考；参考书；提及，涉及；证明人，介绍人 *v.* 引用；参照

refinement [riˈfainmənt] *n.* 精炼，提纯，净化；细微的改良

refuge [ˈrefjuːdʒ] *n.* 避难；避难所；庇护者 *vt.* 给予……庇护；接纳……避难 *vi.* 避难

refurbish [ˈriːˈfəːbiʃ] *vt.* 刷新；使重新干净

rehabilitation [ˈriːəˈbiliˈteiʃn] *n.* 修复；复兴；复职；恢复名誉

rehearse [riˈhəːs] *vt.* 排练，排演 *vt.* 详述；排演；复述，背诵

reject [riˈdʒekt] *vt.* 拒绝；排斥；抛弃 *n.* 被拒绝或被抛弃的人或事物

relief [riˈliːf] *n.* 宽慰，安心；免除，减轻；浮雕

religious [riˈlidʒəs] *adj.* 宗教的；虔诚的；谨慎的 *n.* 修士，修女，出家人

reminiscent [ˈremiˈnisnt] *adj.* 怀旧的；回忆往事的；使人联想……的 *n.* 回忆录作者

remnant [ˈremnənt] *n.* 残余；剩余部分；幸存者 *adj.* 残留的；剩余的

renaissance [riˈneisns] *n.* 文艺复兴；复兴

renovate [ˈrenəveit] *vt.* 翻新，修复，整修；更新；恢复 *adj.* 恢复的，革新的，翻新的

replicate [ˈreplikeit] *vt.* 复制，复写；重复，反复；折转

reputedly [riˈpjuːtidli] *adv.* 据说，根据风评

resent [riˈzent] *vt.* 怨恨；愤恨；厌恶；对……感到愤怒

reservoir [ˈrezəvwɑː(r)] *n.* 蓄水池；储液器；储藏；蓄积

resistance [riˈzistəns] *n.* 电阻；阻力；抵抗；抗力

respectively [riˈspektivli] *adv.* 各自地；各个地；分别地

resplendent [riˈsplendənt] *adj.* 华丽的，辉煌的；鲜艳夺目的

restrain [riˈstrein] *vt.* 制止；抑制；限定，限制

resurrection [ˈrezəˈrekʃn] *n.* 耶稣复活；最后审判日所有死者的复活 *adj.* 复活的，复兴的

retail [ˈriːteil] *n*. 零售 *vt*. 零售；传播 *adj*. 零售的
retard [riˈtɑːd] *vt*. 使减速，妨碍，推迟 *vi*. 减慢，受到阻滞 *n*. 减速，阻滞，延迟
retardant [riˈtɑːdənt] *n*. 延缓剂 *adj*. 延缓的
retreat [riˈtriːt] *vi*. 撤退；撤销；凹进 *n*. 撤回；静居处；引退期间；静思，静修
retrofit [ˈretrəʊfit] *n*. 式样翻新 *vt*. 翻新，改型
rhombus [ˈrɔmbəs] *n*. 菱形
rib [rib] *n*. 肋骨；肋拱；肋骨状的东西 *vt*. 嘲笑，开(某人的)玩笑
ridged [ridʒd] *adj*. 翘起的
ripple [ˈripl] *vt*. 使泛起涟漪；使作潺潺声 *vt*. 在……上形成波痕 *vi*. 发出潺潺声
ritual [ˈritʃuəl] *n*. 典礼；宗教仪式；例行公事，老规矩 *adj*. 礼节性的；例行公事的
Romanesque [ˈrəʊməˈnesk] *n*. 罗马式建筑 *adj*. 罗马风格的
romanticism [rəʊˈmæntisizəm] *n*. 浪漫主义；浪漫的思想感情；传奇小说体裁；传奇性
roofline [ˈruːflain] *n*. 屋顶轮廓线
rotunda [rəʊˈtʌndə] *n*. 圆形建筑，圆形大厅
rugged [ˈrʌgid] *adj*. 崎岖的；凹凸不平的；结实的
runoff [ˈrʌnˈɔːf] *n*. 径流；决胜投票
rupee [ruːˈpiː] *n*. 印度等国的货币单位卢比
rusticated [ˈrʌstikeitid] *adj*. 粗毛石砌筑

S

sack [sæk] *n*. 麻袋；洗劫 *vt*. 解雇；把……装进袋里；掠夺
sacrament [ˈsækrəmənt] *n*. 圣礼；天主教的圣事；神圣的东西；庄严的誓言
sacred [ˈseikrid] *adj*. 神圣的；宗教的；受崇敬的，值得崇敬的；有宗教性质的
sacrificial [ˈsækriˈfiʃl] *adj*. 献祭的，牺牲的
saga [ˈsɑːgə] *n*. 讲述许多年间发生的事情的长篇故事；传说，冒险故事，英雄事迹
sanction [ˈsæŋkʃn] *n*. 约束力；制裁，处罚 *vt*. 批准；鼓励，容忍
sanctuary [ˈsæŋktʃuəri] *n*. 圣所；避难所；庇护所；庇护
sapphire [ˈsæfaiə(r)] *n*. 蓝宝石；蔚蓝色 *adj*. 蔚蓝色的
sarcophagi [sɑːˈkɔfəgai] *n*. 石棺的复数
scaffold [ˈskæfəʊld] *n*. 脚手架
scarce [skeəs] *adj*. 缺乏的，罕见的 *adv*. 勉强；仅仅；几乎不
sconce [skɔns] *n*. 壁突式烛台 *vt*. 筑保垒防卫
scotia [ˈskəʊʃə] *n*. 凹形边饰，深凹面线脚
scroll [skrəʊl] *n*. 纸卷；书卷，画卷，卷轴；涡卷形装饰 *vt*. 使成卷形
secular [ˈsekjələ(r)] *adj*. 现世的，俗界的；长期的，长久的 *n*. 牧师；俗人
secluded [siˈkluːdid] *adj*. 与世隔绝的；偏僻的；隐退的
semispherical [semisˈferikl] *n*. 半球形，半球形顶盖
sequential [siˈkwenʃl] *adj*. 序贯；时序；按次序的，相继的
sermon [ˈsəːmən] *n*. 布道；讲道；讲道文章

setback [ˈsetbæk] *n.* 挫折；退步；阻碍；逆流
shaft [ʃɑːft] *n.* 柄，轴；柱身 *vt.* 给……装上杆柄
shaman [ˈʃeimən] *n.* 萨满，巫师
shed [ʃed] *n.* 棚，库；分水岭 *vt.* 流出；树叶脱落 *vi.* 脱落；流出；散布
shutter [ˈʃʌtə(r)] *n.* 快门；百叶窗 *vt.* 使……停止运行，关闭；装上百叶窗
siege [siːdʒ] *n.* 围攻，围困，围城；不断袭击；长期努力
simultaneously [ˈsiməlˈteiniəsli] *adv.* 同时地；一壁；一齐
single-eave hip gable roof 单檐歇山屋顶
skene [skiːn] *n.* 舞台；短剑，匕首
skew [skjuː] *adj.* 斜的，歪的；不对称的 *vt.* 歪曲；曲解；使歪斜 *vi.* 偏离，歪斜
skyline [ˈskailain] *n.* 地平线；天际线，轮廓线 *vt.* 天空映衬出……的轮廓
skylight [ˈskailait] *n.* 天窗
slate roof [sleit ruːf] 板岩顶板
slay [slei] *vt.* 残杀
slit [slit] *vt.* 切开，撕开；使有狭缝 *n.* 裂缝；狭长切口；投币口
sniper [ˈsnaipə(r)] *n.* 狙击手；狙击兵
sophisticate [səˈfistikeit] *n.* 老于世故的人；见多识广的人
sovereignty [ˈsɔvrənti] *n.* 主权国家；国家的主权；君权，最高统治权
spandrel [ˈspændrəl] *n.* 拱肩；拱脊
spatially [ˈspeiʃəli] *adv.* 空间地，存在于空间地
specimen [ˈspesimən] *n.* 样品；范例
spectacle [ˈspektəkl] *n.* 眼镜；奇观，壮观；景象；表演，场面
spectacular [spekˈtækjələ(r)] *adj.* 场面壮观的；引人注意的 *n.* 壮观的场面，精彩的表演
spices [sˈpaisiz] *n.* 调味品；香料；趣味；情趣
spiritual [ˈspiritʃuəl] *adj.* 精神的；心灵的；宗教的 *n.* 圣歌；教会事物
spotlight [ˈspɔtlait] *n.* 聚光灯；公众注意或突出显著；探照灯 *vt.* 聚光照明
spout [spaʊt] *n.* 喷嘴；喷流 *vt.* 喷出；喷射；滔滔不绝地讲 *vi.* 喷出；喷射
sprinkler [ˈspriŋklə(r)] *n.* 洒水器，喷洒器；自动喷水灭火装置
squat [skwɔt] *vi.* 蹲，蹲伏 *vt.* 使蹲坐 *n.* 蹲坐
squinch [skwintʃ] *n.* 突角拱
stagnation [stægˈneiʃn] *n.* 滞止；淤塞，停滞；不景气
stalactite [ˈstæləktait] *n.* 钟乳石，蜂窝状装饰
stall [stɔːl] *n.* 货摊；畜栏；小隔间 *vt.* （使）熄火，（使）停止转动 *vi.* 拖延
statuary [ˈstætʃuəri] *adj.* 雕塑的 *n.* 塑像；制造雕塑
steeple [ˈstiːpl] *n.* 尖塔，尖顶
stilt [stilt] *n.* 高跷；支撑建筑物高出地面或水面的桩子，支柱
stoa [ˈstəʊə] *n.* 古希腊建筑的拱廊，柱廊
stockade [stɔˈkeid] *n.* 防御用的栅栏，围桩 *vt.* 用栅栏防护

stucco [ˈstʌkəʊ] *n.* 粉饰灰泥　*vt.* 用拉毛粉饰法粉饰
stratified [ˈstrætifaid] *adj.* 成层了的，层积了的，分层的
straddl [ˈstrædl] *v.* 跨坐；交界；不表明态度
strain [strein] *vt.* 拉紧，拉伤　*n.* 血统，家族　*vt.* 拉紧；尽量使力；扭伤；歪曲
streamlining [ˈstriːmlainiŋ] *n.* 流线型；流线
strew [struː] *vt.* 点缀；撒满；撒在……上；散落于
triad [ˈtraiæd] *n.* 三个一组；三幅一组
stronghold [ˈstrɔŋhəʊld] *n.* 据点；要塞；根据地
stylobate [ˈstailəˈbeit] *n.* 柱座，一列柱子下面的柱基；台基
stuck [stʌk] *v.* 刺(stick 的过去式及过去分词)　*adj.* 动不了的；被卡住的；被……缠住的
stupa [ˈstuːpə] *n.* 佛塔
subbase [ˈsʌbbeis] *n.* 子基，底基，底基层
subordinate [səˈbɔːdinət] *adj.* 下级的；次要的；附属的　*n.* 部属；下级　*vt.* 使……居下位
subservient [səbˈsəːviənt] *adj.* 卑躬屈节的；有帮助的；充当下手的
submerge [səbˈməːdʒ] *vt.* 淹没；把……浸入；沉没；使沉浸　*vt.* 淹没；把……浸入　*vi.* 淹没
subterranean [ˈsʌbtəˈreiniən] *adj.* 地表下面的，地下的　*n.* 地下室；生活在地下的人
Sumeru pedestal 须弥座
sultan [ˈsʌltən] *n.* 苏丹；巨头，强人
superimpose [ˈsuːpərimˈpəʊz] *vt.* 添加；附加
superintendent [ˈsuːpərinˈtendənt] *n.* 主管；监督人，管理人；厂长，所长
superstition [ˈsuːpəˈstiʃn] *n.* 迷信；迷信行为
surmount [səˈmaʊnt] *vt.* 战胜，克服；登上，攀登；居于……之上；顶上覆盖着
survive [səˈvaiv] *vi.* 幸存，活下来　*vt.* 幸存；比……活得长　*vt.* 幸免于难；挺过
suspend [səˈspend] *v.* 暂停；悬；挂；延缓
sustain [səˈstein] *vt.* 维持；供养；支撑，支持；遭受，忍受
sutra [ˈsuːtrə] *n.* 佛经，经典
swag [swæg] *n.* 赃物；摇晃；水潭　*vi.* 摇晃，垂下
sweeping [ˈswiːpiŋ] *adj.* 彻底的；影响广泛的　*n.* 扫除；垃圾
swelling [ˈsweliŋ] *n.* 肿胀；膨胀；增大　*v.* 肿胀；增强
syllabary [ˈsiləbəri] *n.* 字音表，音节文字表
symmetrically [siˈmetrikli] *adv.* 对称性地，对称地，平衡地
synagogue [ˈsinəgɔg] *n.* 犹太教堂；犹太人集会
synonymous [siˈnɔniməs] *adj.* 同义词的；同义的，类义的

T

talisman [ˈtælizmən] *n.* 护身符；法宝；驱邪物；有不可思议的力量之物
tangible [ˈtændʒəbl] *adj.* 可触知的；确实的，真实的　*n.* 有形资产
taper [ˈteipə(r)] *vt.* 逐渐变细，变尖　*n.* 细蜡烛；灯芯；锥形物

tavern ['tævən] *n.* 酒馆；小旅馆，客栈
tegulae ['tegjələˌliː] *n.* 翅基片
tenant ['tenənt] *n.* 房客；佃户；占有者
tenon ['tenən] *n.* 木工的榫，凸榫，榫舌 *vt.* 在……上制榫；用榫接合；牢固接合
terminate ['təːmineit] *vt.* 结束；使终结；解雇；到达终点站 *adj.* 结束的
terraced ['terəst] *adj.* 台地的，阶地的；有平台的，有露台的；沿斜坡建造的
tessellate ['tesileit] *vt.* 把……镶嵌成棋盘花纹
tessellation ['tesi'leiʃən] *n.* 镶嵌成小方格，棋盘形布置，镶嵌式铺装；嵌石装饰
thatch [θætʃ] *n.* 茅草屋顶 *vt.* 用茅草盖房子的屋顶，用茅草覆盖
theologian ['θiːə'ləudʒən] *n.* 神学家，宗教研究家
thoroughfare ['θʌrəfeə(r)] *n.* 通行；大道，大街
threefold ['θriːfəuld] *adj.* 三倍的；有三部分的，三重的
thrust [θrʌst] *vt.* 猛推；逼迫 *n.* 侧推力 *vi.* 插入；用力向某人刺去；猛然或用力推
tie beam [tai biːm] 系梁；(水平)拉杆
tier [tiə(r)] *n.* 等级；阶梯座位的一排，一行 *vt.* 层层排列 *vi.* 成递升排列
tolerably ['tɔlərəbli] *adv.* 颇；可容忍地
torus ['tɔːrəs] *n.* 花托，花床，圆环面
tracery ['treisəri] *n.* 花饰窗格，窗饰
tragedy ['trædʒədi] *n.* 悲剧，惨剧；悲剧文学；悲剧表演艺术
translucent [træns'luːsnt] *adj.* 半透明的；透亮的，有光泽的
transome [t'rænsəm] *n.* 横档
trapdoor ['træpdɔː(r)] *n.* 舞台的地板门，活板门，活盖
trapezoid ['træpəzɔid] *n.* 梯形，不等边四边形
travertine ['trævətiːn] *n.* 石灰华
treacherous ['tretʃərəs] *adj.* 奸诈的；不可信的 *adv.* 背信弃义地 *n.* 背叛
trellised ['trelist] *adj.* 成格状的，有格子架支撑的
tribune ['tribjuːn] *n.* 论坛；古罗马护民官；民众领袖；赛马场的看台
triglyph ['traiglif] *n.* 三竖线花纹装饰，三陇板
trim [trim] *vt.* 装饰；修剪；整理 *adj.* 整齐的，整洁的；修长的 *n.* 修剪；整齐
triumphal [trai'ʌmfl] *adj.* 胜利的，用于胜利的，庆祝胜利的
trilateral ['trai'lætərəl] *adj.* 三边的 *n.* 三边形
trophy ['trəufi] *n.* 纪念品，战利品；奖品 *adj.* 显示身份或地位的；有威望的
truss [trʌs] *vt.* 捆绑 *n.* 干草的一捆，一束，构架，梁架
trustee [trʌ'stiː] *n.* 受托人；信托公司；慈善事业或其他机构受托人 *vt.* 移交财产或管理权给
tulip ['tjuːlip] *n.* 郁金香；郁金香的花朵或球根
turquoise ['təːkwɔiz] *n.* 绿松石；青绿色，天蓝色 *adj.* 蓝绿色的
turret ['tʌrət] *n.* 炮塔，塔楼，角楼

tympana ['timpənə] *n.* 鼓室；窗洞口之间的墙面

U

unabated ['ʌnə'beitid] *adj.* 不减弱的，不减退的
unanimous [ju'næniməs] *adj.* 全体一致的；一致同意的，无异议的
underneath ['ʌndə'ni:θ] *adj.* 下面的；较低的 *adv.* 在下面，在底下 *prep.* 在下面
unification ['ju:nifi'keiʃn] *n.* 统一，联合；一致
unrivaled [ʌn'raivəld] *adj.* 无敌的；无双的；无比的；无对手的
upkeep ['ʌpki:p] *n.* 维持；保养，维修；保养费，维修费
unkempt ['ʌn'kempt] *adj.* 邋遢；蓬乱的；粗野的

V

vandalism ['vændəlizəm] *n.* 故意破坏，捣毁
vagrant ['veigrənt] *n.* 流浪者；乞丐；无赖 *adj.* 流浪的；无定向的
vandalize ['vændəlaiz] *vt.* 肆意破坏尤指公共财产
vasque（法语）*n.*（喷泉的）承水盘，间歇泉口小池；放在桌上作摆设的浅口盆
vault [vɔ:lt] *n.* 墓穴；拱顶，穹窿；地下室 *vt.* 做成圆拱形 *vi.* 跳跃；成穹状弯曲
vendor ['vendə(r)] *n.* 卖主；摊贩，小贩；自动售货机；供应商
venerate ['venəreit] *vt.* 崇敬，尊敬
venue ['venju:] *n.* 会场；案发地点；体育比赛场所
ventilation [venti'leiʃn] *n.* 通风设备；空气流通；通风方法；公开讨论
veranda [və'rændə] *n.* 阳台，走廊
vermilion [və'miliən] *n.* 朱红色，鲜红色
vernacular [və'nækjələ(r)] *adj.* 白话的；方言的；本土的；本国的
vestigial [ve'stidʒiəl] *adj.* 残留的，残余的，退化的
viceroy ['vaisrɔi] *n.* 代表国王管辖行省或殖民地等的总督
vicinity [və'sinəti] *n.* 附近地区；附近，邻近
vigorous ['vigərəs] *adj.* 有力的；精力充沛的；充满活力的
viper ['vaipə(r)] *n.* 毒蛇；蝰蛇；阴险恶毒的人
virtue ['və:tʃu:] *n.* 美德；德行；价值；长处
virtuosity ['və:tʃu'ɔsəti] *n.* 精湛技艺；对艺术品的爱好；艺术爱好者；古董收藏家
vivid ['vivid] *adj.* 生动的记忆、描述；清晰的；丰富的；鲜艳的，耀眼的
voluntary ['vɔləntri] *adj.* 志愿的；自愿的，自发的；无偿的 *n.* 自愿者；自愿行动
volute [və'lju:t] *n.* 涡形，涡形花样，柱头上的 *adj.* 向上卷的，涡形的，螺旋形的
vulgar ['vʌlgə(r)] *adj.* 庸俗的，粗俗的；一般大众的，粗野的 *n.* 平民，百姓
vulnerable ['vʌlnərəbl] *adj.*（地方）易受攻击的；易受伤的；易受批评的

W

warehouse ['weəhaʊs] *n.* 仓库，货栈 *vt.* 把……放入或存入仓库
wat [wɑ:t] *n.* 泰国或高棉的佛教寺或僧院

whistle [ˈwisl] *vi.* 吹口哨，鸣汽笛 *vt.* 吹口哨召唤 *n.* 汽笛；口哨

wholesale [ˈhəʊlseil] *n.* 批发；大规模买卖 *adj.* 大规模的；整批卖的 *adv.* 大量地，大批地

winch [wintʃ] *n.* 绞车；摇柄，曲柄 *vt.* 用绞车拉

wrath [rɔθ] *n.* 愤怒；激怒

wrought-iron [rɔːt ˈaiən] *n.* 熟铁

Z

zeal [ziːl] *n.* 热情；热心；奋发

zenith [ˈzeniθ] *n.* 顶点，极点

zig-zag [zigˈzæg] *adj.* 曲折的

References

参考文献

[1] Micheal Fazio, Marian Moffett, Lawrence Wodehouse. A Wolrd History of Architecture. London: Laurence King Publishing, 2009.

[2] Duly Colin. The Houses of Mankind. London:Thames and Hudson,1979.

[3] Rudofsky Bernard. Architecture without architects. New York: Doubleday, 1964.

[4] Lawrence Arnold Walter. Greek and Roman Architecture. New Haven: Yale University Press, 1996.

[5] Huntington Susan L. The art of Ancient India: Buddhist, Hindu, Jain. New York: Weatherhill, 1985.

[6] Boyd Andrew. Chinese Architecture and Town Planning, 1500 B. C. —1911 A. D. Chicago:University of Chicago, 1962.

[7] Hillenbrand Robert. Islamic Architecture. New York : Henry and Abrams, 1977.

[8] Burckhardt J. The architecture of the Italian Renaissance. London: John Murray, 1985.

[9] Blunt A. Art and architecture in France, 1500—1700. New Haven: Yale University Press, 1999.

[10] Summerson J. Architecture in Britain, 1530—1830. New Haven:Yale University Press, 1993.

[11] Torsten Olaf Enge. Garden Architecture in Europe, 1450—1800. London: Mark Thomson, 1992.

[12] Suga Hirofumi . Japanese Garden. Tokoyo: The Images Publishing Group, 2005.

[13] Fairchild Ruggles D. Islamic Gardens and Landscapes. Philadelphia: University of Pennsylvania Press, 2008.

[14] 梁思成. 图像中国建筑史. 北京:生活·读书·新知 三联书店,2011.

[15] 贾珺. 北京四合院. 北京:清华大学出版社,2009.

[16] 黄汉民. 福建土楼. 北京:生活·读书·新知三联书店,2009.

[17] Wang Zhixian. Piled Dwellings. Guiyang: Guizhou Publishing Group, 2010.

[18] Wang Zhixian. Drum Towers and Roofed Bridges. Guiyang: Guizhou Publishing Group, 2010.